Frontiers in Computational Genomics

Functional Genomics Series Volume 3

www.functional-genomics.org

Edited by

Michael Y. Galperin
and Eugene V. Koonin
National Center for Biotechnology Information,
Bethesda, USA

Series Editor: **Milton H. Saier, Jr.**
University of California at San Diego
La Jolla, USA

CAISTER
ACADEMIC PRESS

Caister Academic Press
32 Hewitts Lane
Wymondham
Norfolk NR18 0JA
U.K.

www.caister.com

British Library Cataloguing-in-Publication Data

A catalogue record for this book is available from the British Library

ISBN: 0-9542464-4-6

Printed and bound in Great Britain by the Cromwell Press Limited

Contents

Contributors

Alex Bateman
Wellcome Trust Sanger Institute
Hinxton
Cambridge CB10 1SA
UK
agb@sanger.ac.uk

Faina S. Berezovskaya
Department of Mathematics
Howard University
2400 Sixth Str.
Washington
DC 20059
USA

Jürgen Brosius
Institute of Experimental Pathology
ZMBE
University of Münster
Von-Esmarch-Str. 56
Münster
Germany
RNA.world@uni-muenster.de

Frank Eisenhaber
Research Institute of Molecular Pathology
The Vienna Biocenter
Dr. Bohr-Gasse 7
A-1030 Vienna
Austria
Frank.Eisenhaber@imp.univie.ac.at

Natalie D. Fedorova
National Center for Biotechnology Information
National Library of Medicine
National Institutes of Health
Building 38A
Bethesda
Maryland 20894
USA
fedorova@ncbi.nlm.nih.gov

Michael Y. Galperin
National Center for Biotechnology Information
National Library of Medicine
National Institutes of Health
Building 38A
Bethesda
Maryland 20894
USA
galperin@ncbi.nlm.nih.gov

Mikhail S. Gelfand
IntegratedGenomics-Moscow
P.O.Box 348
Moscow
117333
Russia
gelfand@integratedgenomics.ru

Adam Godzik
The Burnham Institute
10901 N. Torrey Pines Road
La Jolla
California 92037
and Joint Center for Structural Genomics
San Diego Supercomputer Center
University of California San Diego
9500 Gilman Drive
La Jolla
California 92093-0527
USA
adam@sdsc.edu

Nick V. Grishin
Howard Hughes Medical Institute and
Department of Biochemistry
University of Texas Southwestern Medical Center
5323 Harry Hines Boulevard
Dallas
Texas 75390-9050
USA
grishin@chop.swmed.edu

Roderic Guigo
Grup de Recerca en Informàtica Biomèdica
Institut Municipal d'Investigació Mèdica
Universitat Pompeu Fabra
C/ Dr. Aiguader 80
E-08003 Barcelona
Spain
rguigo@imim.es

Alexander Hüttenhofer
Institute of Experimental Pathology
ZMBE
University of Münster
Von-Esmarch-Str. 56
Münster
Germany
huttenh@uni-muenster.de

Martijn A. Huynen
Nijmegen Center for Molecular Life Sciences
P/a Center for Molecular and Biomolecular Informatics
Toernooiveld 1
Nijmegen
Netherlands
Huynen@cmbi.kun.nl

Georgy P. Karev
National Center for Biotechnology Information
National Library of Medicine
National Institutes of Health
Building 38A
Bethesda
Maryland 20894
USA
karev@ncbi.nlm.nih.gov

Eugene V. Koonin
National Center for Biotechnology Information
National Library of Medicine
National Institutes of Health
Building 38A
Bethesda
Maryland 20894
USA
koonin@ncbi.nlm.nih.gov

Jürgen Kopp
Biozentrum der Universität Basel and
Swiss Institute of Bioinformatics (SIB)
Klingelbergstr. 50-70
CH-4056 Basel
Switzerland
Juergen.Kopp@unibas.ch

Eugene N. Kuznetsov
Institute of Control Sciences
Russian Academy of Sciences
Profsoyuznaya Street 65
Moscow 117806
Russia

Olga N. Laikova.
State Scientific Center GosNIIGenetika
1st Dorozhny pr. 1
Moscow 113545
Russia
laikova@mail.ru

Anastasia N. Nikolskaya
National Center for Biotechnology Information
National Library of Medicine
National Institutes of Health
Building 38A
Bethesda
Maryland 20894
USA
nikolska@ncbi.nlm.nih.gov

Chris P. Ponting
MRC Functional Genetics Unit
Department of Human Anatomy and Genetics
University of Oxford
South Parks Road
Oxford OX1 3QX
UK
Chris.Ponting@human-anatomy.oxford.ac.uk

Manuel C. Peitsch
Novartis Pharma AG and Swiss Institute of Bioinformatics (SIB)
WKL-490.4
CH-4002 Basel
Switzerland
manuel.peitsch@pharma.novartis.com

Igor B. Rogozin
National Center for Biotechnology Information
National Library of Medicine
National Institutes of Health
Building 38A
Bethesda
Maryland 20894
USA
rogozin@ncbi.nlm.nih.gov

Andrey Y. Rzhetsky
Columbia Genome Center
Columbia University
1150 St. Nicholas Avenue
Unit 109
New York
NY 10032
USA
ar345@columbia.edu

Torsten Schwede
Biozentrum der Universität Basel and Swiss Institute of Bioinformatics (SIB)
Klingelbergstr. 50-70
CH-4056 Basel
Switzerland
Torsten.Schwede@unibas.ch

Berend Snel
European Molecular Biology Laboratory
Meyherhofstrasse 1
D-69117 Heidelberg
Germany
snel@embl-heidelberg.de

Shamil R. Sunyaev
Division of Genetics
Department of Medicine
Brigham and Women's Hospital and Harvard Medical School
Thorn Bldg. Rm. 914
20 Shattuck St.
Boston
Massachusetts 02115-6195
USA
ssunyaev@rics.bwh.harvard.edu

Vladimir G. Tumanyan
V.A. Engelhardt Institute of Molecular Biology
Russian Academy of Sciences
Vavilov Street 32
117984 Moscow
Russia
tuman@imb.imb.ac.ru

Thomas Wiehe
Working Group Sequence Annotation
Freie Universität Berlin and Berlin Center for Genome Based Bioinformatics
Arnimallee 22
D-14195 Berlin
Germany
twiehe@zedat.fu-berlin.de

Yuri I. Wolf
National Center for Biotechnology Information
National Library of Medicine
National Institutes of Health
Building 38A
Bethesda
Maryland 20894
USA
wolf@ncbi.nlm.nih.gov

Other Titles of Interest

Preface

Computational analysis is a major, integral part of genomics. From sequence fragment assembly, which becomes a particularly formidable problem in the increasingly popular shotgun sequencing, to gene prediction (an especially challenging and unsolved problem when genes of higher eukaryotes are involved), to prediction of protein functions and structure on the basis of comparisons to homologs, to identification of regulatory signals in DNA, to building phylogenetic trees, it would not be an exaggeration to state that genomics is primarily computational. Quite simply, if the entire body of computational-genomic methods and established procedures disappears tomorrow, genome sequencing itself will immediately become not only meaningless, but outright impossible. In contrast, if it so happened that sequencing of new genomes stopped, that certainly would be a great loss, but computational genomics would have things to chew on for many years to come. Thus, the importance of computational approaches in genomics is obvious.

However, there seems to be a considerable confusion as to the status of Computational Biology (often used interchangeably with Bioinformatics) as a scientific discipline. Indeed, a popular (even if not so often articulated) view holds that Bioinformatics (the preferred designation for the activity under this view) is a compendium of computational methods whose function is largely to process, store in a readily accessible form, and visualize the basic information extracted from genomes. Sophisticated and even innovative algorithms may underlie such methods, but, according to this vision, Bioinformatics does not have its own coherent theoretical framework and is not a major source of independent discoveries. The opposite position holds that Computational Biology is a full-fledged research field capable of producing fundamental, biologically important results without recourse to experiment. As long-time practitioners of Computational Biology, the Editors of this book might be expected to be staunch adherents of the latter view. In an attempt to be objective, however, we would like to posit that the two visions of Computational Biology are complementary and mutually enriching, rather than contradictory. Straightforward methods for gene identification and sequence comparison probably are the greatest contributions of Bioinformatics to the progress of genomics and biology in general and improvement of these methods remains extremely important. However, as the field is maturing, two other aspects of Computational Biology are becoming increasingly prominent: (i) its potential for actively directing experimental research through non-trivial functional predictions and prioritization of targets (the burgeoning field of structural genomics is the most obvious illustration of this trend), and (ii) its ability to reconstruct major events of life's history and possibly even deduce the laws of genome evolution. We hope and believe that the latter could become the foundation of theoretical biology of the 21st century. Undoubtedly, to understand life, one must understand how it evolved, and the potential for experiment in this area is limited at best.

We attempted to shape this book such that it reflects the different faces of Computational Genomics. In the opening chapter, Guigo and Wiehe describe their long-term effort to objectively assess the accuracy of gene prediction. Accurate gene prediction is the first, elementary task of genomics, but the problem is far from being solved, at least for eukaryotes where exons are spliced together in a process, the complexity of which we are only starting to appreciate. Importantly, Guigo and Wiehe show how genome comparisons help increasing the accuracy of gene prediction, which illustrates the crosstalk between the methodological and evolutionary aspects of computational genomics. In the next chapter, Godzik discusses a new generation of protein sequence alignment methods, which allow one to transfer information between proteins that previously were deemed too distantly (if at all) related for comparisons to be meaningful. The chapter by Eisenhaber and colleagues presents an in-depth statistical analysis of so-called knowledge-based potentials that are derived from known protein structures and applied to predict structures of uncharacterized proteins from their sequences in the increasingly popular threading methods. Their analysis suggests that unresolved problems lurk in the very foundations of these approaches. Kopp and colleagues address the task of homology modeling in genome-wide protein sets and its functional applications. This directly relates to the objective of structural genomics: solve enough strategically chosen structures to ensure that the structure of each protein encoded in the sequenced genomes can be accurately modeled on the structure of a sufficiently close homolog. The chapter by Ponting and Bateman deals with the notions of domain, repeat and motif, which are crucial in functional interpretation of protein sequences. They show how careful and critical analysis of protein domains and motifs (two types of entities that need to be clearly distinguished but are improperly considered to be interchangeable all too often) opens new opportunities in functional genomics. Martijn Huynen and Berend Snel are pioneers of context analysis, the new buzzword in functional genomics. Here, they take this notion further than before by analyzing the correlation between genomic and functional association of proteins and by building an explicit evolutionary model of the evolution of a pathway. The chapter by Fedorova and coworkers addresses related issues by analyzing domain rearrangements and emergence of new functions in the two most common multi-component complexes in bacteria, the ABC-type transporters and two-component signal transduction proteins. The chapter by Gelfand and Laikova is a preliminary take on a truly grand goal, reconstruction of the evolution of prokaryotic regulatory networks. The available information is certainly insufficient to get the big picture right now, but some glimpses are emerging through case stories of co-evolution of regulators and regulatees presented in this chapter. Hüttenhofer and Brosius introduce the new discipline of "RNomics" (one cannot help wondering, is the name largely tongue-in-cheek or is it really here to stay?) and show how a combination of computational and experimental approaches works towards the identification of the complete ...well, RNome of a cell. The final two chapters of the book directly take on the goal of building evolutionary genomics. Wolf and co-workers explore the notion of genomic trees and review the work of the last two-three years that revealed that potential of genome-scale comparisons to enhance the resolution of phylogenetic trees. Karev and co-workers develop a mathematical model of genome evolution that explains the emergence of power laws in the size distributions of protein families, a subject that recently caused much excitement in the context of the brave new world of scale-free networks. Although it remains unclear just how far this particular approach can take us, this type of analysis seems to show the way to a new level of abstraction in genomics.

Computational biology is on the rise; there is a clear need in books explaining the basics of computational approaches in genomics and presenting the principal methods, and several such books have been recently published (*e.g.* Baxevanis and Ouellette, 2001; Kanehisa, 2000; Mount, 2001). The aim of this volume was different: we tried to highlight research on the emerging, perhaps somewhat controversial, topics in genomics and show the feedback between the new developments and the established methods of computational biology. If the readers share some of the enthusiasm that the editors and authors of this book feel about the future of Computational Genomics, we will consider its goal fulfilled.

Michael Galperin and Eugene Koonin

References

Baxevanis, A.D., and Ouellette, B.F.F. 2001. Bioinformatics: A Practical Guide to the Analysis of Genes and Proteins. 2nd Edition. John Wiley and Sons, New York.

Kanehisa, M. 2000. Post-Genome Informatics. Oxford University Press, Oxford-New York

Mount, D.W. 2001. Bioinformatics: Sequence and Genome Analysis. Cold Spring Harbor Laboratory Press, Cold Spring Harbor, New York

From: *Frontiers in Computational Genomics*

1

Gene Prediction Accuracy in Large DNA Sequences

Roderic Guigó and Thomas Wiehe

Abstract

Essential to our ability of converting the nucleotide sequence of the genomes into valuable biological knowledge of the organisms is the identification of the genes encoded in these genomes. Despite substantial progress during the nineties, computational gene identification remains an open problem in higher eukaryotic organisms: a year after the publication of the draft sequence of the human genome, the exact number of human genes remains highly controversial. In this chapter we review the accuracy of computational gene finding methods. We emphasize recently developed methods, based on the comparison of genomic sequences. These methods show promise in identifying novel genes, not well represented in the databases of known coding sequences.

1. Introduction

1.1. How Many Genes are in the Human Genome?

With the publication of the draft human genome sequence (Lander *et al.*, 2001; Venter *et al.*, 2001), one of the surprises was the apparent low number of human genes. While previous estimates had put this number in the range of 70,000 to 100,000 (Antequera and Bird, 1993; Fields *et al.*, 1994), the initial analysis of the draft sequence provided evidence for only about 35,000 genes.

This number, however, has been much disputed and while the human genome sequence deposited in the databases becomes more and more accurate, gene number estimations remain highly variable. Chromosome 22 - the smallest human chromosome and the first one sequenced - is exemplary. According to the initial analysis (Dunham *et al.*, 1999), it codes for 545 genes and 134 pseudogenes. This number was the result of a combination of human inspection and software procedures. Gene predictions were obtained with GENSCAN (Burge and Karlin, 1997), but were only accepted when they had additional support from similarity to already known genes or matches with so-called expressed sequence tags (ESTs), cDNA fragments deposited in special databases. The total number of GENSCAN predicted genes on chromosome 22 ranges from 817 to 1128, depending on the greediness of repeat-masking programs used to pre-process the sequence. Several groups investigated whether the additional *ab initio* gene predictions, which were not included in the initial annotation, could be real genes. Thus, Das *et al.* (2001) have used a sensitive RT-PCR assay in 17 tissues and one cell line, and found consistent evidence of transcription for about 30% of the additional *ab initio* predictions, suggesting that chromosome 22 has at least 690 genes. On the other hand, Shoemaker *et al.* (2001) have designed microarrays consisting of oligonucleotide probes derived from predicted exons. When hybridized with flourescently labeled cDNAs, the intensities can be used to identify those probes that represent authentic exons. They applied this 'exon array' approach to a detailed analysis of human chromosome 22 under 69 pairs of experimental conditions, and found experimental support for about 57% of those GENSCAN predictions which were not supported by the initial analysis. This would lead to at least 730 genes in chromosome 22. The number, however, could be higher, because only about 85% of the already confirmed genes tested positive in the exon array.

The gene number estimates for the whole genome are even more variable. The two papers by Lander *et al.* (2001) and Venter *et al.* (2001) give similar figures, and claim the number of protein coding genes to be in the range of 30,000 to 40,000, and of 27,000 to 38,000, respectively. In contrast, later analyses gave substantially higher estimates. For instance, Wright *et al.* (2001) compared the human genome draft sequence to a large number of databases of coding sequences, and found evidence for 65,000 to 75,000 genes, among them 854 on chromosome 22. Hogenesch *et al.* (2001) compared the Consortium (as in the ENSEMBL site, see http://www.ensembl.org) and the Celera (Venter *et al.*, 2001) gene annotations. They found that most of the known genes from the REFSEQ database (Pruitt and Maglott, 2001) were included in both annotations, but that there was little overlap between the sets of novel genes predicted by both groups: nearly 80% of the novel genes were predicted by only one group. Hogenesch *et al.* (2001) then used RNA expression profiling on a bank of 13 diverse human tissues and found support for about 80% of the genes in both sets, the one of known genes and the one of novel genes. The authors conclude that the novel transcripts predicted by both groups may encode differentially expressed mRNAs. In such a case, the number of human genes would be at least 50,000.

Previous to the publication of the draft genome sequence the gene number estimates were even more discrepant, ranging from 30,000 to 120,000 (Ewing and Green, 2000; Liang *et al.*, 2000; Crollius *et al.*, 2000). In any case, these highly variable numbers - even after the genome sequence became available - illustrate the difficulties of computational gene prediction in higher eukaryotes.

1.2. Computational Gene Prediction

Gene prediction has a long history in bioinformatics. The first programs to predict coding regions in genomic DNA sequences were developed as early as 1982 (Fickett, 1982). During the nineties, and prompted by the genome sequencing projects, the problem attracted more and more interest (for reviews see Claverie, 1997; Fickett, 1996; Haussler, 1998; Burge and Karlin, 1998; Guigó, 1997a). Substantial progress was made both in terms of computational efficiency and accuracy of the predictions. While the first gene structure prediction programs available in the early nineties (for instance, GENEMODELER (Fields and Soderlund, 1990), GENEID (Guigó *et al.*, 1992), GENEPARSER (Snyder and Stormo, 1993)) were able to analyze only sequences shorter than a few kilobases, assuming a single gene in one strand, by the end of the decade programs were available which were able to analyze chromosome size sequences in a few minutes - predicting full exonic structures of multiple genes in both strands, such as GENSCAN (Burge and Karlin, 1997).

Despite the progress, computational gene prediction is still an open problem, as the discrepant gene number estimates make clear, and the accurate identification of every gene in the human genome by computational methods remains a distant goal.

Only the determination of the mRNA sequence of a gene guarantees the correctness of its predicted exonic structure. But even in such a case a number of important issues remain to be solved: the completeness of the 3' and 5' ends of the gene, in particular if only a partial cDNA sequence has been obtained, the possibility of alternative primary transcripts, the determination for each of these of the complete catalogue of splice isoforms - of which the sequenced mRNA may correspond to only one among multiple instances, the usage of non-canonical splice sites, the precise site (or the alternative sites) at which the translation starts, the possibility of the gene being encoded within an intron, or nested within a gene, or the possibility of the gene not fully conforming to the standard genetic code, such as in the case of selenoproteins. All these 'exceptions' add difficulties to the already poorly solved gene prediction problem.

1.3. Comparative Methods for Gene Prediction

Soon after the human genome project had been started the importance of sequence comparisons to understand the human genome was realized (Brenner *et al.*, 1993). Today, nearly completed genome sequences are available for a variety of model organisms, most notably *Mus musculus*. The expectation that interspecific comparisons help to locate functional domains, protein coding among them, was fully confirmed. Through the analysis of syntenic regions in human and mouse or human and fugu gene predictions can be improved, and new, otherwise unidentified, genes or gene elements can be revealed (Galgoczy *et al.*, 2001; Onyango *et al.*, 2000; Reichwald *et al.*, 2000). New bioinformatics methodologies have been developed to infer protein coding genes from sequence comparisons of two or more different species (Korf *et al.*, 2001; Batzoglou *et al.*, 2000b; Wiehe *et al.*, 2001; Rinner and Morgenstern, 2002; Bafna and Huson, 2000; Novichkov *et al.*, 2001), which lead to highly accurate results. The rationale is that functional regions, in particular protein-coding exons, retained more sequence similarity through evolution than did non-functional parts of the genome (Figure 1).

The analysis of the recently published complete sequence of the human chromosome 20 (Deloukas *et al.*, 2001) has benefited greatly from the availability of the draft mouse genome, as well as of that of other model

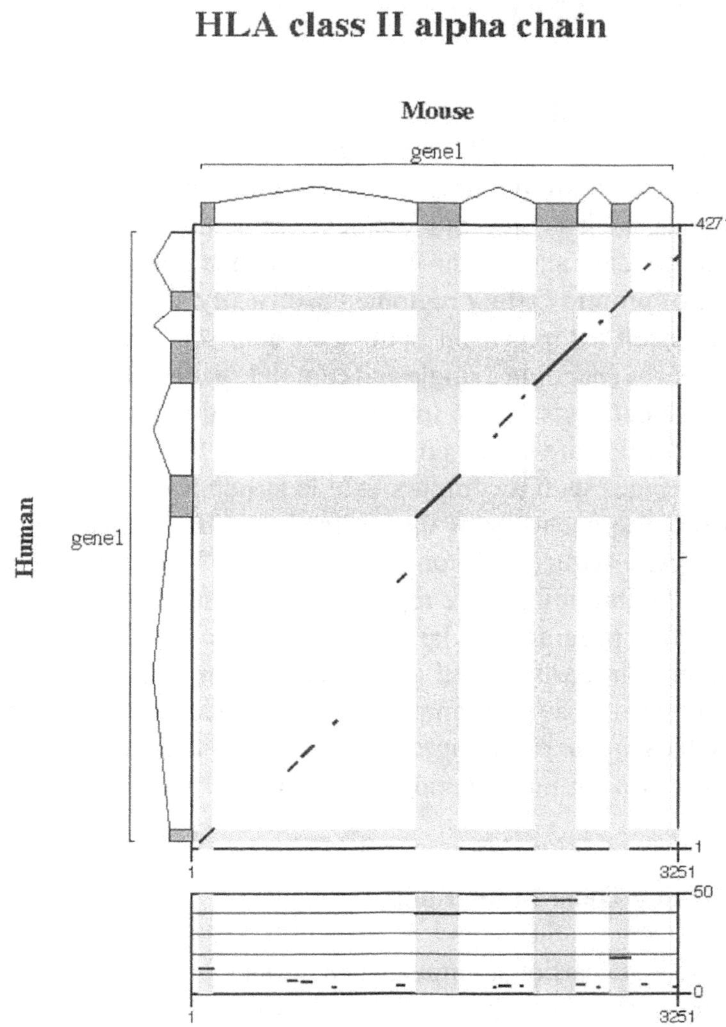

Figure 1. Pairwise comparison using TBLASTX of the human and mouse genomic sequences coding for the HLA class II alpha chain. Red boxes indicate the coding exons, while black diagonals indicate the conserved alignments. The score of the conserved alignments (divided by 10) is given in the lower panel. While conserved regions between human and mouse fully include the coding exons, a substantial fraction of intronic regions is also conserved. TBLASTX output has been post-processed to produce a maximal-scoring, monotonic path. The figure has been obtained using the program GFF2APLOT (http://www1.imim.es/software/gfftools/GFF2APLOT.html). ***Colour plate at back of book.***

vertebrate genomic sequences, for example from the puffer fish *Tetraodon nigroviridis*. In particular, evidence for at least 60 additional genes on chromosome 20, not detected through the traditional gene finding methods, *i.e.* a combination of *ab initio* gene finders, database searches and human inspection, was obtained from sequence comparisons with these two vertebrates.

In the following, we will first describe some standard metrics to evaluate gene predictions. Then, we will review *ab initio* gene prediction methods. *Ab initio* gene finders are the programs which rely on the identification of sequence signals, such as promoter elements, splicing sites, and translation start and stop codons, and on the bias in sequence composition which is characteristic of protein coding regions. Historically, the accuracy of these programs has been estimated on controlled data sets consisting of short genomic sequences encoding a single and complete gene with a simple exonic structure. These data sets are no longer representative of the chromosome size sequences, which are produced now. They have low coding density, and carry multiple genes with a complex exonic structure. In particular, it has been noted that the accuracy of the *ab initio* methods degrades with the length of the gene sequence (Wong *et al.*, 2001). Thus, we will describe efforts aimed at obtaining a more representative estimation of the accuracy of gene prediction programs on large genomic sequences. Finally, we will review sequence similarity based gene prediction methods. Here, we will distinguish between those programs that use sequence comparisons between the query DNA sequence and target protein sequences, and those that use comparisons of two (or more) genomic sequences.

2. Benchmarking Gene Predictions

Although there were earlier attempts to benchmark gene prediction methods or their components - notably the assessment of protein coding measures carried out by Fickett and Tung (1992) - the first comprehensive comparative analysis of gene prediction programs was published by Burset and Guigó (1996). In this work, a number of gene prediction programs were run on a large set of vertebrate genomic sequences coding for single genes, and a number of performance metrics were introduced to evaluate the accuracy of the predictions. Some of the accuracy measures introduced in this article became the de facto standard in the gene prediction field. Later, other measures have been introduced (Pavy *et al.*, 1999; Reese *et al.*, 2000; Rogic *et al.*, 2001). In this section, we will briefly review some of the most common measures. For a more detailed discussion see the papers by Bajic (2000) and Baldi *et al.* (2000).

Table 1. Basic definitions for accuracy measures

Nucleotide			Exon			Gene		
			AE	Actual Exon	number of actual exons	**MG**	Missing Genes	number of actual genes which are completely missing in the prediction
			PE	Predicted Exon	number of predicted exons			
TP	True Positives	number of coding nucleotides predicted as coding	**TE**	True Exons	number of correctly predicted exons	**WG**	Wrong Genes	number of predicted genes without overlap with any actual gene
TN	True Negatives	number of non-coding nucleotides predicted as non-coding	**ME**	Missed Exons	number of exons missing in the prediction	**SG**	Split Genes	number of predicted genes overlapping real genes divided by the number of real genes that were split
FP	False Positives	number of non-coding nucleotides predicted as coding						
FN	False Negatives	number of coding nucleotides predicted as non-coding	**WE**	Wrong Exons	number of wrongly predicted exons	**JG**	Joined Genes	number of real genes that overlap predicted genes divided by the number of predicted genes that were joined
S_n	Sensitivity: TP/(TP+FN)	proportion of coding nucleotides correctly predicted as coding	S_n	Sensitivity: TE/AE	proportion of correctly predicted exons among actual exons	S_n	Sensitivity	proportion of completely correctly predicted genes among actual genes
S_p	Specificity: TP/(TP+FP)	proportion of nucleotides predicted as coding and which are actually coding	S_p	Specificity: TE/PE	proportion of correctly predicted among all predicted exons	S_p	Specificity	proportion of completely correctly predicted among all predicted genes

2.1. Measures of Prediction Accuracy

To evaluate the accuracy of a gene prediction program on a test sequence, we compare the gene structure predicted by the program with the actual gene structure of the sequence, as, for instance, established with the help of an experimentally validated mRNA. The accuracy can be evaluated at different levels of resolution. Typically, these are the nucleotide, exon, and gene levels, see Table 1. These three levels offer complementary views of the accuracy of the program. At each level, there are two basic measures: sensitivity and specificity, which essentially measure prediction errors of the first and second kind. Briefly, sensitivity is the proportion of real elements (coding nucleotides, exons or genes) that have been correctly predicted, while specificity is the proportion of predicted elements that are correct.

Both, sensitivity and specificity, take values from 0 to 1, with perfect prediction when both measures are equal to 1. Neither S_n nor S_p alone constitute good measures of global accuracy, since one can have high sensitivity with little specificity and vice versa. It is desirable to use a single scalar value to summarize both of them. In the gene finding literature, the preferred such measure on **nucleotide level** is the correlation coefficient defined as:

$$CC = \frac{(TP \times TN) - (FN \times FP)}{\sqrt{(TP+FN) \times (TN+FP) \times (TP+FP) \times (TN+FN)}}$$

CC ranges from -1 to 1, with 1 corresponding to a perfect prediction, and -1 to a prediction in which each coding nucleotide is predicted as non-coding and vice versa.

At **exon level**, an exon is considered correctly predicted (TE) only if the predicted exon is identical to the true one, in particular both 5' and 3' exon boundaries have to be correct. A predicted exon is considered wrong (WE), if it has no overlap with any real exon. A summary measure on exon level is simply the average of sensitivity and specificity.

At **gene level**, a gene is correctly predicted if all of the coding exons are identified, every intron-exon boundary is correct, and all of the exons are included in the proper gene. In addition, we compute the missed genes, MG: it is the fraction of those true genes for which none of its exons is overlapped by any predicted gene. WG (wrong genes) is the fraction of those predicted genes for which none of the exons is overlapped by any real gene. As already mentioned, *ab initio* gene finders predict the initial and terminal exons very poorly. This often leads to so-called chimeric predictions - one predicted gene encompassing more than one real gene - or to split predictions - one real gene split in multiple predicted genes. Reese *et al.* (2000) proposed the two measures, split genes, SG, and joined genes, JG, to account for these tendencies. SG is the total number of predicted genes overlapping real genes divided by the number of genes that were split. Similarly, JG is the total number of real genes that overlap predicted genes divided by the number of predicted genes that were joined.

3. Accuracy of *ab initio* Gene Predictions

3.1. Accuracy in Single Gene Sequences

Table 2 reproduces the results in Burset and Guigó (1996). These authors evaluated seven programs in a set of 570 vertebrate single gene genomic sequences deposited in GenBank after January 1993 (the ALLSEQ dataset).

Table 2. Accuracy of gene prediction programs on single gene vertebrate sequences. Adapted from Burset and Guigó (1996)

		Nucleotide			Exon				
Program	Dataset	S_n	S_p	CC	S_n	S_p	$\frac{Sn+Sp}{2}$	ME	WE
FGENEH	ALLSEQ	0.77	0.88	0.80	0.61	0.64	0.64	0.15	0.12
	NEWSEQ	0.70	0.83	0.73	0.51	0.54	0.54	0.22	0.18
GENEID	ALLSEQ	0.63	0.81	0.65	0.44	0.46	0.45	0.28	0.24
	NEWSEQ	0.58	0.78	0.60	0.41	0.43	0.42	0.34	0.27
GENEPARSER2	ALLSEQ	0.66	0.79	0.65	0.35	0.40	0.37	0.29	0.17
	NEWSEQ	0.63	0.76	0.62	0.33	0.39	0.36	0.32	0.20
GENLANG	ALLSEQ	0.72	0.79	0.71	0.51	0.52	0.52	0.21	0.22
	NEWSEQ	0.63	0.73	0.63	0.39	0.44	0.43	0.29	0.25
GRAIL 2	ALLSEQ	0.72	0.87	0.76	0.36	0.43	0.40	0.25	0.11
	NEWSEQ	0.69	0.85	0.72	0.34	0.41	0.38	0.30	0.13
SORFIND	ALLSEQ	0.71	0.85	0.72	0.42	0.47	0.45	0.24	0.14
	NEWSEQ	0.65	0.79	0.65	0.36	0.39	0.38	0.29	0.19
XPOUND	ALLSEQ	0.61	0.87	0.69	0.15	0.18	0.17	0.33	0.13
	NEWSEQ	0.58	0.83	0.64	0.12	0.15	0.14	0.36	0.16
GENEID+	ALLSEQ	0.91	0.91	0.88	0.73	0.70	0.71	0.07	0.13
	NEWSEQ	0.88	0.87	0.84	0.68	0.64	0.66	0.10	0.15
GENEPARSER3	ALLSEQ	0.86	0.91	0.85	0.56	0.58	0.57	0.14	0.09
	NEWSEQ	0.83	0.89	0.82	0.50	0.53	0.51	0.17	0.09

This was done to minimize the overlap between this test set and the sets of sequences where the programs had been trained on. In addition, the accuracy was also measured in the subset (the NEWSEQ dataset) of 196 sequences that did not show sequence similarity to sequences deposited in GenBank prior to January 1993. This set of 'novel' sequences was used to investigate to what extent the programs were dependent on the particularities of the examples they learn from, or were able to generalize characteristic features of gene sequences.

Table 2 shows the results of the accuracy measures averaged over the set of sequences analyzed. Averages are given separately for all of the sequences in the test set, ALLSEQ, and for the NEWSEQ subset.

Average CC for the programs analyzed in the ALLSEQ data set ranged from 0.65 to 0.78 at the nucleotide level, while the average exon prediction accuracy ($(S_n+S_p)/2$) ranged from 0.37 to 0.64. Accuracy was lower and less variable between programs in the NEWSEQ dataset: average CC ranged here from 0.60 to 0.70, and average exon accuracy ranged from 0.36 to 0.54. The authors of the study also found that the programs showed some dependency - usually only slight - on the phylogenetic group within vertebrates in which they had been trained and that they tended to perform worse on low G+C content sequences, but this was not a general trend. Programs were also sensitive to high rates of sequence errors, but their robustness varied widely.

After the publication of this study, and during the second half of the 1990s, there were substantial developments in the field of computational gene identification. A new generation of programs - notably those based on Hidden Markov Models - showed substantially increased prediction accuracy and overcame some of the technical limitations of earlier programs: these programs were able to analyze larger sequences and to predict multiple genes in both strands. Accuracy of most such programs was still being evaluated in the test dataset described in Burset and Guigó (1996). While this allowed for effective comparison between programs, with the years this was also making it increasingly difficult to guarantee a clear separation between the sequence data sets in which the programs had been trained, and the sequence set in which they were being tested.

To address this concern Rogic *et al.* (2001) published recently a new independent comparative analysis of seven gene prediction programs. The programs were again tested in single gene sequences from human and rodent species. In order to avoid overlap with the training sets of the programs, only sequences were selected that had been entered in GenBank, after the programs were developed and trained. Their extraction process resulted in 195 genes, the HMR195 data set. Table 3 shows the accuracy measures averaged over the set of sequences analyzed for each of the tested programs. Obviously, the programs tested by Rogic *et al.* (2001) showed substantially higher accuracy than the programs tested by Burset and Guigó (1996). Now,

Table 3. Accuracy of gene prediction programs on single gene mammalian sequences. Adapted from Rogic *et al.* (2001)

	Nucleotide			Exon				
Program	S_n	S_p	CC	S_n	S_p	$\frac{Sn+Sp}{2}$	ME	WE
FGENES	0.86	0.88	0.83	0.67	0.67	0.67	0.12	0.09
GENEMARK.hmm	0.87	0.89	0.83	0.53	0.54	0.54	0.13	0.11
GENIE	0.91	0.90	0.88	0.71	0.70	0.71	0.19	0.11
GENSCAN	0.95	0.90	0.91	0.70	0.70	0.70	0.08	0.09
HMMGENE	0.93	0.93	0.91	0.76	0.77	0.76	0.12	0.07
MORGAN	0.75	0.74	0.69	0.46	0.41	0.43	0.20	0.28
MZEF	0.70	0.73	0.66	0.58	0.59	0.59	0.32	0.23

average *CC* at the nucleotide level ranged from 0.66 to 0.91, while average exon prediction accuracy ranged from 0.43 to 0.76. The authors also found that while earlier gene finders were tuned to have a low false-positive rate which resulted in, on average, 20% higher specificity than sensitivity, programs of the new generation have more equilibrated sensitivity vs. specificity. Still, as in the study by Burset and Guigó (1996), Rogic *et al.* (2001) found a large gap between accuracy at the nucleotide level and at the exon level. This indicates that, while the programs correctly predict a large proportion of the coding region of the genes, they have trouble identifying the exact exon boundaries. In this regard, the gene prediction problem - formulated as the problem of identifying the nucleic acid signals involved in gene specification, and delineating only from them the corresponding gene structure - is still far from being solved.

Other findings of Rogic *et al.* (2001) include dependency of some programs on G+C content, but little on phylogenetic specific training within mammals, and the general trend among the programs tested to have a very low proportion of correctly predicted short exons or very long exons. This, together with the fact that splice sites seem to be better predicted than start and stop codons, could explain the tendency, consistent among the programs, of better predicting internal exons than initial or terminal exons. This leads to rather poor predictions of the gene termini.

Burset and Guigó (1996) discussed that it may be beneficial to combine outputs of several gene-finding programs. They found that among the 2649 exons of the ALLSEQ dataset, 174 were predicted by all the programs. 172 of them corresponded exactly to actual exons annotated in GenBank, while the remaining two had a large overlap with actual exons. Only 33 exons (around 1%) were missed completely by all programs. This idea of combining several *ab initio* gene finders has been explored a number of times. Murakami and Takagi (1998), for instance, combined four programs. They achieved a 5% increase in accuracy at the nucleotide level in comparison with the best single gene finding program.

3.2. Accuracy in Large Genomic Sequences

While the paper by Rogic *et al.* (2001) represented a valuable update on the accuracy of gene finders, it suffered from the same limitation as did the previous work by Burset and Guigó (1996) and others: gene finders were tested in controlled data sets made of short genomic sequences encoding a single gene with a simple gene structure. These data sets are not representative of the genome sequences produced today: large sequences of low coding density, encoding several genes and/or incomplete genes, with complex gene structure. Thus, accuracy measures inferred from them are likely to be substantial overestimations of the actual accuracy of the programs in real genomic sequences. Indeed, the initial analysis on chromosome 22 suggests that the *ab initio* gene finders predict a large number of false positive genes, and while they tend to predict most of the known genes, only a small fraction of the genes have all exons correctly predicted. For instance, GENSCAN predicts 94% of the known genes in chromosome 22, but only about 20% of them had all the exons correctly predicted (Dunham *et al.*, 1999).

Ideally, one would like to benchmark the gene identification programs in real genomic sequences. The main problem is that until very recently on very few of these sequences the structure of the genes has been exhaustively verified by experimental means, making it almost impossible to construct large enough test sets of well characterized and large enough genomic sequences with which the accuracy of the gene predictions can be calibrated.

A number of authors have recently addressed this concern. Pavy *et al.* (1999) have evaluated a number of gene prediction programs in a set of multigenic sequences from *Arabidopsis thaliana*, and Reese *et al.* (2000) tested gene prediction programs in a large well characterized contig from the *Adh* region of the *Drosophila melanogaster* genome. Because of the lack of well characterized human sequences, and because genes or exons may remain undetected even if a detailed experimental analysis is performed, Guigó *et al.* (2000a) investigated the accuracy of *ab initio* gene finders by constructing a set of semi-artificial sequences to emulate large well annotated genomic sequences. In these semi-artificial sequences, known genic sequences have been embedded in simulated intergenic DNA. Therefore, the location of all coding exons is known.

3.2.1. Accuracy in *Arabidopsis thaliana*

Pavy *et al.* (1999) constructed a data set of contigs of validated genes, and measured prediction accuracy not only of gene models, but also of the gene defining sequence signals. Table 4 summarizes some of the results obtained by Pavy *et al.* (1999). Gene prediction in *A. thaliana* appears to be more accurate than in vertebrate organisms, with average exon accuracy ranging from 0.41 to 0.79. The maximum value is higher in this multigene sequence

Table 4. Accuracy of *ab initio* gene prediction programs in multigene *A. thaliana* sequences. Adapted from Pavy *et al.* (1999)

	Exon				Gene			
Program	S_n	S_p	$\frac{Sn+Sp}{2}$	WE	S_n	S_p	MG	WG
FEX	0.55	0.32	0.44	0.40				
FGENES	0.55	0.54	0.55	0.18				
FGENESP	0.42	0.59	0.51	0.15	0.06	0.11	0.30	0.02
GENEMARK.hmm	0.82	0.77	0.79	0.08	0.40	0.32	0.01	0.16
GENSCAN	0.63	0.70	0.67	0.09	0.17	0.19	0.01	0.08
GRAIL	0.44	0.38	0.41	0.19				
MZEF (prior p=0.01)	0.39	0.63	0.51	0.14				
MZEF (prior p=0.04)	0.45	0.54	0.49	0.18				
MZEF (prior p=0.10)	0.48	0.49	0.48	0.21				

dataset than that found by Rogic *et al.* (2001) in a single gene sequences human dataset. This is probably due to the simpler exonic structure of the *A. thaliana* genes and to the higher coding density of the genome of this plant. Another reason might be that introns are generally shorter and their length is less variable in *A. thaliana* than in human (Deutsch and Long, 1999). The work by Pavy *et al.* (1999) highlights the difficulties of gene finders of correctly predicting complete genes, and gene boundaries. Although accuracy at the exon level is quite high, accuracy at the gene level remains low. The most accurate program has an average accuracy at the gene level of only 0.36, that is only 40% of the genes in the data set are completely correctly predicted, and only 32% of the predicted genes are completely correct. In addition, 16% of the predicted genes do not overlap real genes. The authors suggest that these could correspond to novel, yet undetected, genes.

3.2.2. Accuracy in *D. melanogaster*: the GASP Experiment

Taking the already traditional CASP (Critical Assessment of techniques for protein Structure Prediction, Venclovas *et al.*, 2001) as a paradigm, Reese *et al.* (2000) organized the GASP experiment (Genome Annotation Assessment Project). In this experiment, a large well-characterized sequence contig of 2.9 Mb from the *Adh* region of *D. melanogaster* was provided to a number of research groups participating in the experiment. After the groups had submitted their predictions on this region, these predictions were then compared with the standard annotation under the supervision of a team of independent assessors.

Table 5. The GASP experiment. Accuracy of gene prediction programs on a large genomic contig from *D. melanogaster*. Adapted from Reese *et al.* (2000)

	Nucleotide		Exon					Gene						
Program	S_n	S_p	S_n	S_p	$\frac{Sn+Sp}{2}$	ME	WE	S_n	S_p	MG	WG	SG	JG
FGENES	0.89	0.77	0.65	0.49	0.57	0.10	0.32	0.30	0.27	0.09	0.24	1.10	1.06
GENEID	0.86	0.83	0.58	0.34	0.46	0.21	0.47	0.26	0.10	0.14	0.30	1.06	1.11
GENIE	0.96	0.92	0.70	0.57	0.63	0.08	0.17	0.40	0.29	0.05	0.11	1.17	1.08
GENIE est	0.97	0.91	0.77	0.55	0.66	0.05	0.20	0.44	0.28	0.05	0.13	1.15	1.09
GENIE esthom	0.97	0.83	0.79	0.52	0.65	0.03	0.23	0.44	0.26	0.05	0.15	1.16	1.09
GRAIL	0.81	0.86	0.42	0.41	0.41	0.24	0.29	0.14	0.12	0.16	0.24	1.23	1.08
HMMGENE	0.97	0.91	0.68	0.53	0.60	0.05	0.20	0.35	0.30	0.07	0.15	1.04	1.12
MAGPIE exon	0.96	0.63	0.63	0.41	0.52	0.12	0.50	0.33	0.21	0.05	0.55	1.22	1.06

Two standard annotations were used to evaluate the predictions: one based on previously unreleased high-quality full-length cDNA sequences and a second based on the set of annotations generated as part of an in-depth study of the region by a group of fly experts. The first standard is likely to be correct in the details of the genes that it describes, though only includes a small portion of the genes in the region. It has been thus used to calculate sensitivity measures. The second standard is as complete as possible, but it does not have rigorous evidence for all of its annotations, and it has been used to compute specificity measures.

Six groups submitted *ab initio* gene predictions. The results are summarized in Table 5. Accuracy of the programs in this large genomic sequence from *D. melanogaster* is comparable to the accuracy of the programs in the set of single gene sequences by Rogic *et al.* (2001), and slightly lower than the accuracy on the multigenic *A. thaliana* sequences with average exon accuracy ranging from 0.41 to 0.66. As with the work by Pavy *et al.* (1999), the GASP experiment highlighted the difficulties of predicting complete genes correctly. The average gene prediction accuracy at the gene level of the most accurate program is only 0.36, exactly as in *A. thaliana*.

3.2.3. Accuracy in Semi-artificial Genomic Sequences

To overcome the lack of large well annotated genomic sequences in the human genome - comparable to those existing in *A. thaliana* and *D. melanogaster* - Guigó *et al.* (2000a) constructed a set of semi-artificial genomic sequences. They proceeded in the following way. First, a typical benchmark set was extracted made of sequences from the EMBL database, release 50 (1997). These are 178 human genomic sequences coding for single complete genes for which both the mRNA and the coding exons are known. The procedure used to extract the sequences is described in Burset and Guigó (1996) and Guigó (1997b). We will refer to this set here as h178. The coding density of this set is 21%, slightly higher than the coding density in Rogic *et al.* (2001), 14%, and Burset and Guigó (1996), 15%. Second, to simulate large sequence contigs, the sequences in h178 were randomly placed in a background of intergenic DNA, randomly generated using a Markov model of order 5 as described in Guigó and Fickett (1995). The length of each artificial contig is generated randomly according to a normal distribution. Genomic fragments containing genes and random-sized segments of intergenic sequence are then concatenated until their combined lengths exceed the predetermined contig target length. The strands are also chosen at random for each genic subsequence. After this process, the 178 genic sequences collapsed into 42 semiartificial genomic sequences (SAG). We will refer to this set as Gen178. The coding density on this set was only 2.3%, very close to the coding density estimated for the whole human genome.

Table 6 shows the accuracies of GENSCAN - as the most accurate representative of *ab initio* gene finders - in the set of single gene sequences (h178) and in the set of large artificial contigs (Gen178). While the accuracy of GENSCAN in the single gene sequences is similar to that estimated by Rogic *et al.* (2001), it is substantially lower in the set of larger multiple gene SAG sequences: specificity at the nucleotide level drops to 0.64 from 0.92, and the proportion of wrong exons increases to 0.41 from 0.09 in the single gene sequences. In addition, a significant decrease in sensitivity is also observed, with GENSCAN failing to predict exons that were correctly identified in the single gene sequences. For instance, the proportion of missing exons increases from 0.09 to 0.14. Only about 30% of the GENSCAN genes

Table 6. Accuracy of gene prediction programs on semiartificial multigene human sequences. Adapted from Guigó *et al.* (2000a)

		Nucleotide			Exon					Gene	
Program	Dataset	S_n	S_p	CC	S_n	S_p	$\frac{Sn+Sp}{2}$	ME	WE	MG	WG
GENSCAN	Gen178	0.89	0.64	0.76	0.64	0.44	0.54	0.14	0.41	0.03	0.28
	h178	0.92	0.92	0.91	0.76	0.76	0.76	0.09	0.09		
BLASTX nr	h178	0.90	0.92	0.90	0.10	0.12	0.11	0.12	0.07		
GENEWISE	Gen178	0.98	0.98	0.97	0.88	0.91	0.89	0.06	0.02		
	h178	0.98	0.98	0.97	0.88	0.91	0.89	0.06	0.02		
PROCRUSTES	Gen178	0.93	0.94	0.93	0.80	0.75	0.77	0.10	0.16		
	h178	0.93	0.95	0.93	0.76	0.82	0.79	0.11	0.04		

are predicted in the simulated-intergenic DNA. Guigó *et al.* (2000a) concluded that for *ab initio* gene finders, these accuracy values (on SAG sequences) are more representative of their true accuracy on large genomic sequences than those obtained in the typical single gene benchmark experiments.

3.3. Accuracy in Chromosome 22

The recent availability of the complete sequence of the first human chromosomes allows one to obtain a more realistic estimation of the accuracy of the gene finders in human genome sequences. Of the three chromosomes (20, 21 and 22) completely sequenced so far, the annotation of chromosome 22 is likely to be the most complete. We have evaluated the accuracy of two *ab initio* gene finders, GENSCAN and GENEID, in the complete sequence of chromosome 22. The analysis has some circularity, since the annotation of chromosome 22 is already based on GENSCAN predictions. In that respect, the results here could overestimate the actual accuracy of gene finders. We used the sequence of chromosome 22 as in the April 2001 release of the UCSC assembly (http://genome.cse.ucsc.edu/). The GENSCAN predictions were taken from the same site. The GENEID predictions were obtained after masking the sequence with RepeatMasker (A. Smith and P. Green, unpublished). We used version 1.0 of GENEID (http://www1.imim.es/software/geneid.html). The accuracy of the predictions was measured with respect to the set of 554 annotated genes (see http://www.cs.columbia.edu/~vic/sanger2gbd/ (CDS set)). Results appear in Table 7. GENSCAN sensitivity in chromosome 22 is quite similar to that estimated by Guigó *et al.* (2000a) in the set of artificial genomic sequences, but specificity is somewhat lower. A reason could be the existence of un-annotated genes in the chromosome 22 sequence, and need not necessarily be due to poorer GENSCAN performance. This is consistent with recent updates of the number of genes in chromosome 22, which could be as high as 700 or more, instead of the 554 used here to evaluate the accuracy of the predictions. In any case, it seems unlikely that the overall exon accuracy of GENSCAN and GENEID on the chromosome 22 sequence will be much higher than 60%.

Table 7. Accuracy of gene prediction programs on the complete chromosome 22 sequence

	Nucleotide			Exon				
Program	S_n	S_p	CC	S_n	S_p	$\frac{Sn + Sp}{2}$	ME	WE
GENEID	0.80	0.55	0.64	0.60	0.54	0.57	0.21	0.30
GENSCAN	0.89	0.46	0.61	0.69	0.36	0.52	0.13	0.54

4. Accuracy of Sequence Similarity Based Gene Prediction Tools

In contrast to *ab initio* methods, similarity-based methods seek to extract the information in genomic sequence data by comparing the sequence with other sequences (known to be coding or not), and to look for segments that are conserved over evolutionary time. Essentially, there are two main classes of similarity-based approaches for gene identification: the comparison of the DNA query sequence with a protein or cDNA sequence, or a database of such sequences, and the comparison of two or more genomic sequences. In both approaches query and target sequences may, but need not, be from different species. In the latter approach, one may even compare a sequence with itself and search for self-similarity. This would be a reasonable strategy for those DNA regions that have experienced local duplication events in their evolutionary history. In particular, for many plant species gene duplication is an important evolutionary mechanism (Bancroft, 2001). The more typical case, however, is to compare sequences from two or more different species which are evolutionarily sufficiently close to each other, so that conserved fragments can be detected and distinguished from non-conserved fragments.

4.1. Genomic Query Against Protein or cDNA Target

The backbone of similarity-aided or similarity-based gene structure determination is constituted by those methods that rely on a comparison of the query sequence with protein, or cDNA sequences. Although mostly known as a database search program, BLASTX (Altschul *et al.*, 1990; Gish and States, 1993) illustrates the rationale behind this approach. With BLASTX a genomic query is translated into a set of amino acid sequences in the six possible frames and compared against a database of known protein sequences. The assumption is that those segments in the genomic query which are similar to database proteins are likely to correspond to coding exons. A similar assumption is behind the comparison of the genomic query against a database of cDNA sequences (such as ESTs), using BLASTN (Altschul *et al.*, 1990), FASTA (Pearson, 1999), or similar programs.

Such database search programs, however, are not dedicated gene prediction tools; they report exclusively matching sequences, but are not capable to automatically identify start and stop codons or splice sites. Thus, after a database search and the identification of potential targets, additional tools are required to define exonic structures.

One approach is to use the top database match as target sequence and obtain a so-called spliced alignment between this and the genomic query. In such an alignment, large gaps - likely to correspond to introns - are only allowed at legal splice junctions. A popular program (SIM4) to calculate such a spliced pairwise alignment has been developed by Florea *et al.* (1998). One of the first programs for the specific task to map ESTs onto a genomic sequence is EST_GENOME (Mott, 1997). When aligning a genomic query with ESTs, several high scoring database matches may map to the same genomic locus. Some programs, therefore, first try to assemble overlapping database matches into an artificial target and then use this to calculate a spliced alignment.

Splicing alignment algorithms with protein targets require a conceptual translation of the query sequence, computation of the alignment and post-processing, which includes the combinatorial problem of finding the best fit of a multi-exon structure to a related protein. Most such splicing alignment tools employ for this task dynamic programming techniques. Therefore, gene prediction in large scale sequences may become extremely time- and space-demanding. This investment, however, is usually at the benefit of prediction accuracy. PROCRUSTES (Gelfand *et al.*, 1996) and GENEWISE (Birney and Durbin, 1997) are powerful programs to predict genes based on a comparison of a genomic query with protein targets.

Guigó *et al.* (2000a) tested the accuracy of GENEWISE and PROCRUSTES in the set of single gene sequences h178, and in the set of semiartificial sequences Gen178 (Table 6). Accuracy values for these programs are obtained when the target amino acid sequence is the top scoring protein after a BLASTX search of the genomic sequence against the non-redundant protein sequence database. For comparison, Table 6 also contains the accuracy values for the raw BLASTX search in the set of single gene sequences. Obviously, GENEWISE and PROCRUSTES have very high accuracy. Unlike *ab initio* gene finders, they are not disturbed by the inclusion of real genes in the artificial intergenic context.

Usuka and Brendel (2000) (see also Usuka *et al.*, 2000) tested their splicing alignment program GENESEQER on a set of confirmed *A. thaliana* genes for which target proteins were available, and compared them to those obtained with GENSCAN. The results confirm that *ab initio* gene prediction can be substantially improved by the spliced alignment method (see Table 8). The authors, however, note that a spliced alignment algorithm has often difficulties in detecting, for instance, unusually large introns.

Table 8. Accuracy of GENSCAN, GENEGENERATOR (Kleffe *et al.*, 1998) and GENESEQER (Usuka and Brendel, 2000) in a set of 50 genes from *A. thaliana* (adapted from Usuka and Brendel (2000)). For GENESEQER the performance values are averaged over *Number* spliced alignments with target proteins of a similarity score *Sim*.

			Nucleotide			Exon			
Program	*Sim*	*Number*	S_n	S_p	*CC*	S_n	S_p	ME	WE
GENSCAN		50	0.90	0.96	0.89	0.74	0.81	0.10	0.02
GENEGENERATOR		50	0.91	0.95	0.89	0.64	0.57	0.05	0.17
GENESEQER	10-19%	14	0.60	0.93	0.54	0.06	0.04	0.19	0.17
GENESEQER	50-59%	84	0.88	0.97	0.88	0.52	0.53	0.05	0.05
GENESEQER	100%	119	1.00	1.00	1.00	0.98	0.98	0.00	0.00

Also, both sets of results agree in the finding that little sequence similarity between target protein and query sequence does not improve the prediction accuracy compared to that obtained with an *ab initio* method, in particular on exon level (see Tables 8 and 9 and also Figure 1 in Guigó *et al.*, 2000a). Unfortunately, such a situation is not infrequent in practice: a cognate target protein or EST, *i.e.* from the same or a very closely related sister species as the query, may not be available.

In a totally different approach, the results of a database search can be integrated in a more or less ad-hoc way into the framework of a typical *ab initio* gene prediction program. In essence, these methods promote candidate exons in the query sequence for which similar known coding sequences exist. Indeed, the score of the candidate exon - initially a function of the score of the splice (start, stop) sites and of the coding potential of the exon sequence - is increased as a function of the similarity between the candidate

Table 9. Accuracy of gene prediction tools on semi-artificial multigene human sequences, when either strongly or moderately similar sequences are used as target for GENEWISE and PROCRUSTES. Adapted from Guigó *et al.* (2000a).

	Very similar target p-value < 10^{-50} 17 SAG sequences						Moderately similar target $10^{-50} < p$-value < 10^{-6} 26 SAG sequences					
	Nucleotide			Exon			Nucleotide			Exon		
Program	S_n	S_p	*CC*	S_n	S_p	$\frac{Sn+Sp}{2}$	S_n	S_p	*CC*	S_n	S_p	$\frac{Sn+Sp}{2}$
GENSCAN	0.91	0.66	0.77	0.67	0.46	0.56	0.91	0.61	0.74	0.67	0.43	0.55
GENEWISE	0.99	0.99	0.99	0.90	0.93	0.91	0.68	0.98	0.81	0.46	0.63	0.54
PROCRUSTES	0.92	0.96	0.94	0.80	0.75	0.77	0.66	0.79	0.72	0.48	0.32	0.40

exon and the known coding sequences. In this way, candidate exons showing similarity to known coding sequences, are more likely to be included in the final gene prediction. In theory, this approach should produce predictions as accurate as pure *ab initio* programs when no similar target sequences exist, but more accurate ones (ideally, as accurate as those from splicing alignment tools) when such target sequences do exist.

The integration of database search results into the prediction scheme was proposed since the early 1990s. Table 2 shows the accuracy of two such programs, GENEID+ and GENEPARSER3. Both programs exhibit substantially increased accuracy compared to their pure *ab initio* counterparts GENEID (Guigó *et al.*, 1992) and GENEPARSER (Snyder and Stormo, 1993). Since then, a large number of *ab initio* gene prediction programs have been extended to integrate database search results: GENIE (Kulp *et al.*, 1997), GRAILEXP (Xu and Uberbacher, 1997), AAT (Huang *et al.*, 1997), EbEST (Jiang and Jacob, 1998), GENEBUILDER (Milanesi *et al.*, 1999), GIN (Cai and Bork, 1998), HMMGENE (Krogh, 2000), EUGENE (Schiex *et al.*, 2001), and GENOMESCAN (Yeh *et al.*, 2001) are only some of the published methods. All these methods report increased accuracy when incorporating database sequence searches. GENOMESCAN, an extension of GENSCAN, for instance, was tested in the h178 set of single gene sequences. Results in this set (see Figure 2 in Yeh *et al.*, 2001) show a steady increase in accuracy as similarity to the target protein increases. When only a very weakly similar protein is available, GENOMESCAN has only a slight advantage over GENSCAN. It predicts about 80% of the annotated exons in this set correctly, with similar levels of specificity. The improvement with respect to GENSCAN increases with similarity, and more than 90% of exons are exactly predicted (with comparable specificity) when a very similar protein is available, rivaling GENEWISE accuracy.

Despite these encouraging results, in the practice of large scale sequence analysis, pure *ab initio* gene prediction programs appear to be preferred over their extended versions which incorporate sequence similarity searches. Difficulty of usage and substantially longer execution time may be part of the reasons. Difficulty in training and, given the incompleteness of reference databases, in extrapolating the results to sequence sets other than those in which the programs have been trained, may be another reason. Also, the temporal dynamics of databases - their exponential growth and also the accumulation of annotation errors - may present a problem for sequence similarity based gene prediction methods. For instance, a simultaneous search in December 2001 of the nucleotide section of the GenBank and EMBL databases with the query word 'GBF2' returned 5 entries in GenBank and 7 entries in EMBL. Repeating the same search in March 2002 resulted in 7 entries in GenBank, and 8 entries in EMBL (J. Kleffe, pers. comm). Among them, there are two genomic contigs (Acc No. AL161491 and AF007269) of the same chromosomal region in *A. thaliana*, but with a highly contradictory annotation. Even worse, the more recent annotations, often done automatically, need not necessarily be more reliable. For a discussion and analysis of the problems associated with sequence similarity-based gene prediction, see Guigó *et al.* (2000b).

4.2. Genomic Query Against Genomic Target

As noted above, genome-genome comparison of sufficiently closely related organisms was recently established as a new approach for gene identification. Unfortunately, it is not possible to make a generally valid statement about what 'sufficiently close' means. It depends on various aspects: whether the researcher is interested in protein-coding genes or in conserved non-coding DNA, whether evolution in the particular region of interest proceeded slowly or quickly, whether the species studied diverged slowly or quickly, on the sensitivity of the employed alignment algorithm and so on. Despite this problem, recent studies (Crollius *et al.*, 2000; Wiehe *et al.*, 2000; Rinner and Morgenstern, 2002) suggest that similarity-based gene prediction can be very reliable over a fairly wide spectrum of species and evolutionary divergence times, ranging from a few million to several hundred million years. In any case, the underlying rationale is that a random mutation in a functional, not necessarily protein-coding, region is usually deleterious to the organism, and hence unlikely to become fixed in the species, whereas mutations in non-functional regions are not, or at least much less, surveyed by natural selection. Thus, they are more likely to be fixed and thus generate the sequence divergence between species that we observe today. This fact is exploited in order to localize genes, determine gene structures and regulatory regions and to infer gene function.

A typical partner species for gene prediction in human sequences is *Mus musculus*, one of the most important mammalian model organisms. Its divergence time from human is about 80 million years ago. However, conservation of functionally important sequences can also be detected for much more distantly related taxa (see Table 10 and Figure 2), such as human and fugu (Brenner *et al.*, 1993), human and drosophila (Kornberg and Krasnow, 2000; Rubin *et al.*, 2000), or even human and yeast (Gavin *et al.*, 2002).

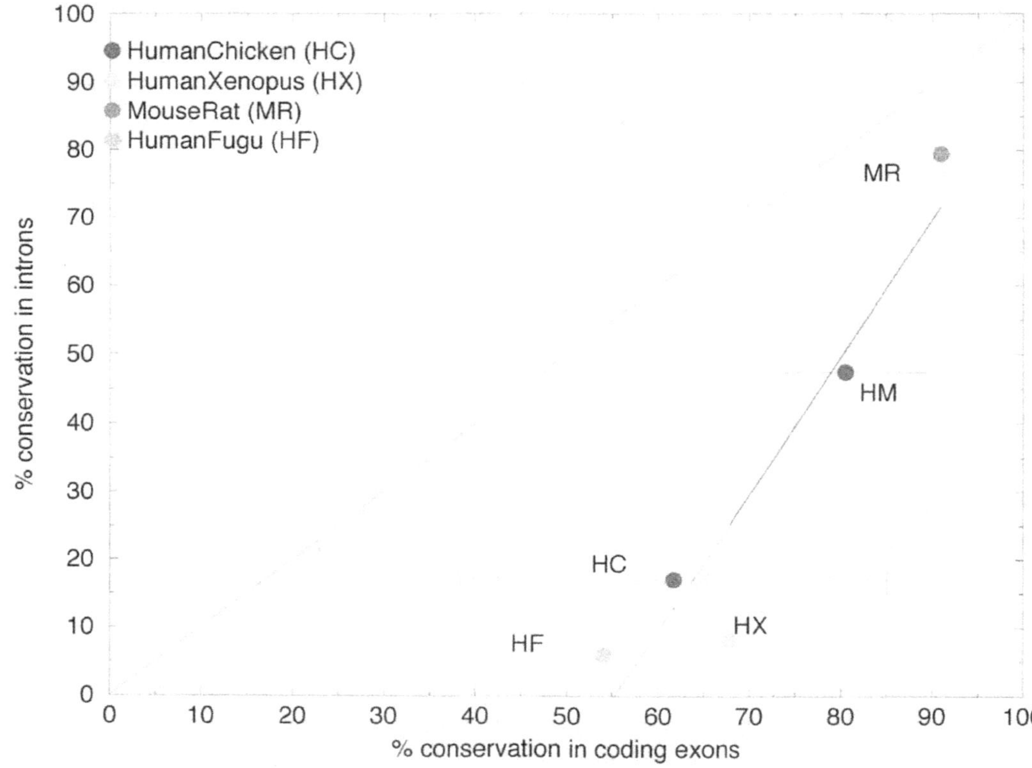

Figure 2. Conservation in coding exons versus introns. The reported values are averages obtained from pairwise comparisons of a set of orthologous genes (kindly provided by L. Duret). Pairs of some vertebrate species (see the inlet) at varying evolutionary distances have been compared. ***Colour plate at back of book.***

In plant comparative genomics promising pairs to compare are maize and rice or species in the genera *Arabis* and *Brassica*. For instance, it has been shown that *Arabidopsis thaliana* and other, closely related Brassicaceae share gene structures and regulatory units that can be detected based on their sequence similarity (Paterson *et al.*, 2001; Koch *et al.*, 2001).

Generally, the potential to infer complete gene structures from their similarity with genes in related species decreases with divergence time, leading to an increased number of false-negative predictions (*i.e.* missing genes; see Table 10). On the other hand, recent divergence times, usually correlated with high sequence similarity, may increase the false-positive rate (see the results for mouse/rat comparisons in Table 10). However, the rate of evolution varies among clades and species. Therefore, a divergence time that is uniformly best suited for comparative gene prediction cannot be given in absolute time. For instance, while human-avian comparisons (about 300 Myr divergence)

Table 10. Nucleotide level accuracy of TBLASTX applied to orthologous sequences from species with varying degrees of divergence

	H. sapiens					*R. norvegicus*				
	Myr	matrix	S_n	S_p	*CC*	Myr	matrix	S_n	S_p	*CC*
M. musculus	80	blosum85	0.95	0.68	0.77	10	blosum90	0.99	0.53	0.66
G. gallus	300	blosum75	0.85	0.87	0.84					
F. rubripes	450	blosum70	0.81	0.88	0.84					
D.melanogaster	670	blosum62	0.48	0.96	0.68					

may work well, Brendel *et al.* (2002) raise concerns about the feasibility of comparative gene prediction using *A. thaliana* and *O. sativa*, although the divergence time of monocots and eudicots is estimated to be 150 to 200 Myr (Bennetzen, 2002). Relatively recently diverged plants, such as *Arabidopsis thaliana* and *Brassica oleracea* (divergence time less than 20 Myr, Paterson *et al.*, 2001), have been shown to lead to good results in plant comparative gene prediction.

For the prediction of gene regulatory elements, which are often short motifs and functionally little constrained on sequence level, other pairs or sets of species may be suited than for the prediction of coding genes. Hardison (2000) advocates the comparison of very closely related species. Dubchak *et al.* (2000) could show that a three-way comparison of human, mouse and dog sequences can help to pinpoint putative regulatory elements in non-coding DNA.

Unfortunately, varying rates of evolution among species is not the only problem. The rates of evolution may be different also among genes. Therefore prediction accuracy can vary a great deal even within a given species pair, depending on the chromosomal regions compared. For instance, in human/mouse comparisons and in highly conserved regions, such as the HOX cluster, the false-positive rate may be considerably higher than in less conserved regions, such as the ERCC locus (see Table 2 in Wiehe *et al.* (2000)). Rate variation is due to the nature of the evolutionary process with natural selection as only one of the components to cause deviation from a neutral molecular clock.

Large genomic sequences are produced for a growing number of species. Among the most important vertebrates to aid annotation of the human genome are *Mus musculus*, *Rattus norvegicus*, and *Fugu rubripes*. Their genome projects (http://fugu.hgmp.mrc.ac.uk/, http://mrcseq.har.mrc.ac.uk/, http://

www.celera.com) are about to be completed. Also, *Gallus gallus* genomic sequences may play an increasingly important role (http://poultry.mph.msu.edu/).

The program ExoFish (Crollius *et al.*, 2000) predicts human exons based on a comparison with a database of random sequences from *Tetraodon nigroviridis*, another puffer fish species. Bafna and Huson (2000) and Batzoglou *et al.* (2000b) have developed programs (CEM and ROSETTA, respectively) for gene prediction based on a pairwise comparison of human and mouse homologous sequences. These programs have been tested on benchmark data sets of pairs of homologous genes in human and mouse (see Table 11). The authors of these programs showed that similarity based gene prediction with homologous genomic sequences from human and mouse significantly increases the fraction of exactly identified coding exons. However, alignment calculation of large genomic sequences is often the computational bottleneck. The program by Batzoglou *et al.* also includes a newly designed alignment algorithm (GLASS) (Batzoglou *et al.*, 2000a), which permits the user to analyze BAC-size sequences in reasonable time.

A problem with some gene prediction tools is species-specific training. In this case, they cannot easily be applied to arbitrary species. This is a problem not only for *ab initio* programs, but to some extent also for similarity-based ones. The reason is that programs often employ coding usage tables and other species specific sequence properties to enhance their prediction accuracy. It has been pointed out (Usuka and Brendel, 2000; Wiehe *et al.*, 2001) that a scoring system for exon prediction which relies on splice site quality together with sequence similarity can help to circumvent the problem of species-specificity of gene prediction programs. For instance, SGP-1 (Wiehe *et al.*, 2001) explicitly avoids species-specific pre-processing of the query sequences, such as repeat-masking, as well as recurring on codon usage. Among related organisms, for instance among vertebrates or angiosperms, splice site profiles, *i.e.* the sequence composition around splice sites, are less variable than is codon usage and therefore similarity-based gene prediction can be extended to a fairly wide range of related species without the need of specific training. SGP-1 was shown to produce good results in test sets of homologous genes in plants as well as in vertebrates (Table 11).

Other recently developed similarity-based gene prediction programs working with homologous genomic sequences are SLAM (Pachter *et al.*, 2001) (http://baboon.math.berkeley.edu/~syntenic/slam.html), and TWINSCAN (Korf *et al.*, 2001) (http://genes.cs.wustl.edu/). These programs are currently used to

Table 11. Accuracy for similarity based gene prediction programs (adapted from Wiehe *et al.*, 2001).

	nucleotide			exon			
	S_n	S_p	CC	S_n	S_p	ME	WE
Human/Rodent orthologous sequences with single genes[1]:							
SGP-1	0.94	0.96	0.94	0.70	0.76	0.12	0.04
ROSETTA	0.95	0.97	0.96	0.84	0.84	0.04	0.05
GENSCAN	0.97	0.89	0.92	0.79	0.74	0.06	0.13
Human/Mouse orthologous sequences with multiple genes[2]:							
SGP-1	0.89	0.94	0.91	0.72	0.75	0.15	0.11
GENSCAN	0.96	0.74	0.83	0.75	0.58	0.08	0.29
A.thaliana/Brassicacea orthologous sequences with single genes[2]:							
SGP-1	0.93	0.99	0.94	0.57	0.63	0.12	0.00
GENSCAN	0.92	0.90	0.87	0.61	0.57	0.09	0.15

(1) The test set was kindly provided by S. Batzoglou; see also Batzoglou *et al.* (2000b)
(2) The data sets are described in Wiehe *et al.* (2001)

cross-annotate human and mouse genomes. SGP-2 (Guigó *et al.*, unpublished) is also being used with this purpose, with genome-wide predictions available at http://www1.imim.es/genepredictions/index.html. SLAM is an application of generalized pair Hidden Markov Models (HMM), while TWINSCAN is also an extension of GENSCAN (Burge and Karlin, 1997) and, as its predecessor, incorporates a pseudo-HMM algorithm to generate gene models. As the authors note, TWINSCAN has in the past been applied to human/ mouse gene prediction but it should be applicable to a range of other species as well. Korf *et al.* (2001) measured the accuracy of TWINSCAN on two data sets of human/mouse orthologs. They found a considerable improvement of TWINSCAN results compared to GENSCAN on exon level, and, most notably, on gene level. Here, accuracy has improved by more than 60% compared to the GENSCAN results (see their Figure 4 in Korf *et al.* (2001)).

Although similarity-based gene prediction with homologous genomic sequences may produce high quality results (Miller, 2001), an obvious shortcoming is the need for two homologous sequences. Also, genes without a homolog in the partner sequence will escape detection. This is particularly problematic if species are compared where gene order or synteny is not preserved. Given only a single query sequence, it is therefore desirable to automatically search for homologous sequences or syntenic chromosome stretches in other species that are suited to be applied to similarity-based programs. This is a realistic strategy, given that large parts of the nuclear genome of nine eukaryotic species are already publicly available. The TWINSCAN web server already allows the user to specify only a single query sequence and let the server do the search for a suitable homologous partner within the human or mouse genome. Another advantage of this approach is that several partners of the query sequence, possibly from different species, maybe used for gene prediction in a single run. This will be of particular interest for studies on gene structure evolution.

Fostered by commercial genome sequencing, we are today in a situation where for the next future primarily sequence drafts-instead of assembled genomes-will be available and where analysis programs need to cope with low sequence quality. Both, SGP-2 and TWINSCAN are able to annotate the sequence of a genome when the sequence of the target genome is only in shotgun form.

5. Discussion and Outlook

The survey here illustrates the difficulties of converting genomic data into valuable biological knowledge. Even the first and one of the simplest, but essential steps, namely the identification of the genes encoded in a genome, is still difficult and not satisfactorily solved. Today, a truly reliable "prediction" of the exonic structure of a gene can only be obtained if a full length cDNA of this gene has been experimentally established. Although a number of public and private projects exist to create a comprehensive database of human (and other species) full length cDNAs (see, for instance, http://www.nedo.go.jp/bio-e, http://mips2.gsf.de/proj/cDNA, http://www.resgen.com/full_length, Kawai *et al.* (2001)), it is still unclear how correct and complete these collections are. Because low-level background transcription of almost the whole genome has been recently reported (Wong *et al.*, 2001), it is possible that a fraction of so-called full-length cDNAs never end up in translated protein products. Because of technical difficulties, it is often almost impossible to guarantee that the 5' end of a gene has been correctly sequenced. Furthermore, because a number of genes, probably even a substantial fraction of all genes, may be expressed only during particular stages of the developmental process, in particular and very specialized tissues or cells, or in response to particular environmental conditions, it is currently impossible to assess how complete these collections of full-length cDNAs are. Nevertheless, even with incomplete collections, it is clear (Haas *et al.*, 2002) that full length cDNAs are valuable resources to improve gene anntations.

Only through a combination of experimental and computational approaches can the complete catalogue of human genes be fully elucidated. Today, two problems stand up as particularly relevant toward this end: the characterization of the splicing pattern, or alternative patterns, of each gene and the correct determination of the gene boundaries. The latter is linked to a correct characterization of the gene promoter regions. In both cases, comparative methods can substantially contribute. Indeed, assuming that the splicing pattern of homologous genes is at least partially conserved between sufficiently close species, methods could be developed that exploit the conservation of homologous splice signals to reconstruct such a pattern.

Despite recent progress (Davuluri *et al.*, 2001), the reliable prediction of the transcript boundaries remains to be a hard problem. Again, based on a comparative approach, and using genomically aligned ESTs, Kan *et al.* (2001) recently published a method (termed TAP) that promises to reduce the false-positive predictions. A comparative approach has also been introduced to the analysis of regulatory regions (Ohler and Niemann, 2001). However, regulatory regions can tolerate much more sequence divergence and still retain the original function than can coding sequences. That is one source of the notorious difficulties encountered by promoter prediction programs, and it is also a major hurdle for a comparison of promoter sequences. Because regulatory regions function in part via structural properties of their constituent DNA, one must consider the similarity of structural parameters, *i.e.* physico-chemical properties, as well as sequence similarity. It is hoped that a better understanding of the role of structural properties of DNA will also lead to more accurate gene prediction results in the near future.

Acknowledgements

This work was supported by a grant (Förderkennzeichen 0312705A) from the German Ministry for Education and Research to TW, and by a grant (BIO98-0443-C02-01) from the Spanish Ministry of Education and Science to RG. We would like to thank G. Parra for the data on chromosome 22, and J. Kleffe for comments on the manuscript.

References

Altschul, S.F., Gish, W., Miller, W., Myers, E.W., and Lipman, D.J. 1990. Basic local alignment search tool. J. Mol. Biol. 215: 403-410.

Antequera, F. and Bird, A. 1993. Number of CpG islands and genes in human and mouse. Proc. Natl. Acad. Sci. USA. 90: 11995-11999.

Bafna, V. and Huson, D.H. 2000. The conserved exon method. Proceedings of the eighth international conference on Intelligent Systems in Molecular Biology (ISMB), pages 3-12.

Bajic, V. 2000. Comparing the success of different prediction software in sequence analysis: A review. Briefings in Bioinformatics. 1: 214-228.

Baldi, P., Brunak, S., Chauvin, Y., Andersen, C., and Nielsen, H. 2000. Assessing the accuracy of predicition algorithms for classification: an overview. Bioinformatics. 16: 412-424.

Bancroft, I. 2001. Duplicate and diverge: the evolution of plant genome microstructure. Trends Genet. 17: 89-93.

Batzoglou, S., Pachter, L., Mesirov, J., Berger, B., and Lander, E. 2000a. Human and mouse gene structure: Comparative analysis and application to exon prediction. In: Press, A., editor, Proceedings of the fourth annual conference on Computational Molecular Biology (RECOMB 2000), pages 46-53, New York.

Batzoglou, S., Pachter, L., Mesirov, J.P., Berger, B., and Lander, E.S. 2000b. Human and mouse gene structure: Comparative analysis and application to exon prediction. Genome Res. 10: 950-958.

Bennetzen, J. 2002. Opening the door to comparative plant biology. Science. 296: 60-63.

Birney, E. and Durbin, R. 1997. Dynamite: a flexible code generating language for dynamic programming methods used in sequence comparison. Proceedings of the fifth international conference on Intelligent Systems in Molecular Biology (ISMB), pages 56-64.

Brendel, V., Kurtz, S., and Walbot, V. 2002. Comparative genomics of arabidopsis and maize: prospects and limitations. Genome Biol. 3: Reviews1005.1-1005.6.

Brenner, S., Elgar, G., Sandford, R., Macrae, A., Venkatesh, B., and Aparicio, S. 1993. Characterization of the pufferfish (Fugu) genome as a compact model vertebrate genome. Nature. 366: 265-268.

Burge, C.B., and Karlin, S. 1997. Prediction of complete gene structures in human genomic DNA. J. Mol. Biol. 268: 78-94.

Burge, C.B., and Karlin, S. 1998. Finding the genes in genomic DNA. Curr. Opin. Struct. Biol. 8: 346-354.

Burset, M., and Guigó, R. 1996. Evaluation of gene structure prediction programs. Genomics. 34: 353-357.

Cai, Y., and Bork, P. 1998. Homology-based gene prediction using neural nets. Anal. Biochem. 265: 269-274.

Claverie, J.-M. 1997. Computational methods for the identification of genes in vertebrate genomic sequences. Hum. Mol. Genet. 6: 1735-1744.

Crollius, H.R., Jaillon, O., Bernot, A., Dasilva, C., Bouneau, L., Fischer, C., Fizames, C., Wincker, P., Brottier, P., Quetier, F., Saurin, W., and Weissenbach, J. 2000. Estimate of human gene number provided by genome-wide analysis using Tetraodon nigroviridis DNA sequence. Nat. Genet. 25: 235-238.

Das, M., Burge, C.B., Park, E., Colinas, J., and Pelletier, J. 2001. Assessment of the total number of human transcription units. Genomics. 77: 71-78.

Davuluri, R.V., Grosse, I., and Zhang, M.Q. 2001. Computational identification of promoters and first exons in the human genome. Nat. Genet. 29: 412-417.

Deloukas, P. *et al.* 2001. The DNA sequence and comparative analysis of human chromosome 20. Nature. 414: 865-871.

Deutsch, M. and Long, M. 1999. Intron-exon structures of eukaryotic model organisms. Nucleic Acids Res. 27: 3219-3228.

Dubchak, I., Brudno, M., Loots, G.G., Pachter, L., Mayor, C., Rubin, E.M., and Frazer, K.A. 2000. Active conservation of noncoding sequences revealed by three-way species comparisons. Genome Res. 10: 1304-1306.

Dunham, I. *et al.* 1999. The DNA sequence of human chromosome 22. Nature. 402: 489-495.

Ewing, B., and Green, P. 2000. Analysis of expressed sequence tags indicates 35,000 genes. Nat. Genet. 25: 232-234.

Fickett, J.W. 1982. Recognition of protein coding regions in DNA sequences. Nucleic Acids Res. 10: 5303-5318.

Fickett, J.W. 1996. Finding genes by computer: The state of the art. Trends Genet. 12: 316-320.

Fickett, J.W. and Tung, C.-S. 1992. Assessment of protein coding measures. Nucleic Acids Res. 20: 6441-6450.

Fields, C., Adams, M., White, O., and Venter, J. 1994. How many genes in the human genome? Nat. Genet. 7: 345-346.

Fields, C.A., and Soderlund, C.A. 1990. GM: a practical tool for automating DNA sequence analysis. Comp. Appl. Biosci. 6: 263-270.

Florea, L., Hartzell, G., Zhang, Z., Rubin, G.M., and Miller, W. 1998. A computer program for aligning a cDNA sequence with a genomic DNA sequence. Genome Res. 8: 967-974.

Galgoczy, P., Rosenthal, A., and Platzer, M. 2001. Human-mouse comparative sequence analysis of the NEMO gene reveals an alternative promoter within the neighboring G6PD gene. Gene. 271: 93-98.

Gavin, A. *et al.* 2002. Functional organization of the yeast proteome by systematic analysis of protein complexes. Nature. 415: 141-147.

Gelfand, M.S., Mironov, A.A., and Pevzner, P.A. 1996. Gene recognition via spliced alignment. Proc. Natl. Acad. Sci. USA. 93: 9061-9066.

Gish, W., and States, D. 1993. Identification of protein coding regions by database similarity search. Nat. Genet. 3: 266-272.

Guigó, R. 1997a. Computational gene identification. J. Mol. Med. 75: 389-393.

Guigó, R. 1997b. Computational gene identification: An open problem. Comp. Chem. 21: 215-222.

Guigó, R., Agarwal, P., Abril, J.F., Burset, M., and Fickett, J.W. 2000a. Gene prediction accuracy in large DNA sequences. Genome Res. 10: 1631-1642.

Guigó, R., Burset, M., Agarwal, P., Abril, J.F., Smith, R.F., and Fickett, J.W. 2000b. Sequence similarity based gene prediction. In: Suhai, S., editor, Genomics and Proteomics: Functional and Computational Aspects, pages 95-105. Kluwer Academic / Plenum Publishing.

Guigó, R. and Fickett, J.W. 1995. Distinctive sequence features in protein coding, genic non-coding, and inter-genic human DNA. J. Mol. Biol. 253: 51-60.

Guigó, R., Knudsen, S., Drake, N., and Smith, T.F. 1992. Prediction of gene structure. J. Mol. Biol. 226: 141-157.

Haas, B., Volfovsky, N., Town, D., Troukhan, M., Alexandrov, N., Feldmann, K., Flavell, R., White, O., and Salzberg, S. 2002. Full-length messenger RNA sequences greatly improve genome annotation. Genome Biol. 3: 29.1-29.12.

Hardison, R.C. 2000. Conserved noncoding sequences are reliable guides to regulatory elements. Trends Genet. 16: 369-372.

Haussler, D. 1998. Computational gene finding. Trends Biochem. Sci. Supplementary Guide to Bioinformatics, pages 12-15.

Hogenesch, J.B., Ching, K.A., Batalov, S., Su, A.I., Walker, J.R., Zhou, Y., Kay, S.A., Schultz, P.G., and Cooke, M.P. 2001. A comparison of the Celera and Ensembl predicted gene sets reveals little overlap in novel genes. Cell. 106: 413-415.

Huang, X., Adams, M.D., Zhou, H., and Kerlavage, A.R. 1997. A tool for analyzing and annotating genomic sequences. Genomics. 46: 37-45.

Jiang, J., and Jacob, H.J. 1998. EbEST: An automatic tool using expressed sequence tags to delineate gene structure. Genome Res. 5: 681-702.

Kan, Z., Rouchka, E.C., Gish, W., and States, D. 2001. Gene structure prediction and alternative splicing analysis using genomically aligned ESTs. Genome Res. 11: 889-900.

Kawai, J. *et al.* 2001. Functional annotation of a full-length mouse cDNA collection. Nature. 409: 685-690.

Kleffe, J., Hermann, K., Vahrson, W., Wittig, B., and Brendel, V. 1998. GeneGenerator: A flexible algorithm for gene prediction and its application to maize sequences. Bioinformatics. 14: 232-243.

Koch, M.A., Weisshaar, B., Kroymann, J., Haubold, B., and Mitchell-Olds, T. 2001. Comparative genomics and regulatory evolution: conservation and function of the Chs and Apetala3 promoters. Mol. Biol. Evol. 18: 1882-1891.

Korf, I., Flicek, P., Duan, D., and Brent, M.R. 2001. Integrating genomic homology into gene structure prediction. Bioinformatics. 17: S140-S148.

Kornberg, T.B. and Krasnow, M.A. 2000. The drosophila genome sequence: implications for biology and medicine. Science. 287: 2218-2220.

Krogh, A. 2000. Using database matches with HMMGene for automated gene detection in drosophila. Genome Res. 10: 523-528.

Kulp, D., Haussler, D., Reese, M.G., and Eeckman, F.H. 1997. Integrating database homology in a probabilistic gene structure model. In: Altman, R.B., Dunke, A.K., Hunter, L., and Klein, T.E., editors, Biocomputing: Proceedings of the 1997 Pacific Symposium, pages 232-244. World Scientific, New York.

Lander, E.S. et al. 2001. Initial sequencing and analysis of the human genome. Nature. 409: 860-921.

Liang, F., Holt, I., Pertea, G., Karamycheva, S., Salzberg, S., and Quackenbush, J. 2000. Gene index analysis of the human genome estimates approximately 120,000 genes. Nat. Genet. 25: 239-240.

Milanesi, L., D'Angelo, D., and Rogozin, I.B. 1999. GeneBuilder: interactive *in silico* prediction of gene structure. Bioinformatics. 15: 612-621.

Miller, W. 2001. Comparison of genomic DNA sequences: solved and unsolved problems. Bioinformatics. 17: 391-397.

Mott, R. 1997. EST_GENOME: a program to align spliced DNA sequences to unspliced genomic DNA. Comp. Appl. Biosci. 13: 477-478.

Murakami, K. and Takagi, T. 1998. Gene recognition by combination of several gene-finding programs. Bioinformatics. 14: 665-675.

Novichkov, P., Gelfand, M.S., and Mironov, A. 2001. Gene recognition in eukaryotic DNA by comparison of genomic sequences. Bioinformatics. 17: 1011-1018.

Ohler, U. and Niemann, H. 2001. Identification and analysis of eukaryotic promoters: recent computational approaches. Trends Genet. 17: 56-60.

Onyango, P., Miller, W., Lehoczky, J., Leung, C.T., Birren, B., Wheelan, S., Dewar, K., and Feinberg, A.P. 2000. Sequence and comparative analysis of the mouse 1-megabase region orthologous to the human 11p15 imprinted domain. Genome Res. 10: 1697-1710.

Pachter, L., Alexandersson, M., and Cawley, S. 2001. Applications of generalized pair Hidden Markov Models to alignment and gene finding problems. In: Press, A., editor, Proceedings of the fifth annual conference on Computational Molecular Biology (RECOMB 2001), pages 241-248, New York.

Paterson, A.H., Lan, T., Amasino, R., Osborn, T.C., and Quiros, C. 2001. Brassica genomics: a complement to, and early beneficiary of, the arabidopsis sequence. Genome Biol. 2: reviews1011.1-1011.4.

Pavy, N., Rombauts, S., Déhais, P., Mathé, C., Ramana, D., Leroy, P., and Rouzé, P. 1999. Evaluation of gene prediction software using a genome data set. Bioinformatics. 15: 887-899.

Pearson, W.R. 1999. Flexible similarity searching with the Fasta3 program package. In Misener, S. and Krawetz, S.A., editors, Bioinformatics Methods and Protocols, pages 185-219, Totowa, New Jersey. Humana Press.

Pruitt, K. and Maglott, D. 2001. Refseq and LocusLink: NCBI gene-centered resources. Nucleic Acids Res. 29: 137-140.

Reese, M.G., Hartzell, G., Harris, N.L., Ohler, U., Abril, J.F., and Lewis, S.E. 2000. Genome annotation assessment in Drosophila melanogaster. Genome Res. 10: 483-501.

Reichwald, K., Thiesen, J., Wiehe, T., Weitzel, J., Strätling, W.H., Kioschis, P., Poustka, A., Rosenthal, A., and Platzer, M. 2000. Comparative sequence analysis of the MECP2-locus in man and mouse reveals novel transcribed regions. Mammalian Genome. 11: 182-190.

Rinner, O. and Morgenstern, B. 2002. Agenda: Gene prediction by comparative sequence analysis. In silico Biology, 2: 0018.

Rogic, S., Mackworth, A.K., and Ouellette, F. 2001. Evaluation of gene-finding programs on mammalian sequences. Genome Res. 11: 817-832.

Rubin, G.M. *et al.* 2000. Comparative genomics of the eukaryotes. Science. 287: 2204-2215.

Schiex, T., Moisan, A., and Rouzé, P. 2001. Eugene: an eukaryotic gene finder that combines several sources of evidence. In: Gascuel, O. and Sagot, M.-F., editors, Lecture Notes in Computer Science, volume 2066 (Computational Biology).

Shoemaker, D.D. *et al.* 2001. Experimental annotation of the human genome using microarray technology. Nature. 409: 922-927.

Snyder, E.E. and Stormo, G.D. 1993. Identification of coding regions in genomic DNA sequences: an application of dynamic programming and neural networks. Nucleic Acids Res. 21: 607-613.

Usuka, J. and Brendel, V. 2000. Gene structure prediction by spliced alignment of genomic DNA with protein sequences: increased accuracy by differential splice site scoring. J. Mol. Biol. 297: 1075-1085.

Usuka, J., Zhu, W., and Brendel, V. 2000. Optimal spliced alignment of homologous cDNA to a genomic DNA template. Bioinformatics. 16: 203-211.

Venclovas, C., Zemla, A., Fidelis, K., and Moult, J. 2001. Comparison of performance in successive CASP experiments. Proteins: Structure, Function, and Genetics, Suppl. 5: 163-170.

Venter, J.C. *et al.* 2001. The Sequence of the Human Genome. Science. 291: 1304-1351.

Wiehe, T., Gebauer-Jung, S., Mitchell-Olds, T., and Guigó, R. 2001. SGP-1: Prediction and validation of homologous genes based on sequence alignments. Genome Res. 11: 1574-1583.

Wiehe, T., Guigó, R., and Miller, W. 2000. Genome sequence comparisons: Hurdles in the fast lane to functional genomics. Briefings in Bioinformatics. 1: 381-388.

Wong, G. K.-S., Passey, D.A., and Yu, J. 2001. Most of the human genome is transcribed. Genome Res. 11: 1975-1977.

Wright, F.A., Lemon, W.J., Zhao, W.D., Sears, R., Zhuo, D., Wang, J.-P., Yang, H.-Y., Baer, T., Stredney, D., Spitzner, J., Stutz, A., Krahe, R., and Yuan, B. 2001. A draft annotation and overview of the human genome. Genome Biol. 2: 1-18.

Xu, Y., and Uberbacher, E.C. 1997. Automated gene identification in large-scale genomic sequences. J. Comput. Biol. 4: 325-338.

Yeh, R., Lim, L., and Burge, C. 2001. Computational inference of homologous gene structures in the human genome. Genome Res. 11: 803-816.

From: *Frontiers in Computational Genomics*

2

Sensitive Protein Alignment Algorithms

Adam Godzik

Abstract

The recent advances in sensitive protein alignment algorithms come mostly from exploring additional information available either from the analysis of substitution patterns in homologous proteins (profile or PSSM methods) or from the requirements of the stability of the final protein structure (threading). Many new methods bring in new types of information beyond structure and sequence variation among homologs and thus hold significant promise for further advancing the sensitivity of discovering distant homologs. Methods such as ISS (intermediate sequence search) or other neighborhood methods explore features of protein sequence space searching for "bridging proteins" to connect divergent families by a series of comparisons. Context analysis implemented in systems use position in a genome and function of genome neighbors to identify function and indirectly recognizes homologies. Other methods use keyword and literature searches to build a "literature profile" of a protein, discovering distant relations between proteins.

1. Introduction

Evolution is a central concept in biology and many, some even claim that all, important questions in biology involve relations between organisms, but also individual genes, processes or pathways. DNA and protein sequences offer a molecular picture of evolution and no wonder that questions of evolutionary relations between genes dominate their study on all levels. As genes evolve by mutations, deletions, swapping, inversions and other molecular processes, their relations are documented in similarities between their sequences. Similarity between closely related sequences, connected by a small number of molecular rearrangements, could be recognized by eye and simple counting of a number of identical residues in both sequences is still one of the most popular measures of describing sequence similarity. As the number of known sequences grew, more refined analysis tools become necessary. The first natural step was to generalize naïve sequence comparison by introducing statistical measure of similarity between amino acids, so a less than perfect conservation could be differentiated from a random similarity. First described over 20 years ago, algorithms using mutation matrices to give a numerical measure of protein similarity and dynamic programming and its variants to find an alignment that optimizes the similarity (or minimized the distance) still dominate the field of sequence analysis.

Thanks to the success of genomic projects, the rapid growth of the number of known protein sequences seems to accelerate faster and faster and so is the need for faster and more sensitive comparison algorithms. In all genomic projects, the computer analysis of their sequences is the only tool that can keep pace with the speed of sequencing. Speed of algorithms such as BLAST or FASTA helped deal with the rapidly growing databases. At the same time, only a part of the newly sequenced proteins were closely homologous to well known proteins – a majority were not and tools based on traditionally defined sequence similarity were not sufficient to reliably recognize their similarity to other proteins. Many of such proteins, thought to represent novel fold and function, were later found to be similar to already known proteins after all, with the obvious conclusion that some hints of the similarity of their sequences were not recognized in the initial analysis. A growing number of such cases prompted a vigorous research effort into the development of new, sensitive algorithms for the recognition of distant homologies. In last few years it became a vibrant research field with hundreds of novel algorithms being introduced every year. In this short review we would like to describe main classes of the new methods and concentrate on assumptions underlying various classes of methods rather than on direct comparison of specific methods.

With the increasing evolutionary distance between proteins their sequences diverge, and at certain point their similarity becomes difficult to recognize. To solve this problem, the similarity measure must be redefined, possibly taking into account additional sources of information. With the bewildering variety of comparison methods and algorithms developed to deal with this issue, it is important to organize the discussion around their most fundamental features. This perspective is necessary to understand different methods, assumptions they are built on and the fundamental reasons for their improved sensitivity in recognizing remotely homologous proteins. Without such understanding it is impossible to critically evaluate results obtained by different methods and thus fully utilize results of the genomic projects.

Every method that aims at comparing two sequences consists of three major elements:

- **Scoring system** for comparing two sequences. In most computer-based algorithms this is a numerical score, but some programs are based on visual analysis of the output
- **comparison threshold** or an implicit or explicit mathematical model of the distribution of scores between unrelated proteins, so the statistical significance of the specific alignment score could be compared to the value expected by chance
- **alignment algorithm** for finding the alignment with an optimal score

Similarities and differences between any sequence alignment methods can be described by comparing them in these three areas. Consequently, we will organize this review around these points, concentrating mostly on the first two, since this is where most of the development is happening now. However, we start from a summary of the standard approach, introduced in classical papers (Needleman and Wunsch, 1970; Smith and Waterman, 1981) and implemented in hundreds of programs, for instance GAP or BestFit from the GCG software package (GCG, 1991). This would set the stage for introducing and describing more sophisticated techniques. Most of the alignment algorithms can be used to align either DNA or protein sequences, but almost all super-sensitive alignment tools focus on aligning protein sequences and consequently in this review we will focus on protein sequence alignment.

2. Standard Approach

Since the alignment between two sequences usually involves not only mutations, but also gaps and insertions in both sequences, finding an optimal alignment is based on testing various combinations of gaps/insertions to optimize score of the alignment of the remaining parts of the sequence. This seems to be a combinatorial problem, but the most successful and widely used dynamic programming algorithm uses an optimization technique that finds the optimal solution including gaps in time proportional to the product of lengths of both sequences being compared. This algorithm is described in many introductory texts in bioinformatics and in various variants is used in most of the protein alignment programs. A very simplified description of the dynamic programming algorithms is that the optimal solution is built recursively by extending the best solution found in previous iterations. Depending on the choice of the mutation matrix and the scoring system, two variants of this approach – one maximizing the similarity, the second minimizing the difference between the two sequences – are used. Dynamic programming is an O(nm) algorithm, *i.e.* it's running time is proportional to the product of the length of the two sequences being aligned, which is one of the reasons for it's popularity. Another important variant of the dynamic programming concerns boundary conditions for the alignment, with the alignment constrained to encompass complete proteins (global alignment) or free to start and end at arbitrary position along the sequence (local alignment). A mixed model (local-global) alignment is also possible, when sequences are not treated in a symmetric way and the alignment is global in respect to one, but not the second sequences.

The speed and ease of implementation of the dynamic programming algorithm resulted in its wide popularity. A "standard" sequence comparison program, including a dynamic programming alignment between two protein sequences using a mutation matrix have been described many times in computational biology textbooks (Waterman, 1995) and implemented in hundreds of programs.

Another popular alignment method is based on identification of uninterrupted high scoring fragments of similarity in two sequences and subsequently connecting them by introducing gaps/insertions (Karlin and Altschul, 1990; Karlin and Altschul, 1993). One of the biggest advantages of the HSF approach is that the significance estimates are easier for alignments without gaps. A particular implementation of the HSF algorithm used in BLAST (Altschul *et al.*, 1990; Altschul *et al.*, 1997) also includes a specific hashing algorithm for a quick estimate of a highest possible score between a pair of proteins, resulting in a complicated alignment algorithm. This allows BLAST to avoid aligning proteins that are not likely to produce a high scoring alignment and gives it a phenomenal speed and allows comparison of hundreds of thousands of sequences in a matter of seconds. Thanks to its speed and easy availability at the NCBI WEB site, BLAST has become the single most popular sequence alignment program.

In both dynamic programming and HSF alignments, the alignment between two sequences can be broken down into a series of comparisons between single amino acids, including situations where a residue from one protein will be matched to a gap in the other protein. Then, the score of the entire alignment is calculated as a sum of individual scores of pairs of aligned residues. There are 21*21 possibilities of residue-residue pairing (including gaps), or half of that if we assume that the comparison is symmetric. The matrix with scores for aligning pairs of residues is usually called a mutations or similarity matrix and starting from the pioneering work of Dayhoff (Dayhoff *et al.*, 1978) several different mutation matrices were developed, based on various assumptions about the mutation process and/or differences and similarities between amino acid features (Tomii and Kanehisa, 1996). This type of scoring is called local, to highlight the fact that the comparison of any given pair of residues doesn't depend on other positions in the alignment. This feature of the scoring function is necessary if the dynamic programming alignment algorithm is to be used.

It is generally accepted that if two sequences can be aligned in a way that the alignment score is significantly larger than a score of a random alignment, it can be reasonably argued that these two sequences evolved from a common ancestor by mutations, deletions and insertions – all known to occur at the DNA level as part of normal duplication process. Thus similarity between two sequences is interpreted as a valid argument for homology between the two genes or proteins coded by these sequences. In fact, these two concepts, sequence similarity and protein homology become so intertwined that often one is used as a synonym of the other.

The key statement in the previous paragraph is "significant". How to decide if a given score is significant or not? The simplest answer is to calculate the score of the best alignment and compare it to the "null hypothesis" that these two sequences are not related and the observed similarity is a result of a random event. If a large percentage of positions along the sequence remained unchanged, the naïve reasoning "strong similarity cannot be random" is sufficient. This applies to comparisons between closely related proteins, and as a result, very similar sequences. When reporting similarities of 50% identical residues or higher, most researches don't bother with significance analysis. However, with weaker similarities and more distant relations, more sophisticated analysis is necessary.

When using an alignment score or some features of the alignment to decide whether or not a given pair of proteins is homologous, we are in fact comparing two hypotheses – one that the proteins can be connected by a common ancestor and the second is that they can not. The way to test this hypothesis is to calculate the distribution of scores (or other features) for unrelated proteins and calculate the probability of a specific score belonging to this distribution.

Distribution of scores between non-homologous sequences is assumed to follow an extreme value distribution (Evans *et al.*, 1993), which was rigorously proven only for alignments of randomly generated sequences with no gaps (Waterman, 1995), but is generally believed to be applicable in more general cases (Altschul *et al.*, 2001). However, empirically calculated distributions of scores between biological sequences differ from the extreme value distribution. Biological sequences are not truly random and display hidden regularities that are not well understood, but probably related to foldability requirements. One of the most interesting, but seldom discussed discrepancies between theory and practice of sequence alignments is that the distribution of scores for non-homologous sequences could be calculated in two ways – by repeatedly aligning randomly permuted sequences or by aligning large number of sequences in the database and assuming that the vast majority of them could not be related. Unfortunately these two distributions are often very different and empirically the latter gives the better reliability estimates.

Standard methods, loosely defined here as tools comparable to GAP, BLAST or FASTA algorithm are very efficient in identifying close homologs. Typically proteins with more than 40% sequence identity are easy to identify and align using almost any alignment algorithm and their homology is unquestionable. With increasing evolutionary distance the sequence similarity decreases and around 25% of sequence identity it becomes difficult to recognize using standard algorithms. For instance, there are many examples of spurious sequence similarities on the level of 25% sequence identity and above. This level of sequence similarity was historically called the twilight zone.

Figure 1. Distant homologs share significant structural similarity with barely detectable sequence similarity: flavodoxin (1fla), mannose permease (1pdo) and CheY (1ab5). Sequence similarity is smaller than 15% of identical residues for each pair, structural similarity better than 3Å RMSD over more than 50% of the structure. ***Colour plate at back of book.***

At the same time, it is obvious that many homologous proteins have sequence similarities well below the twilight zone. Classical examples, such as hemoglobins, have clearly recognizable functional similarity as defined both by the conserved cofactor (heme), activity (oxygen binding) and function (oxygen transport). Some features, such as type of a complex they form (tetramer, dimer or monomer) changes, but there is little doubt that despite low sequence similarity that can reach even less than 10% of identical residues, hemoglobins from such organisms as human and worms are indeed homologous. The strongest argument in this and many other cases comes from structural studies, where independently solved structures of many putative homologs proved to be remarkably similar, essentially settling the case in favor of distant relation between such proteins despite weak conservation of sequence. In many cases, such homology was recognized only after the structures of both proteins were solved (Petsko, 1991).

The growing number of experimentally determined protein structures contributed decisively to the wider acceptance of the existence of distant homologues. When two proteins generally thought to be unrelated are found to have very similar three dimensional structures, it often motivates researchers to reevaluate other features of both proteins and arrive at realization that there is a distant homology between them. Sometimes such analysis even leads to discoveries of new mechanisms and functions in supposedly well studied protein families (Babbitt *et al.*, 1995). Lists and databases classifying proteins based on their structural similarity, such as SCOP or DALI are also critical in the development of super-sensitive sequence alignment programs, providing training sets and quantitative

A

- phosphoribosyltransferase and viral coat protein, identity: 42%, different folds, different functions

```
 99 IRLKSYCNDQSTGDIKVIGGDDLSTLTGKNVLIVEDIIDTGKTMQTLLSLVRQY.NPKMVKVASLLVKRTPRSVGY 173
    : ||.  |||  ||          |.    || |  : |   |  |  | || |  || |:|      | ||.| |
214 VPLKTDANDQ.IGDSLY....SAMTVDDFGVLAVRVVNDHNPTKVT..SKVRIYMKPKHVRV...WCPRPPRAVPY 279
```

B

- Histon H5 and transcription factor E2F4, identity 7%, similar fold, similar function (DNA binding)

```
PTYSEMIAAAIRAEKSRGGSSRQSIQKYIKSHYKVGHNADLQIKLSIRRLLAAGVLKQTKGVGASGSFRL
              |  |                        |          |            |
GLLTTKFVSLLQEAKD-GVLDLKLAADTLA------VRQKRRIYDITNVLEGIGLIEKKS----KNSIQW
```

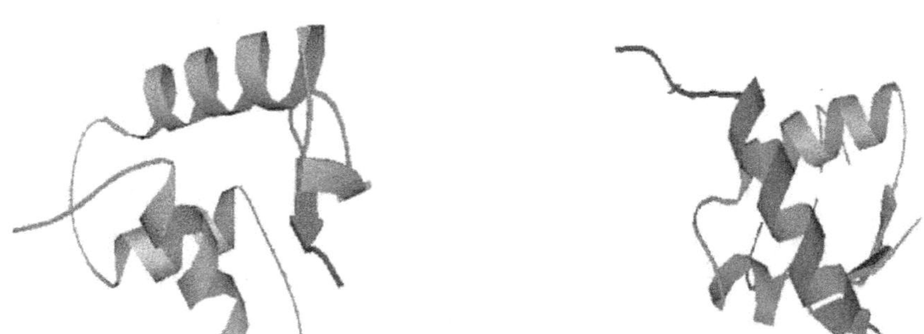

Figure 2. Sequence similarity is not equivalent to homology. A) Two proteins similar on the sequence level, but not homologous, as shown by their function and structure differences. B). Two proteins, homologous as shown by their function and fold similarity, but with no apparent sequence similarity (7% sequence similarity). ***Colour plate at back of book.***

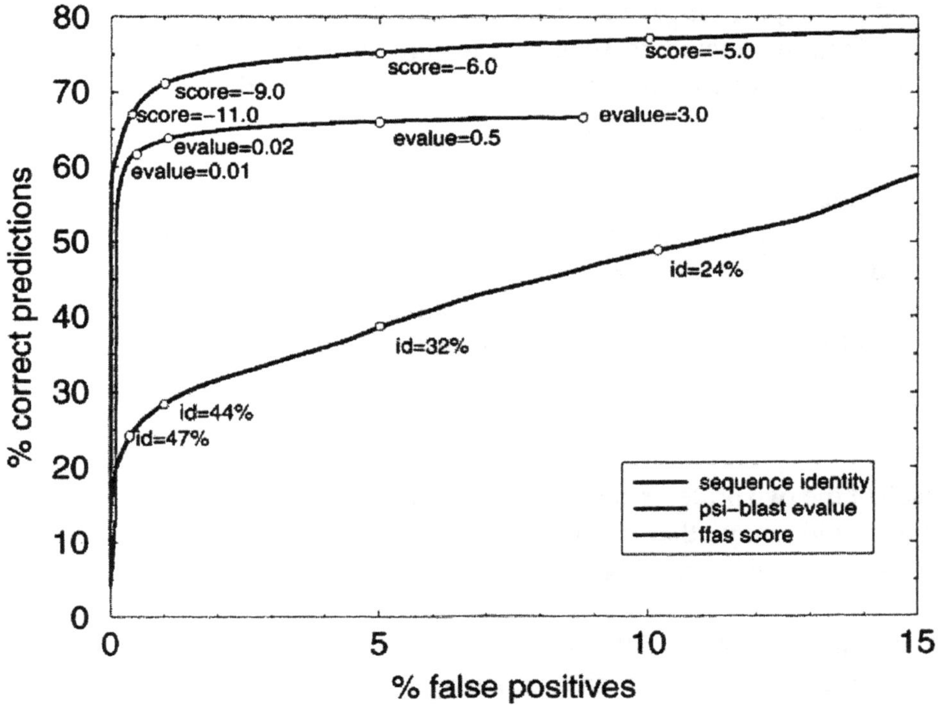

Figure 3. Comparison of homology recognition programs on a fold recognition benchmark, consisting of several hundred pairs of proteins with similar structures but varying sequence identity. Percentage of correct prediction is given as a function of incorrect ones for every level of statistical significance. (Both numbers do not add to 100% as there are cases where there is no prediction with a given significance level). ***Colour plate at back of book.***

validation criteria for new algorithms. For instance, a typical test for a new algorithm would involve its application to a set of proteins whose distant homologs were identified by structure comparison in an attempt at finding these homologs using only sequence information.

Such tests can, of course, be performed only on a subset of proteins with known three dimensional structures, but we can expect that the same relative performance of different algorithms on other sets of proteins.

3. Scoring Methods

As discussed earlier, to quantify the similarity between two proteins, it is necessary to introduce a numerical score that can be calculated for every alignment. In most alignments it is the sequence of both proteins that is being displayed, but in fact each position can be characterized by many other features, including sequence features of a given protein and/or its homologs, or structural features, either known or predicted. For instance, for each position we may list:

1. amino acid sequence (one amino acid)
2. list of amino acids present at a given position in a family of closely related homologs
3. secondary structure (real or predicted) at this position
4. environment of a given position in the final folded structure (real or predicted)
5. list of interactions of the residue at this position

Each of the two proteins can be described in any of the representations listed above (and many other possible representations) and any combination can be used to define a scoring system.

The standard scoring method uses representations 1) from a list above for both proteins, and a mutation or similarity matrix, providing a score for every pair of positions in the alignment. Several hundred different similarity matrices had been developed using a variety of assumptions, starting from a pioneering work of Dayhoff (Dayhoff *et al.*, 1978). One of the recent reviews compiled and analyzed a collection of 167 mutation matrices (Tomii and Kanehisa, 1996) and the updated list is available at the web site http://www.genome.ad.jp/dbget/aaindex.html. A prime rational to develop new mutation matrices was to be able to recognize and properly align distantly homologous proteins (Tomii and Kanehisa, 1996). Indeed, there is a significant difference between various mutation matrices (Vogt *et al.*, 1995), however, even the best ones fail to significantly penetrate the twilight zone of borderline similarities.

Mutation matrix lists amino acid – amino acid scores, assuming that identical mutations happening on various positions along the sequence should have identical score. While DNA-level mutations result from random errors during the DNA duplication phase and are generally believed to be universal (although some disagree with this statement), the probability of accepting a given mutation depends on the role of the appropriate position in the final protein structure. Thus a mutation changing an active site residue will not be accepted since the resulting protein will not be active and the organism expressing it would be deprived of its function. Protein sequences of orthologs from various species document series of successful *in vivo* experiments which

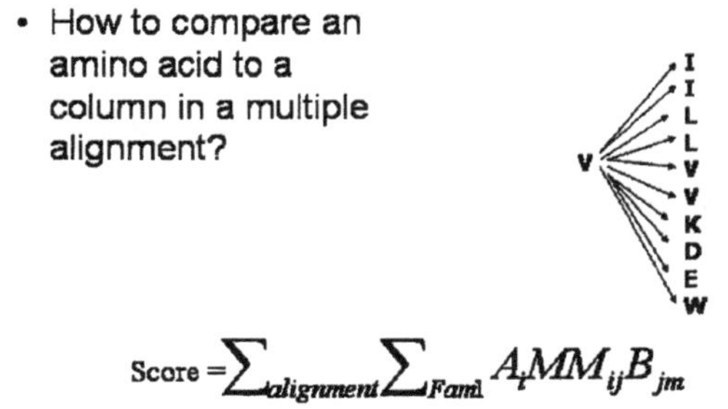

Figure 4. A simple variant of a scoring system in a profile method. First summation defines a position specific similarity matrix by averaging a regular scoring matrix over all amino acids found at a given position within a homologous family.

all lead to viable organisms. Failed mutations are not present in our databases as organisms that tested them died as a result of this experiment. Therefore we cannot expect that the mutation rules at different positions would be identical. In contrast, we can expect very different patterns of accepted mutations, for instance, in the hydrophobic core of the protein, the surface accessible loop that is not directly involved in the activity of the protein or in the active site. All this information can be derived from a multiple alignment of the proteins in the already known homologous family. Thus, using representation 2) from the list in the beginning of this paragraph on one or both sides of the alignment leads to profile class of methods. Various algorithms differ in what formalism is used to describe this information and how it is extracted from the multiple alignments.

For instance, a very natural way of describing the different rules for accepting mutations of different positions along the sequence is by a position-specific similarity matrix (PSSM), where position-specific considerations are expressed by changing scores for various mutations in different positions along the sequence. Thus a similarity matrix for a position that in the final structure forms a hydrophobic core would allow easy mutations from one hydrophobic residue to another, while introducing stiff penalties for mutations to a polar or charged residue. The standard mutation or similarity matrix is universal and has to be derived once to be used for any sequence. PSSM is protein- and position-specific and must be derived separately for every protein or protein family and for each position along the sequence. However, its application is very easy, as most alignment algorithms can easily be modified to work with a position-specific similarity matrix without any loss of speed. For instance, since PSSM as described above is local in the sense discussed earlier, it can be used in dynamic programming alignment, like for instance in GCG Profile program (Gribskov *et al.*, 1987). It can also be easily applied in scoring high-scoring sequence fragments – an example of such application is the PSI-BLAST algorithm (Altschul *et al.*, 1997). There are many programs

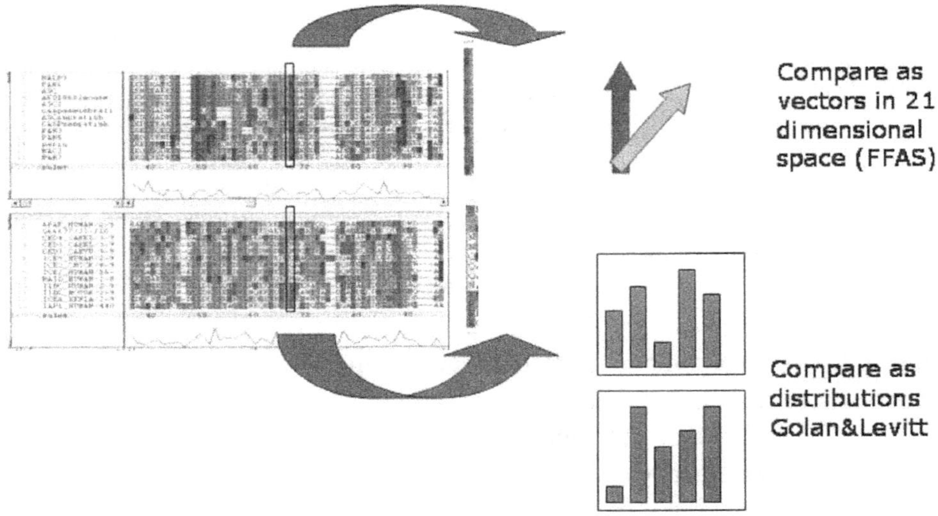

Figure 5. Two possibilities of implementing a profile-profile scoring systems. ***Colour plate at back of book.***

using PSSM concept, often under different names. Despite all their differences, programs such as GCG Profile (GCG, 1991; Gribskov *et al.*, 1987), PSI-BLAST (Altschul *et al.*, 1997) or even to some extent Hidden Markov Models approach (Eddy, 1996) are all based on very similar ideas.

In all such methods, the initial step of deriving a PSSM for a given protein consists of identifying and aligning closely homologous proteins. The result could be presented in a form of multiple alignment (even though some methods do not directly present this as an output), which in turn could be described as a matrix, where rows are representing various members of the homologous family and columns contain amino acids that can be found at specific positions. Similarity matrix for a given position is derived from the analysis of amino acids in the appropriate column. Various methods differ significantly in the steps of this procedure, *i.e.* in how homologous proteins are identified, how they are aligned and how the similarity matrix for each position is derived.

The same idea is carried further in methods that use profile description for both proteins being compared, such as BLOCKS (Pietrokovski *et al.*, 1998), FFAS (Rychlewski *et al.*, 1999) or Yona-Levitt algorithm (Yona and Levitt, 2002).

Representations 3)-5) from the list at the beginning of this paragraph use structure (real or predicted) in the scoring system. Since one of the proteins being compared is usually uncharacterized, in these algorithms the structural information is used for only one protein. Methods from this class are often referred to as threading. Some algorithms from this group still use concepts that could be translated into the PSSM language. For instance, the 3D profiles of Eisenberg and colleagues (Bowie *et al.*, 1991) and following work (Fischer and Eisenberg, 1996), can be viewed as a form of PSSM despite the differences in terminology. Even though positions are described in the language of "environments" or interaction tables, the result is formally equivalent to that of the PSSM, albeit using an intermediate entity where environments at every position are described as a 3D profile and amino acid preference for each environment is derived separately. Another apparently different method which is formally equivalent to PSSM is "interaction fingerprint" threading (Godzik *et al.*, 1992), where each position is described by a list of interactions involving a residue occupying this position in the three-dimensional structure. Again, the list of interactions, coupled with the empirical interaction energy parameters leads to a functional equivalent of PSSM, which contains empirical energies associated with each possible substitution at every position along the sequence.

The interesting feature of methods using information about protein structure to calculate PSSMs is that in these methods it is necessary to introduce additional assumptions to maintain a local character of a score and thus its formal equivalence to PSSM. In particular, interaction environments or list of interacting residues depend on the alignment itself, thus the factorization of the score required by dynamic programming is not fulfilled. In threading algorithm it is necessary to assume that the environment (or the interaction list and interaction partners) doesn't change between distant homologs – to be able to use dynamic programming for alignment. This assumption was called the "frozen approximation" (Godzik *et al.*, 1992) and threading algorithms that do not make this assumption must resort to more complex alignment methods, such as two-dimensional dynamic programming (Jones *et al.*, 1992), Monte Carlo alignment (Bryant and Lawrence, 1993), or branch and bound minimization (Lathrop and Smith, 1996). Algorithms from this group often explicitly aim at recognizing non-homologous proteins with similar structures and are classified as fold or structure prediction and not distant homology recognition algorithms – thus formally they are beyond the scope of this review. However, in reality, most of the protein similarities discovered using such algorithms represent cases for which homology is often disputed, but not unlikely.

4. Score Significance Analysis

When comparing two proteins, whether using a simple similarity or mutation matrix or a position specific-similarity matrix or its logical equivalent, the main result of the comparison is the numerical score of the alignment. To make any statements about homology between the two proteins being compared, we need to assign the relative significance to the score, or in other words compare it to scores of other alignments, including alignments between non-homologous proteins.

In principle, our task is simple – we need to access a probability that a score between a specific protein pair belongs to a distribution of scores between non-homologous proteins. If it is a part of a distribution, the score can be explained as a random event and no arguments about homology of the two proteins can be made. Of course, they still could be related, but the particular alignment method has failed to recognize and properly score the similarity. Even the best existing alignment methods fail to recognize a significant number of distantly related proteins whose homology can be established based on the strong sequence similarity. On the other hand, if a given alignment score is unlikely to be found by chance, one of the possible explanations might be that the two proteins are indeed related. Unfortunately, there might be other valid explanations for the alignment score not to be part of the random distribution. For instance, all alignment scores are strongly composition-biased and two unrelated proteins with similarly biased amino acid compositions may get a very strong alignment score despite lack of any relation between them.

There are further problems in obtaining parameters of the distribution of scores for non-related proteins, most of which are related to the fact that we are dealing with real, biological sequences. For instance, several approaches that were derived, proved and tested for gapless alignments between computer generated random pseudo-protein sequences often give very different results when applied to real-life biological sequences. There are few approaches to this problem, each with it's own problems and successes

- analytical derivation of the parameters of the distribution
- empirical derivation of the distribution using randomized sequences of the proteins being compared
- empirical derivation of the distribution using large groups of biological sequences to sample a space of non-related sequences
- empirical derivation of the distribution using suboptimal alignments between the proteins being compared (declumping; Waterman, 1995).

The first approach is really applicable only for gapless alignments (Karlin and Altschul, 1990), for alignment with gaps it is widely believed that the distribution has the same functional form, but this conjecture is not proven. The second approach works very well for computer-generated sequences, but generally fails, or at least doesn't perform as well as the third approach, for biological sequences. The reasons for this failure are not clear, but they may be related to the presence of some subtle regularities in naturally occurring protein sequences that are being destroyed in the randomization process. As a result, the scores between randomized sequences have a different distribution from scores of the alignment of the random, unrelated biological sequences, therefore the significance analyses based on the former are not useful. A popular simple variant of this approach is to invert one of the sequences and to compare the alignment score for both runs. If the high score was purely a function of amino acid composition or some specific regularities in the sequence it would be reflected in both scores.

The third approach, *i.e.* empirical distribution of scores for alignments with large group of natural sequences (most of which must be unrelated) suffers from the effect mentioned earlier – the composition dependence of the alignment score. Proteins display significant variations in their composition and scores for proteins with unusual and similar amino acid composition may reflect the composition similarity rather than evolutionary relations. There are several methods to correct the composition dependence of the alignment score, but none of them is really fool-proof. Finally, the fourth method in principle is able to circumvent the limitations of both of the previous methods, but it is not widely used and there are no data for a wide comparison.

As mentioned earlier, the most important source of errors in distant homology recognition is caused by protein pairs whose outstanding score is caused by factors not related to their evolutionary relation. Several steps in significance analysis are useful in identifying some of them. For instance, while the alignment score is symmetric, *i.e.* the alignment score in the comparison of proteins A and B is the same as in the comparison of proteins B and A, the same is not true about distributions. Therefore, to avoid false positives, distributions for both proteins should be calculated, the significance of the score in comparison to both distributions calculated and the lower of the two is used in further considerations. This approach for instance allows distinguishing spurious and significant similarities to large protein families such as colicins and tropomyosins, which often appear as apparently significant similarities to many helical proteins. Another large group of proteins that are often producing spurious high-scoring alignments are low complexity proteins, where composition is dominated by a small number of amino acids. Again, the "two distribution" approach is a good, if not very efficient way of identifying such questionable alignments. This method was for instance implemented in FFAS fold recognition algorithm (Rychlewski *et al.*, 1999).

A variant of this approach is to precompile a list of "promiscuous" proteins that align with high scores to proteins which are known (for instance by comparison of their three dimensional structure) not to be homologous. Such proteins can be removed from the database entirely or any apparently high scoring alignments to proteins from this group are treated with more caution. Another variant of this tactic is to predict and remove from consideration sequence fragments that are almost certain to cause problems with scores and alignments. This approach is used for instance to mask out low-complexity regions in proteins by programs such as BLAST and PSI-BLAST. Both the "black list" and masking out strategies limit the number of false predictions, but effectively make it impossible to recognize genuine similarities from these groups.

Comparing the alignment score against the empirical distribution of scores calculated for the whole database of protein sequences is by far the most popular approach to estimate statistical significance of alignment scores. With some minor variants it is used in such popular programs as BLAST and FASTA, and in detailed analysis of several algorithms for distant homology recognition as applied to fold recognition it proved to be far superior to other approaches (Bryant and Altschul, 1995). In all its variants, this method depends strongly on the database of sequences used to calculate the empirical distribution. In a typical problem, a natural solution is to use the largest possible database, such as the nonredundant set of all available sequences compiled at NCBI (the NR database), which is a standard solution for the BLAST and FASTA families of programs or SwissProt, which is a curated database of protein sequences maintained at Geneva Institute of Bioinformatics and EBI. However, both databases were created with a goal of providing a comprehensive list of all known proteins and this creates problems in using them to calculate score distributions. The content of these databases reflects research interests of general community and is not optimized to provide a balanced group of proteins to estimate distributions of scores. In particular, NR and SwissProt contain large numbers of closely related sequences, such as splicing variants of many human proteins or multiple entries of proteins related to diseases which are a focus of intensive research by many research groups. It is possible to cluster these databases and retain only representative proteins from large groups of closely related proteins. Despite removing some sequences and, consequently, some information from the databases, the smaller, well-balanced databases proved to be superior as reference databases in recognizing distant homologies (Li *et al.*, 2002). There are also other reasons for the improved performance of various distant homology reconition programs on clustered databases (Li *et al.*, 2002), but improved estimates of distribution of random scores also contribute to this effect.

Another way to manipulate the reference database to improve the chances of recognizing distant homologs is to limit it to smaller sets of proteins where from other arguments we are certain that such homologs would be present. For instance, from biochemical considerations we might know that a homolog of a certain protein should be present in a given genome. Using only this genome as a reference database may enable us to recognize a similarity that in a larger database would be lost among spurious alignments. This approach could be called recognition of least dissimilarity, rather then recognition of similarity, but in many tests this approach proved to be remarkably efficient. For instance, in the context of fold recognition we have tested a method we called "PBB-BLAST" where a profile (PSSM) of a protein being analyzed is compared only to sequences of proteins with known structure. To our surprise, this approach improved the recognition of correct structural templates by over 50% as compared to using the same program (PSI-BLAST) with a larger database (NR). Similar approach is often used in identifying orthologs in genomes, where the least dissimilar protein in a given genome is recognized as an ortholog, even if the alignment score significance is not reliable when compared to a score distribution in a larger database (Tatusov *et al.*, 2001).

5. Summary and Discussion

Tracking distantly related homologs is a fascinating subject. In its biological aspect we can discover connections between systems and pathways thought to be unrelated and thus recognize deep relations between organisms and learn how complex systems evolved from simpler ones. On the algorithmic level, its akin to deciphering a message hidden in a noisy transmission.

In this review we focused on the recognition aspect of sensitive alignment algorithms, but for the meaningful analysis of the potential distant homology,

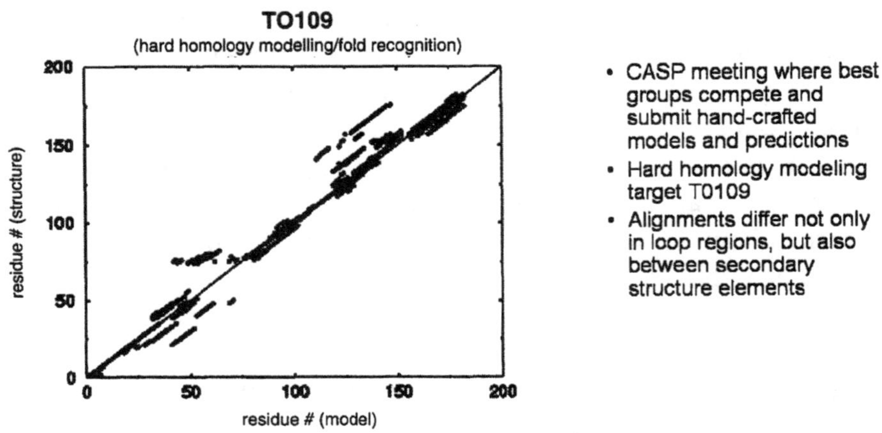

Figure 6. Alignment quality is the next challenge for distant homology analysis.

the alignment must be accurate. Many algorithms correctly recognize distant homology but produce an incorrect alignment. This interesting problem has only recently become a focus of attention (Jaroszewski *et al.*, 2000).

The recent advances in sensitive protein alignment algorithms come mostly from exploring additional information available either from the analysis of substitution patterns in homologous proteins (profile or PSSM methods) or from the requirements of the stability of the final protein structure (threading). Even such a broad review doesn't cover all methods for detection of distant homologs. Many new methods bring in new types of information beyond structure and sequence variation among homologs and thus hold significant promise for further advancing the sensitivity of discovering distant homologs. Methods such as ISS (intermediate sequence search) or other neighborhood methods (Yona *et al.*, 1999) explore features of protein sequence space searching for "bridging proteins" to connect divergent families by a series of comparisons. Context analysis, implemented in systems such as WIT (Overbeek *et al.*, 2000), uses position in a genome and function of genome neighbors to identify function and indirectly recognizes homologies. Other methods use keyword and literature searches to build a "literature profile" of a protein, discovering distant relations between proteins (MacCallum *et al.*, 2000). We can definitely expect new exciting developments in the field of distant homology recognition.

Acknowledgements

Research discussed in this review was supported by NIH GM-60049.

References

Altschul, S.F., Bundschuh, R., Olsen, R., and Hwa, T. 2001. The estimation of statistical parameters for local alignment score distributions. Nucleic Acids Res. 29: 351-361.

Altschul, S.F., Gish, W., Miller, W., Myers, E.W., and Lipman, D.J.1990. Basic local alignment search tool. J. Mol. Biol. 215: 403-410.

Altschul, S.F., Madden, T.L., Schaffer, A.A., Zhang, J., Zhang, Z., Miller, W. and Lipman, D.J. 1997. Gapped BLAST and PSI-BLAST: a new generation of protein database search programs. Nucleic Acids Res. 25: 3389-3402.

Babbitt, P.C., Mrachko, G.T., Hasson, M.S., Hiusman, G.W., Kolter, R., Ringe, D., Petsko, G.A., Kenyon, G.L., and Gerlt, J.A. 1995. Functionally diverse enzyme superfamily that abstracts the a proton of carboxylic acids. Science. 267: 1159-1161.

Bowie, J.U., Luethy, R., and Eisenberg, D. 1991. A method to identify protein sequences that fold into a known three dimensional structure. Science. 253: 164-170.

Bryant, S.H., and Altschul, S.F.1995. Statistics of sequence-structure threading. Curr. Opin. Struct. Biol. 5: 236-244.

Bryant, S.H., and Lawrence, C.E. 1993. An empirical energy function for threading protein sequence through folding motif. Proteins. 16: 92-112.

Dayhoff, M.O., Schwartz, R.M., and Orcutt, B.C. 1978. A model of evolutionary change in protein. Atlas of protein sequence and structure. 5: 345-352.

Eddy, S., 1996. Hidden Markov Models. Curr. Opin. Struct. Biol. 6: 351-355.

Evans, M., Hastings, N., and Peacock B. 1993. Statistical Distributions. Wiley, New York.

Fischer, D., and Eisenberg, D. 1996. Fold recognition using sequence derived properties. Protein Science. 5: 947-955.

GCG. 1991. Program Manual for the GCG Package.

Godzik, A., Kolinski, A., and Skolnick, J. 1992. Topology fingerprint approach to the inverse protein folding problem. J. Mol. Biol. 227: 227-238.

Gribskov, M., McLachlan, M., and Eisenberg, D. 1987. Profile analysis:

Detection of distantly related proteins. Proc. Natl. Acad. Sci. USA. 84: 4355-4358.

Jaroszewski, L., Rychlewski, L., and Godzik, A. 2000. Improving the quality of twilight-zone alignments. Protein Sci. 9: 1487-1496.

Jones, D.T., Taylor, W.R., and Thornton, J.M. 1992. A new approach to protein fold recognition. Nature. 358: 86-89.

Karlin, S., and Altschul, S.F. 1990. Methods for assessing the statistical significance of molecular sequence features by using general scoring schemes. Proc. Natl. Acad. Sci. USA. 87: 2264-2268.

Karlin, S. and Altschul, S.F. 1993. Applications and statistics for multiple high-scoring segments in molecular sequences. Proc. Natl. Acad. Sci. USA. 90: 5873-5877.

Lathrop, R., and Smith, T.F. 1996. Global optimum protein threading with gapped alignment and empirical pair scoring function. J. Mol. Biol. 255: 641-665.

Li, W., Jaroszewski, L., and Godzik, A. 2002. Database clustering improves sensitivity and efficiency in protein fold recognition. Protein Engineering (in press).

MacCallum, R.M., Kelley, L.A., and Sternberg, M.J. 2000. SAWTED: structure assignment with text description—enhanced detection of remote homologues with automated SWISS-PROT annotation comparisons. Bioinformatics. 16: 125-129.

Needleman, S.B., and Wunsch, C.D. 1970. A general method applicable to the search for similarities in the amino acid sequence of two proteins. J, Mol. Biol. 48: 443-453.

Overbeek, R., Larsen, N., Pusch, G.D., D'Souza, M., Selkov, Jr., E., Kyrpides, N., Fonstein, M., Maltsev, N., and Selkov, E. 2000. WIT: integrated system for high-throughput genome sequence analysis and metabolic reconstruction. Nucleic Acids Res. 28: 123-125.

Petsko, G.A., 1991. Déjà vu all over again. Nature. 352: 104-105.

Pietrokovski, S., Henikoff, J., and Henikoff, S. 1998. Exploring protein homology with the BLOCKS server. Trends Genet. 14: 162-163.

Rychlewski, L., Zhang, B., and Godzik, A. 1999. Functional insights from structural predictions: analysis of the Escherichia coli genome. Protein Sci. 8: 614-624.

Smith, T.F., and Waterman, M.S. 1981. Identification of common molecular subsequences. J. Mol. Biol. 147: 195-197.

Tatusov, R.L., Natale, D.A., Garkavtsev, I.V., Tatusova, T.A., Shankavaram, U.T., Rao, B.S., Kiryutin, B., Galperin, M.Y., Fedorova, N.D., and Koonin, E.V. 2001. The COG database: new developments in phylogenetic classification of proteins from complete genomes. Nucleic Acids Res. 29:

22-28.

Tomii, K., and Kanehisa, M. 1996. Analysis of amino acid indices and mutation matrices for sequence comparison and structure prediction of proteins. Prot. Engineer. 9: 27-36.

Vogt, G., Etzold, T., and Argos, P. 1995. An assessment of amino acid exchange matrices in aligning protein sequences: the twilight zone revisited. J. Mol. Biol. 249: 816-831.

Waterman, M.S. 1995. Introduction to computational biology: Maps, sequences and genomes (Interdisciplinary Statistics), Chapman and Hall, New York.

Yona, G., and Levitt, M. 2002. Within the twilight zone: a sensitive profile-profile comparison tool based on information theory. J. Mol. Biol. 315: 1257-1275.

Yona, G., Linial, N., and Linial, M. 1999. ProtoMap: automatic classification of protein sequences, a hierarchy of protein families, and local maps of the protein space. Proteins. 37: 360-378.

From: *Frontiers in Computational Genomics*

3

Measuring Generality of Knowledge-Based Potentials Extracted from Protein Structure Sets

Frank Eisenhaber, Shamil R. Sunyaev, Vladimir G. Tumanyan, and Eugene N. Kuznetsov

Abstract

Sequence-structure matching techniques such as threading with so-called knowledge-based potentials rely on the extraction of amino acid location distributions (*e.g.*, of solvent accessibilities, backbone conformational descriptors or pairwise residue-residue distances) from learning sets of protein structures. These functions are used for the characterization of amino acid type preferences as well as for the description of spatial locations in structural templates in the same physical terms. The transferability of these functions from tertiary structure sets used for their derivation to new proteins outside the learning set is critical for predictor performance. We present the problem in the framework of statistical hypotheses tests that allow quantifying to which extent the extracted knowledge-based functions can be extrapolated to new proteins. Several statistical models for knowledge-based functions are considered in this work: (1) We show that the derivation of knowledge-based potentials within the framework of statistical physics, especially

with Boltzmann's distribution, is not consistent with the observed distributions of residue location parameters. (2) The assumption of amino acid type-specific but protein-independent propensities is often a valid approximation, although many cases of clear disagreement can be found. (3) We propose two models of propensities corrected for the actual amino acid composition in a query protein.

1. Introduction

Although threading techniques rely on additional tertiary structural data for the derivation of so-called knowledge-based potentials, these methods have not been more successful in recognizing distantly related proteins than the much more simple sequence similarity search approaches (Eisenhaber *et al.*, 1995b; Jernigan and Bahar, 1996; Jones and Thornton, 1996; Lemer *et al.*, 1995; Matsuo and Nishikawa, 1994; Smith *et al.*, 1998). This is especially true in the comparison with much more widely used sequence profile-based methods (Bork and Gibson, 1996; Bork and Koonin, 1998; Park *et al.*, 1998) which have also proven to be more powerful in the case of the *M.genitalium* benchmark (Fischer and Eisenberg, 1997; Huynen *et al.*, 1998; Jaroszewski *et al.*, 1998). With ever growing sequence databases due to many genome projects, the sequence space will be even more densely populated, profile techniques will become relatively more powerful and, therefore, this trend will probably continue in the near future. This evaluation of the capacity of threading is unexpected and deserves thorough investigation.

Paradox findings reported in several publications (Crippen, 1996; Furuichi and Koehl, 1998; Sunyaev *et al.*, 1998; Thomas and Dill, 1996) indicate that the methodical problems with threading may be much more deeply rooted than commonly believed and that they are not removed by just learning the knowledge-based potential from more protein structures or by adding a few additional energy terms.

It should be noted that, as a fundamental step, the collection of empirical distributions of amino acid residue properties in tertiary structure sets is part of the derivation of any knowledge-based potential (see below). These empirical distributions are then interpreted as realization of statistical distribution functions (possibly, in form of Boltzmann distribution functions) that describe preferences of amino acid types for spatial sites in structural templates regardless of the structures they are learned from. In their attempt to prove the validity of knowledge-based potentials or propensities, previous authors checked if these propensities converge with growing size of the learning set of tertiary structures. However, observed convergence does not guarantee that the propensity even computed on the basis of a large dataset can be transferred to a particular individual protein. The task does not consist in predicting states of residues in large structure sets but in individual proteins. Thus, the question arises how valid are these functions derived from learning sets if applied to single proteins.

In earlier work (Sunyaev *et al.*, 1998), we have outlined a framework of statistical quality tests for structure dataset-derived parameters applicable independent of and prior to any prediction experiment. We have also shown that, for successful application in predictions, knowledge-based potentials must discriminate amino acid types from each other (or from an averaged amino acid residue) but, surprisingly, energy functions currently in use lack this ability for about a third of all amino acid types.

The main result presented here is a statistical measure for the generality of knowledge-based prediction parameters; *i.e.*, a criterion quantifying the degree of transferability of such parameters to protein structures outside of the learning set. We directly calculate the probability of the compatibility of the statistical model with the available dataset of tertiary structures. Thus, the degree of possible extrapolation can be determined in a pure form, especially prior to a prediction experiment and independent of the influence of its algorithm (in contrast to Furuichi and Koehl (1998)). We apply this methodology on amino acid type-specific potentials based on solvent accessibility, backbone structure, and/or pairwise residue-residue distances. Our results are of general importance for the evaluation of knowledge-based prediction techniques.

2. Fundamentals in the Derivation of Knowledge-Based Potentials

2.1. Necessity of Environment Parameter Transferability for the Quantification of a Sequence-Structure Match

The physical approach to the protein folding problem presumes the characteristics of the native structure being conformational invariants of an ensemble of polypeptide conformations at a state of lowest free energy; *i.e.*, the complete information for protein folding is contained in the amino acid sequence (see section IV.A.1. in Eisenhaber *et al.* 1995b). Thus, the native structure as well as the location of different residues within the structure are determined by the inherent physico-chemical properties of the amino acids and especially of their side chains. The major idea developed by the threading community to describe both amino acid type preferences as well as residue environments in tertiary structures by environment parameters in the same physical terms is fundamental for comparisons of protein sequences and tertiary structures:

amino acid type $\boldsymbol{a}$ ↔ environment parameter x ↔ structural site t;

i.e., the suitability of an amino acid type $\boldsymbol{a}$ for a structural site $\boldsymbol{t}$ (template position) is estimated indirectly via the physical quantity $\boldsymbol{x}$, for example, in form of solvent accessibility, secondary structure, backbone conformation, or pairwise residue-residue distances. This finally allows the possible sequence-structure match or mismatch to be evaluated as, previously, the homology of two sequences since it is decomposed to the evaluation of the match between a single amino acid $\boldsymbol{a}$ (or a combination of a few residues) and a single structural site $\boldsymbol{t}$ (or a combination of a few sites).

For the case of environmental parameters $\boldsymbol{x}$ as functions of only a single structural location $\boldsymbol{t}$ or a single amino acid type $\boldsymbol{a}$, the fundamental assumptions have been most clearly formulated by Bowie *et al.* (1991) (see p. 165 in this reference): A structural position of a residue in a template is characterized by secondary structure, solvent accessibility and accessible area covered by polar protein atoms or water. (1) This combination of properties is **extrapolated to the equivalent residue location in the unknown structure** of the query sequence. (2) Preferences of amino acid types for certain environment classes are calculated from sets of well-refined 3D-structures. These preferences are considered as **inherent properties of the amino acid types**, which are expected to be independent of the structures they are calculated from.

The generalization to environmental parameters $\boldsymbol{x}$ as functions of two spatial locations $\boldsymbol{t}_1$ and $\boldsymbol{t}_2$ or two amino acid types $\boldsymbol{a}_1$ and $\boldsymbol{a}_2$ such as pairwise residue-residue distances is straightforward (Crippen and Viswanadhan, 1985; Koppensteiner and Sippl, 1998; Miyazawa and Jernigan, 1985; Sippl, 1994; Tanaka and Scheraga, 1976). The pairwise distance distributions are considered both (1) as inherent properties of the pair ($\boldsymbol{a}_1$, $\boldsymbol{a}_2$) and (2) as independent of the protein structural dataset they are calculated from. If this condition is indeed fulfilled, then the distribution of $\boldsymbol{x}$ is a two-particle correlation function and the framework of statistical mechanics appears applicable. The environment parameter $\boldsymbol{x}$ is assumed Boltzmann-distributed and mean-force potentials for (a_1, a_2)-interactions are calculated (Koppensteiner and Sippl, 1998; Sippl, 1994).

It was also suggested to normalize amino acid type preferences by a reference state which is basically identical with the value distribution of the environment parameter x in a set of structures regardless of amino type (Crippen and Viswanadhan, 1985; Koppensteiner and Sippl, 1998; Miyazawa and Jernigan, 1985; Sippl, 1994; Tanaka and Scheraga, 1976). This reference state is considered of being independent of the individual structures in the dataset. It is thought that it can also be extrapolated to and applied for comparison with another query protein. As was shown by Koppensteiner and Sippl (1998), this normalization is eventually equivalent to defining an energy difference relatively to some ideal state thought to describe the protein in a generally non-native sequence-structure match.

Thus, the prediction algorithms do not operate directly with the probability distribution $P_a(x)$ ($=P(a \mid x)$=expected occurrence of amino acid type a among template sites with property x in the case of discrete states x) itself but with the propensity $\Pi_a(x)$

$$\Pi_a(x) = \frac{P_a(x)}{P(x)} \tag{1}$$

Here, $P(x)$ is the probability distribution regardless of amino acid type. Especially in the case of continuous distributions, the logarithmic form of (1) is more common and the notion "knowledge-based" potential energy E as component of the total score function is used:

$$E = -RT \ln \frac{f_a(x)}{g(x)} \tag{2}$$

Here, $f_a(x)$ is the probability distribution density of property x for amino acid type a, $g(x)$ is the probability distribution density for any amino acid type. In the case of pairwise properties, the index "a" has to be substituted by the index "a_1, a_2" denoting two amino acid types.

2.2. Paradox Findings Indicating Limited Transferability of Learned Parameters

The very general assumption of environmental parameters being independent of tertiary structure sets they are extracted from has never been directly investigated although all prediction methods using knowledge-based potentials or propensity expressions rely on the possibility to transfer (to extrapolate) such parameters to query proteins without known structure. Although the quality of the potential is just one aspect of a prediction algorithm, structure prediction experiments with known outcome have suggested that the extrapolation may not work well in all cases. Furuichi and Koehl (1998) demonstrated that the prediction accuracy of knowledge-based potentials in threading experiments depends on the learning set of protein structures, especially on the distribution of protein sequence lengths, the amount and the diversity of secondary structures. Earlier, Thomas and Dill (1996) have found that the assumptions used in the derivation of knowledge-based potentials (see Sippl (1990; 1994) and Finkelstein *et al.* (1995b; 1995a) lead to serious contradictions, for example, to greatly different conformational temperatures for the exterior/interior distribution of amino acid types with respect to various protein structure sets.

Considerable attention has been paid to the computation of amino acid type-specific propensities from protein tertiary structural data in the scientific literature. Curiously, the Chou-Fasman parameters (Kyngäs and Valjakka, 1998) and Ponnuswamy's bulk hydrophobic character (Cid *et al.*, 1992) were shown to depend on the learning set of protein structures (with respect to composition of secondary structures) and not only on amino acid type as originally expected. Although these observations question the propensity concept as a whole, they remained without critical consideration.

3. Statistical Model 1: Extraction of Knowledge-Based Potentials from Structural Datasets with the Boltzmann-Distribution Approach

3.1. Statistical Model 1: Theory

The well-known thermodynamic relationship

$$\Delta W_{ab}(x) = -RT\log\left[\frac{f_{ab}(x)}{g(x)}\right] - RT\log\frac{Z_{(ab)}}{Z}, \tag{3}$$

the equation (10) in the paper of Sippl (1990), relates the mean-force potential ΔW_{ab} between two particles of type a and of type b with their two-particle correlation function $f_{ab}(x)$ where x denotes the distance between a and b. The net potential ΔW_{ab} arises from the specific interaction between a and b, therefore, it is taken relative to the reference state defined by a two-particle correlation function $g(x)$ regardless of the type of the interaction partners (Godzik *et al.*, 1995; Ota and Nishikawa, 1995; Rooman and Wodak, 1995; Sippl, 1994). The Boltzmann sums $Z_{(ab)}$ and Z do not depend on x and contribute maximally a sequence-specific constant to ΔW_{ab}. In accordance with equation (25) of Sippl's paper (1990), the total energy of $\Delta W(S,T)$ of the sequence-structure match can then be calculated as sum over all interactions between pairs (a, b) of amino acid residues in the sequence S at distances x defined by the structural template T

$$\Delta W(S,T) = \sum_{(a,b)} \Delta W_{ab}(x) \quad (4)$$

The density distribution functions $f_{ab}(x)$ and $g(x)$ are thought to be estimated from sets of protein tertiary structure. In this way, ΔW_{ab} becomes a "knowledge-based" potential. Obviously, formulas (1) and (2) can also be easily extended to properties x such as solvent accessibility, secondary structure, or backbone conformation (Eisenhaber *et al.*, 1995b; Kocher *et al.*, 1994) which depend only on a single amino acid type a:

$$\Delta W_a(x) = -RT \log\left[\frac{f_a(x)}{g(x)}\right] - RT \log\frac{Z_{(a)}}{Z} \quad (5)$$

and

$$\Delta W(S,T) = \sum_{(a)} \Delta W_a(x) = \sum_{a} \sum_{(x)} f_a(x) \Delta W_a(x) \quad (6)$$

where the summation is carried out over all amino acid residues a of the sequence S occupying a site t with the environment parameter value $x(t)$ of the template T.

Let us summarize: The derivation of the pairwise residue-residue potential ΔW_{ab} rests on the assumption of space homogeneity, interaction isotropy, interaction additivity, and most of all on the fact that $f_{ab}(x)$ and $g(x)$ are distribution density functions in the statistical sense. The latter statement needs some additional explanation: Not every collection of values can be rightly considered as realization of a statistical distribution function. Mathematical statistics operates with data as with samples of random variables; *i.e.*, data samples are considered drawn from one and the same probability distribution function. This can be imagined as if some generator (the random variable $\boldsymbol{x}$) produces the values of some data sample $\{\boldsymbol{x}\}$ in accordance with some fixed probability density distribution function $\boldsymbol{f}(\boldsymbol{x})$. Only in this case is the empirical density distribution function $\boldsymbol{f}^*(\boldsymbol{x})$ calculated from the data sample $\{\boldsymbol{x}\}$ an estimate of the hidden probability distribution function $\boldsymbol{f}(\boldsymbol{x})$. Thus, just a collection of values obtained from a mixed set of various samples (for example, measured in a series of experiments) cannot be immediately treated as a representative set to estimate single distribution function - unless all samples are proven to be drawn from one and the same distribution $\boldsymbol{f}(\boldsymbol{x})$ (condition of homogeneity). There are many methods for testing the latter supposition. For example, the total set can be split into portions and statistical criteria such as χ^2 or Kruskal-Wallis can determine whether the data subsets are drawn from the same statistical distribution.

Since we are worried about the possibility to extrapolate the potential learned from a structure set to a single new protein, it is natural to analyze whether the hidden density distribution function is the same for all proteins; *i.e.*, in the statistical terminology, the density distribution function is called homogeneous over all proteins. Consequently, we have to study the following statistical test (homogeneity test): We consider $\boldsymbol{K}$ sets of observations, in our case, each of the $\boldsymbol{K}$ sets contains data for one protein structure $\boldsymbol{k}$ ($\boldsymbol{k}$=1,...,$\boldsymbol{K}$) from the structural dataset Σ. For each sample set, the values of an environment parameter $\boldsymbol{x}$ (for example, the solvent accessibility of all residues) are calculated. For each of the samples $\{\boldsymbol{x}\}_1$, $\{\boldsymbol{x}\}_2$,..., $\{\boldsymbol{x}\}_{\boldsymbol{K}}$, the empirical distribution density functions $\boldsymbol{f}_{a,*}^{(k)}(\boldsymbol{x})$ for all amino types $\boldsymbol{a}$ ($\boldsymbol{a}$=*A*, *C*, *E*, *F*,..., *Y*; in the case of pairwise residue-residue distances, $\boldsymbol{a}$ denotes a pair of amino acid types) and $\boldsymbol{g}_*^{(k)}(\boldsymbol{x})$ as estimates for the hidden statistical distribution density function $\boldsymbol{f}_a^{(k)}(\boldsymbol{x})$ and $\boldsymbol{g}^{(k)}(\boldsymbol{x})$ of the $\boldsymbol{k}$-th protein are computed. Obviously, the functions defined so far are interrelated via

$$g^{(k)}(x) = \sum_{a=1}^{20} v_a^{(k)} f_a^{(k)}(x) \tag{7}$$

where $v_a^{(k)}$ denotes the fraction of amino acid type $\boldsymbol{a}$ in the $\boldsymbol{k}$-th protein. Accordingly, the empirical distribution density functions $f_{a,*}^{(\Sigma)}(x)$ and $g_*^{(\Sigma)}(x)$ can be determined over the whole learning set Σ together with their hidden counterparts $f_a^{(\Sigma)}(x)$ and $g^{(\Sigma)}(x)$. For the following text, we agree a notation: Whereas the lowercase letter functions $\boldsymbol{f}$ and $\boldsymbol{g}$ denote distribution density functions, the uppercase letters $\boldsymbol{F}$ and $\boldsymbol{G}$ stand for the distributions themselves (frequently called cumulative distribution functions).

The homogeneity hypothesis $\boldsymbol{H}_0$ for $g^{(k)}(x)$ is the condition

$$\boldsymbol{H}_0: \quad \forall x: g^{(1)}(x) = g^{(2)}(x) = \cdots = g^{(K)}(x). \tag{8}$$

The alternative hypothesis $\boldsymbol{H}_1$ assumes that at least one subset $\boldsymbol{l}$ ($1 \le l \le K$) has a different density distribution function $g^{(l)}(x)$. Similarly, the homogeneity condition for $f_a^{(k)}(x)$ reads as

$$\boldsymbol{H}_0: \quad \forall x: f_a^{(1)}(x) = f_a^{(2)}(x) = \cdots = f_a^{(K)}(x). \tag{9}$$

Two standard criteria have been used for this statistical test: the rank test of Kruskal and Wallis and $\boldsymbol{k}$-sample χ^2-homogeneity test (see Appendix). The Kruskal-Wallis test is especially sensitive for mean shifts between distributions whereas the χ^2-homogeneity criterion measures the average deviation between distributions.

3.2. Statistical Model 1: Methods and Results

Formally, a statistical test looks as follows: The functions $f_{a,*}^{(k)}(x)$ and $g_*^{(k)}(x)$ calculated experimentally from the structural learning set enter the formulas for the statistical criteria (see Appendix) which return a value of the statistic KW (Kruskal-Wallis) or χ^2, respectively. Their asymptotic distribution is well studied so that the probability of a statistic larger than or equal to KW (or χ^2) can be determined. This probability is equal to the probability $\boldsymbol{P}$ of an erroneous decision made by the observer (*i.e.*, by us analyzing the structural dataset) if we reject $\boldsymbol{H}_0$ and, at the same time in reality (which is hidden from the observer), $\boldsymbol{H}_0$ is true. The homogeneity hypothesis $\boldsymbol{H}_0$ has to be rejected if $\boldsymbol{P}$, the probability of an erroneous decision, is below a significance threshold (traditionally, 0.05, 0.01, or 0.001).

For the relationship (9), the statistic can be computed for a single amino acid type individually. Alternatively, the statistics obtained for the individual amino acid types can be summed up and a simultaneous test for all amino acid types can be executed.

In the case of the χ^2-homogeneity criterion, we compute also a significance $\boldsymbol{P}_{\text{protein}}$ for an average protein in the learning database in accordance with

$$\chi^2_{\text{protein}} = \frac{\chi^2}{K}. \tag{10}$$

In this case, the number of degrees of freedom is just computed from the number of non-empty bins. For all $\boldsymbol{P}_{\text{protein}}$ values discussed below, the number of data is small for a statistical test since, for an average protein structure with ca. 200 amino acid residues, many residues (and also residue-residue combinations) have a rare occurrence. Thus, a $\boldsymbol{P}_{\text{protein}}$ value above 0.05 does not necessarily indicate homogeneity of the distribution analyzed but a low value <0.001 allows confident rejection of H_0.

In our calculations, a learning set of polypeptide chains with non-homologous sequences (residue identity ≤35%, resolution ≤2.0Å) was used. This set is without NMR structures as well as structures with incomplete backbone or side chains (except for the first or last residue of the polypeptide chain) and contains no oligomeric or membrane-associated proteins (153l, 1aaj, 1acf, 1acx, 1adl, 1ahc, 1alc, 1ald, 1amp, 1arb, 1ast, 1bcx, 1btc, 1btl, 1cbs, 1cdp, 1cfb, 1cgt, 1cie, 1cnr, 1cot, 1cp4, 1crm, 1ctf, 1cyo, 1eas, 1edb, 1enj, 1esl, 1fas, 1fdx, 1fkb, 1flp, 1frd, 1fxd, 1gcd, 1gcs, 1gdi, 1gky, 1glg, 2glt, 1gof, 1gox, 1gpr, 1hbg, 1hfc, 1hoe, 1hpi, 1huw, 1hyp, 1iag, 1icn, 1igd, 1kdb, 1knt, 1l99, 1lis, 1mba, 1mjc, 1nar, 1npc, 1ofv, 1oyb, 1paz, 1pbe, 1php, 1pii, 1poa, 1poc, 1poh, 1ppa, 1ppn, 1ptx, 1r69, 1rbp, 1rbu, 1ris, 1sbp, 1sgt, 1shg, 1srp, 1st3, 1tag, 1tca, 1tdy, 1ten, 1thw, 1tib, 1tml, 1top, 1ubi, 1ukz, 1yma, 21bi, 2acs, 2apr, 2bat, 2che, 2cna, 2ctb, 2cut, 2cy3, 2dri, 2ebn, 2exo, 2fcr, 2fx2, 2hts, 2lhb, 2mcm, 2pia, 2ran, 2sil, 2sn3, 351c, 3bcl, 3blm, 3cms, 3fxn, 3il8, 4enl, 4icb, 5fd1, 6abp, 6ldh, 6pti, and 7pcy.) Residues denoted as ASX or GLX were considered asparagine and glutamine, respectively. The set contains 25638 residues in total.

We analyzed residue location parameters in terms of solvent accessibility surface area (total SASA of the residue in the 3D structure), backbone structural descriptors (α-helix, β-strand, and coil in agreement with DSSP (Kabsch and Sander, 1983) and the 5 conformational states following Matsuo and Nishikawa (1994) and pairwise residue-residue distance functions (as a function of minimal sequence separation with at least 1, 10 and 20 residues between the interacting amino acids, distances are calculated between C_β atoms (except of glycine: C_α) in accordance with Matsuo and Nishikawa (1994). In our previous publication (Sunyaev *et al.*, 1998), we have outlined in detail that these definitions of knowledge-based potentials are reasonable and sufficiently representative for the many variations described in the literature. Area computation of the solvent-accessible surface as defined by Lee and Richards (1971) were made with the program ASC (Eisenhaber *et al.*, 1995a; Eisenhaber and Argos, 1993). The atomic radii from Tab. 1 of Juffer *et al.* (1995) were used. The probe (solvent) radius was set equal to 1.4 Å. The Kabsch and Sander method (Kabsch and Sander, 1983) for secondary structure assignment produces an artificially large number of small helices and strands (Colloc'h *et al.*, 1993). We reassigned to coil all helices shorter than 5 residues and all strands shorter than 3 residues.

In Table 1, we summarized the results of tests in accordance with equation (8). For all residue environment parameters and for all conditions considered, the P-values are far below 10^{-10} and the homogeneity hypothesis can be rejected without risk in all cases. The inhomogeneity affects both average function value (χ^2-test) as well as its median (Kruskal-Wallis test). Nevertheless, the inhomogeneity is different for the residue environment parameters considered. It is improbable to expect even distribution of backbone conformational states among protein structures and this is reflected by the significances. Likewise, the $g^{(k)}$ functions for pairwise distances are highly inhomogeneous. Total SASA is closest to homogeneity among the three types of parameters analyzed. If estimated for an average protein, the significance of the test is in the order of 1%. We have searched for systematic deviations from homogeneity. Such deviation occur, for example, as a result of various domain size (sequence length) or amino acid composition in the structural learning set. If the homogeneity is calculated over subclasses of structures with similar length and amino acid composition, significances increase but are still very low (data not shown).

Table 1. Homogeneity tests for distribution density function $g^{(k)}(x)$. The test is calculated in accordance with hypothesis H_0 in equation (8) for the total solvent accessible surface area (total SASA), backbone conformation (DSSP and 5 conformational states) as well as pairwise residue-residue distances (PD) with 1, 10, and 20 residues minimal sequence separation. We present the value of the statistic (χ^2: χ^2-test, KW- Kruskal-Wallis test, see Appendix) as well as the significance ***P***. In the case of the χ^2-test, we calculated also an average ***P***-value per protein ($P_{protein}$) from the χ^2-statistic divided by the number of proteins in the set. The Kruskal-Wallis test was applied only for the total SASA having continuous real values. The degrees of freedom (column ***df***) were calculated in accordance with the formulas in Appendix but diminished by the number of empty bins. The total number of proteins was 127 (126 – without 3bcl – for pairwise distances). For the SASA evaluation, the results were distributed in 5 bins. Bin width was 1 Å for pairwise distances up to a maximum of 30 Å (in total 28 occupied bins). The total number of distances was 2983513, 3215698 and 3435651 for the sequence separations 20, 10 and 1, respectively. The result "0.000" indicates a positive value smaller than the smallest positive real machine number. "n/a" – non-applicable.

	df	χ^2	***P***	$P_{protein}$	***df***	***KW***	***P***
Total SASA	504	1596.5	1.04E-113	0.014	126	1022.5	2.94E-140
DSSP	252	5790.3	0.000	6.92E-10	n/a	n/a	n/a
5 states	504	6683.4	0.000	1.02E-10	n/a	n/a	n/a
PD (20 res.)	3500	381831.3	0.000	0.000	n/a	n/a	n/a
PD (10 res.)	3500	414661.2	0.000	0.000	n/a	n/a	n/a
PD (1 res.)	3500	465073.9	0.000	0.000	n/a	n/a	n/a

In the upper part of Table 2, we present the results of statistical tests calculated in accordance with equation (9). The statistics (χ^2 or KW) for the $f_a^{(k)}$ functions have been summed up over all amino acid types. In this way, we estimate the significance for all $f_a^{(k)}$ functions together (for all homogeneity tests simultaneously). In average, the $f_a^{(k)}$ functions appear also highly inhomogeneous although slightly less than the $g^{(k)}$ functions as is exemplified by the significances in the upper part of Table 2. Again, the total solvent accessible surface area performs far better than any other parameter in this test. The minimal sequence separation appears to have an influence on the inhomogeneity of pairwise residue-residue distances. The significance is largest for the ≥20 residues separation.

Separate analysis of the $f_a^{(k)}$ functions with regard to amino acid types shows a more heterogeneous picture. The computation results for the total SASA are listed in Table 3. We find significances (***P***-values) above 0.05 for glutamine, proline and tryptophan in the χ^2-test; thus, the homogeneity

Table 2. Homogeneity tests for distribution functions $f_a^{(k)}(x)$ averaged over all amino acid types. The test is calculated in accordance with hypothesis H_0 in equation (9) for the case of model 1 and in equations (12) and (13) for the case of model 2. As environment parameters, the total solvent accessible surface area (total SASA), backbone conformation (DSSP and 5 conformational states) as well as pairwise residue-residue distances (PD) with 1, 10, and 20 residues minimal sequence separation considered. For column headings see legend of Table 1.

	df	χ^2	***P***	P_{protein}	***df***	***KW***	***P***
Model 1							
Total SASA	9064	11526.9	4.29E-64	0.024	2266	4135.4	1.70E-112
DSSP	4532	9833.4	0.000	6.94E-05	n/a	n/a	n/a
5 states	8817	16143.8	0.000	6.34E-05	n/a	n/a	n/a
PD (20 res.)	623143	1.34E+06	0.000	1.90E-307	n/a	n/a	n/a
PD (10 res.)	629563	1.34E+06	0.000	0.000	n/a	n/a	n/a
PD (1 res.)	3500	1.36E+06	0.000	0.000	n/a	n/a	n/a
Model 2							
Total SASA	8988	10288.8	8.88E-21	0.104	n/a	n/a	n/a
DSSP	4272	4892.2	6.94E-11	0.142	n/a	n/a	n/a
5 states	7477	8682.3	3.40E-21	0.413	n/a	n/a	n/a
PD (20 res.)	619195	797855.6	0.000	0.385	n/a	n/a	n/a
PD (10 res.)	626432	719395.2	0.000	0.355	n/a	n/a	n/a
PD (1 res.)	632194	719816.0	0.000	0.380	n/a	n/a	n/a

hypothesis cannot be rejected for them. The ***P***-values calculated in the Kruskal-Wallis test are generally lower but, nevertheless, significances above 0.001 have been obtained for asparagines, glutamine, methionine, proline and tryptophan. For 8 amino acid types in the χ^2-test and 15 in the Kruskal-Wallis test, the significance is below 0.001 and homogeneity can be rejected for them. Apparently, the $f_a^{(k)}$ functions are more homogeneous for rarely occurring amino acid types; thus, the higher significances reflect, at least in some cases, the lack of data in the structural learning set. The significances computed for the average protein (P_{protein}) are all above 17% but, within a single protein, each amino acid does not occur very often (<<100 times) and the high P_{protein} may also be a consequence of the limited data set (naturally, since the typical protein in the learning set contains about 200 amino acid residues).

In the case of both backbone structural descriptors (see Table 4), the significances (***P***-values) are generally below 10^{-6} except for cysteine,

Table 3. Homogeneity tests for amino acid type-specific distribution functions $f_a^{(k)}(x)$: Total surface area. The statistical tests results for the distribution functions F_a have been calculated separately for each amino acid type (homogeneity model 1, equation 9). N_a denotes the occurrence of the amino acid type a in the sequences of the learning set. The number of degrees of freedom df in the χ^2-test is 504 (5 bins) if all bins are occupied. For the Kruskal-Wallis test, df is 126. We present the value of the statistic (χ^2, KW) as well as the P-value. In the case of the χ^2-test, we calculated also an average P-value per protein (P_{protein}) from the χ^2-statistic divided by the number of proteins in the set.

Amino acid type	N_a	χ^2-test				Kruskal-Wallis test		
		df	χ^2	P	P_{protein}	df	KW	P
ALA	2288	504	714.47	1.80E-09	0.229	126	318.01	1.40E-18
ARG	1042	480	575.95	1.68E-03	0.313	120	252.72	1.75E-11
ASN	1327	504	566.59	2.77E-02	0.347	126	174.67	2.70E-03
ASP	1615	500	633.93	4.29E-05	0.284	125	279.67	7.45E-14
CYS	448	376	455.73	3.00E-03	0.309	94	159.30	3.03E-05
GLN	947	488	502.25	3.18E-01	0.395	122	153.11	2.97E-02
GLU	1387	500	608.12	6.40E-04	0.306	125	229.82	3.39E-08
GLY	2151	504	615.63	4.73E-04	0.303	126	237.78	6.93E-09
HIS	488	420	472.64	3.85E-02	0.347	105	173.98	2.69E-05
ILE	1411	496	624.44	7.47E-05	0.288	124	220.87	2.00E-07
LEU	1926	500	618.21	2.33E-04	0.297	125	199.17	2.75E-05
LYS	1559	496	794.96	3.19E-16	0.174	124	271.29	5.20E-13
MET	450	428	477.84	4.80E-02	0.352	107	129.36	6.97E-02
PHE	1011	488	552.13	2.33E-02	0.344	122	179.50	5.49E-04
PRO	1084	496	497.98	4.67E-01	0.408	124	169.05	4.47E-03
SER	1779	500	600.97	1.26E-03	0.312	125	205.44	7.73E-06
THR	1606	492	563.16	1.43E-02	0.338	123	209.74	1.80E-06
TRP	412	420	448.51	1.62E-01	0.376	105	138.48	1.59E-02
TYR	932	472	554.98	4.96E-03	0.324	118	177.11	3.53E-04
VAL	1775	504	648.41	1.36E-05	0.277	126	256.37	6.31E-11

histidine, methionine and tryptophan. The extremely small value for phenylalanine (in the case of 5 conformational states) is caused by a residue with typically excluded backbone dihedrals in 2sn3. The P_{protein} values are generally lower than for total SASA but, in most cases, still above 5%.

For pairwise residue-residue distance functions, all P-values in the statistical test (9) are far below 0.001. At the left side of Table 5, the P_{protein} data is summarized. The number of pair functions involving a specific amino acid type with a significance <0.001 for an average protein (P_{protein}) in the test (7) is presented. In these cases, homogeneity can be ruled out without risk. For distance functions of amino acid pairs including cysteine, histidine,

Table 4. Homogeneity tests (model 1) for amino acid type-specific distribution functions $f_a^{(k)}(x)$: Secondary structures and conformational states. For dataset description, table headings, amino acid type occurrences N_a and calculations of degrees of freedom df, see legend of Table 3. Some amino acid types do not occur in certain proteins or in certain secondary structural states, therefore, the number of degrees of freedom is smaller in these cases. Secondary structure was defined in accordance with DSSP (Kabsch and Sander, 1983); i.e., 3 bins enter the computation of degrees of freedoms. The conformational state of a residue among a classification of 5 states was calculated following Nishikawa and Matsuo (1993); *i.e.*, 5 bins are to be considered.

Amino acid type	DSSP				5 conformational states			
	df	χ^2	P	$P_{protein}$	df	χ^2	P	$P_{protein}$
ALA	252	804.92	4.89E-59	0.042	504	1226.43	4.94E-62	0.047
ARG	240	426.76	1.27E-12	0.171	480	1076.41	9.58E-48	0.064
ASN	252	465.54	6.77E-15	0.160	504	715.04	1.65E-09	0.229
ASP	250	474.33	4.43E-16	0.152	500	724.11	1.91E-10	0.219
CYS	188	269.75	8.57E-05	0.242	376	450.47	4.96E-03	0.315
GLN	244	441.47	1.50E-13	0.166	488	731.42	5.29E-12	0.203
GLU	250	558.55	1.25E-25	0.109	500	914.21	1.13E-26	0.123
GLY	252	654.35	1.55E-37	0.076	504	911.68	6.74E-26	0.127
HIS	210	278.55	1.07E-03	0.269	420	482.29	1.90E-02	0.337
ILE	248	541.74	5.87E-24	0.115	496	764.9	8.39E-14	0.190
LEU	250	677.62	3.89E-41	0.068	500	944	1.09E-29	0.112
LYS	248	586.96	1.60E-29	0.096	496	925.39	2.46E-28	0.116
MET	214	301.87	7.06E-05	0.247	428	478.63	4.56E-02	0.351
PHE	244	403.95	5.07E-10	0.194	488	1529.8	8.48E-108	0.014
PRO	248	387.28	3.53E-08	0.212	372	506.94	3.84E-06	0.399
SER	250	609.7	4.78E-32	0.089	500	891.59	1.86E-24	0.132
THR	246	520.11	9.45E-22	0.123	369	679.55	1.07E-20	0.241
TRP	210	285.75	3.92E-04	0.260	420	534.58	1.24E-04	0.283
TYR	236	479.17	1.08E-18	0.134	472	783.14	8.34E-18	0.160
VAL	252	665.07	5.50E-39	0.073	504	873.19	3.22E-22	0.143

methionine or tryptophan, homogeneity cannot be excluded. Unfortunately, these are also rare amino acid types in globular proteins. Since the relative strength of deviation from homogeneity can be approximately quantified with Bahadur's slope ($=\ln(P)/N$, here is N the number of data, see Sunyaev *et al.* (1998), the small sample size does not allow a final decision. For the remaining amino acid types, the homogeneity can be rejected for a medium (arginine, glutamine, proline) or large number of pair functions with this amino acid. As a tendency, we observe less inhomogeneity for sequence separation ≥20 residues.

Table 5. Homogeneity tests (model 1) and goodness-of-fit test (model 2) for distribution functions $f_{ab}^{(k)}(x)$ specific for amino acid type pairs: Pairwise residue-residue distances. We computed the χ^2-test statistics averaged per protein for all 210 different amino acid type pairs a and b separately (for the learning set except of 3bcl). Since the sequential separation may have an influence on the distance distribution, minimal sequence separations of 1, 10 and 20 residues were independently analyzed. The distance histograms have been calculated with 1 Å bins up to a maximum of 30 Å. In the table, we list the number of residue-residue distance functions involving the given amino acid for which the P-value of the χ^2-test is <0.001 (for model 1, clear cases of absent homogeneity) and >0.05 (for model 2, rejection of homogeneity hypothesis is not possible).

Amino acid type	Model 1			Model 2		
	Sequence separation ≥			Sequence separation ≥		
	20	10	1	20	10	1
ALA	16	16	16	1	1	2
ARG	3	3	4	4	6	9
ASN	11	11	11	1	4	2
ASP	11	10	11	1	1	2
CYS	0	0	0	11	10	8
GLN	6	8	5	4	2	5
GLU	10	8	11	2	2	4
GLY	15	15	16	2	1	1
HIS	0	1	0	6	4	4
ILE	12	13	12	6	4	6
LEU	15	17	16	1	0	2
LYS	15	13	12	1	0	1
MET	1	0	0	1	4	7
PHE	13	12	11	3	4	5
PRO	6	5	5	3	3	1
SER	13	13	13	2	2	1
THR	14	14	14	3	2	3
TRP	0	0	0	4	7	11
TYR	10	10	9	6	7	9
VAL	15	16	16	4	3	6

3.3. Lessons from the Absence of Homogeneity

To conclude, the homogeneity tests (8) and (9) are not at all or very weakly fulfilled for solvent accessibility, backbone conformation or pairwise residue-residue functions. The conditions are best approximated for solvent accessibility and, among the distance functions, for ≥20 residues separation. Typically, the homogeneity hypothesis can be clearly rejected at the level of the test's ***P***-value. In tests for an average protein from the learning set, $\boldsymbol{P}_{\text{protein}}$ is often also small. The tests show that the empirical occurrences of amino acid types calculated for single proteins with respect to physical parameters such as solvent accessibility, backbone conformation and pairwise distances differ in a statistically significant manner from the average empirical occurrences in large protein structure sets. Hence, the distributions $f_a^{(\Sigma)}(x)$ and $g^{(\Sigma)}(x)$ calculated from a structure set cannot be transferred to a new protein. Although this might be not surprising, especially for secondary structure descriptors, this result has consequences for the derivation of knowledge-based potentials. Thus, a statistical distribution of the generally used residue environment parameters describing the behavior of residues regardless of globular protein type does not exist. An interpretation of the empirical distributions observed in tertiary structures in the framework of gas and liquid statistical physics, especially in context of the Boltzmann distribution as in equations (3) and (5), is not justified by the available structural data.

It should be noted that, in the case of predicting a match between a query protein and a structural template, the sequence and the amino acid composition $\left\{v_a^{(k)}\right\}$ of the query protein are known in advance as well as $g^{(k)}(x)$ (*i.e.*, the collection of spatial sites) for the structural template. Thus, the construction of approximations for the functions $f_a^{(k)}(x)$ that are specific for the query protein is essential for the prediction task. The absence of homogeneity both for $g^{(k)}(x)$ and $f_a^{(k)}(x)$ will introduce non-statistical noise in the calculation of energies between native and non-native states with knowledge-based potentials that can go beyond any error limit. Even in the case if their ratio $f_a^{(k)}(x)/g^k(x)$ (*i.e.*, the propensity) was not dependent on an individual protein, the score E calculated in accordance with (6)

$$E = -RT\sum_a \sum_{(x)} f_a^{(k)}(x) \ln\left[f_a^{(k)}(x)/g^{(k)}(x)\right] \qquad (11)$$

depends on the reliability of the model for $f_a^{(k)}(x)$. Thus, it becomes also problematic to introduce thresholds for E associated with defined prediction accuracies (statistical significances).

4. Statistical Model 2: Existence of Amino Acid Type-Specific Propensities

4.1. Statistical Model 2: Theory

Alas, the derivation of the knowledge-based potential in the framework of statistical physics is not supported by our observations, the beauty of applying fundamental relationships is not justified by the context and a residue in a protein structure cannot be considered as an independently interacting particle. But maybe, equations (1) and (2) describe nevertheless useful empirical relationships and can be applied purely pragmatically. For example, propensities for secondary structural preferences of residues ($\boldsymbol{x}$ = helix, sheet or coil) (Garnier *et al.*, 1978; Gibrat *et al.*, 1987) or for interior/exterior distributions with respect to the protein globular surface (Miller *et al.*, 1987) have been used in early work since they were found to vary remarkably little among the protein sets studied at that time.

Within this framework, the existence of amino acid type-specific propensities $\Pi_a(\boldsymbol{x})$ is assumed that are only a function of the residue location parameter $\boldsymbol{x}$ and that are valid for any protein structure and relate the functions $f_a^{(k)}(x)$ and $g^{(k)}(x)$ for the $\boldsymbol{k}$-th protein to each other

$$f_a^{(k)}(x) = \Pi_a(x) \cdot g^{(k)}(x) \cdot \tag{12}$$

In the case of pairwise residue-residue distance functions between two amino acid residues of type $\boldsymbol{a}$ and $\boldsymbol{b}$, the equation (12) reads as

$$f_{ab}^{(k)}(x) = \Pi_{ab}(x) \cdot g^{(k)}(x) \tag{13}$$

and the expression $-RT \ln \Pi_a(x)$ (following equation 2) would enter the total score function. Typically, the propensities are estimated from as large as possible structural sets. In our case, they can be calculated as

$$\Pi_a(x) = \Pi_a^{(\Sigma)}(x) \approx \Pi_{a,*}^{(\Sigma)}(x) = \frac{f_{a,*}^{(\Sigma)}(x)}{g_*^{(\Sigma)}(x)} \tag{14}$$

using the empirical distribution density functions $f_{a,*}^{(\Sigma)}(x)$ and $g_*^{(\Sigma)}(x)$ from the whole learning set.

4.2. Statistical Model 2: Methods and Results

Relationships (12) and (13) can also be verified with statistical criteria evaluating the so-called goodness of fit. The hypothesis $\boldsymbol{H}_0$ involves that the equation (12) (or 13, respectively) holds for all proteins $\boldsymbol{k}$ with $1 \leq k \leq K$ and all values $\boldsymbol{x}$. The alternative hypothesis assumes that this is not true for at least one protein. The χ^2-criterion is a useful measure for the deviation between the two sides of equations (12) and (13). We checked the same residue environment parameters as in the case of model 1.

In the lower part of Table 2, we have summarized the calculation results for the simultaneous tests for all amino acid types together. Except for pairwise distances, the $\boldsymbol{P}$-values for the test are dramatically increased but still much lower than 0.001 so that the hypothesis of a good fit in (12) or (13) can be rejected without risk. Although the expected average number of instances per average protein is too small for a statistical test, the improvement of P_{protein} (>0.100 for all parameters) indicates that model 2 matches the protein structure reality apparently much better than model 1. Surprisingly, the assumption of protein-independent propensities appears slightly worse for sequence separation ≥10 residues among the three variants of distance functions as judged by the P_{protein} values.

The tendency towards higher significances is also followed in the results for the specific amino acid types. In Table 6, the computation results for the total SASA are listed. We find that 14 amino acid types exhibit significances above 0.05; *i.e.*, the hypothesis of a good fit in equation (12) cannot be rejected. Only for two types (leucine and lysine), the significance is below 0.001 and the hypothesis $\boldsymbol{H}_0$ can be rightly rejected. Apparently, a few buried lysines and highly accessible leucines in the structure sample are responsible. Surprisingly, the $\boldsymbol{P}$-value in this test for lysine (10^{-22}) is worse than that for model 1 (10^{-16}). For asparagines, serine, tryptophane and valine, the significance does not allow a clear distinction. The significances calculated per average protein (P_{protein}) are all above 0.300 except for leucine and lysine.

The goodness of fit with secondary structural propensities apparently depends to a certain extent on the coarseness of backbone conformation description (see Table 7). In the case of DSSP, the significances for 14 amino acid types

Table 6. Evaluation of the goodness of fit (model 2) for amino acid type-specific distribution functions $f_a^{(k)}(x)$: Total surface area. For dataset description, table headings, amino acid type occurrences N_a and calculations of degrees of freedom df (note the change in degrees of freedom as a result of the linear fit), see legend of Table 3.

Amino acid type	N_a	χ^2-test			
		df	χ^2	*P*	$P_{protein}$
ALA	2288	500	504.33	4.37E-01	0.410
ARG	1042	476	463.82	6.47E-01	0.429
ASN	1327	500	528.42	1.83E-01	0.385
ASP	1615	496	526.01	1.70E-01	0.383
CYS	448	372	429.71	2.07E-02	0.340
GLN	947	484	430.32	9.62E-01	0.478
GLU	1387	496	519.32	2.27E-01	0.390
GLY	2151	500	512.88	3.35E-01	0.401
HIS	488	416	401.11	6.91E-01	0.436
ILE	1411	492	516.00	2.19E-01	0.3890
LEU	1926	496	639.63	1.32E-05	0.280
LYS	1559	492	859.81	1.98E-22	0.142
MET	450	424	425.35	4.72E-01	0.414
PHE	1011	484	506.00	2.36E-01	0.391
PRO	1084	492	431.26	9.77E-01	0.486
SER	1779	496	553.32	3.79E-02	0.356
THR	1606	488	478.22	6.16E-01	0.426
TRP	412	416	467.05	4.23E-02	0.354
TYR	932	468	518.05	5.45E-02	0.360
VAL	1775	500	578.21	8.76E-03	0.336

are all above 0.05 indicating no reason for rejection of a good fit. For asparagine, cysteine, glycine, histidine, proline and serine, the ***P***-values are between 0.05 and 0.001 with the lowest value observed for serine (2.05E-3). In the case of the 5 conformational states defined by Nishikawa and Matsuo (1993), significances above 0.05 have been obtained for 12 amino acid types. For phenylalanine, a good fit can be apparently excluded considering the ***P***-value below 0.001 (one residue with strained backbone conformation in 2sn3 is the major cause). Alanine, arginine, asparagine, histidine, proline, serine and threonine have significances between 0.001 and 0.05 for this test. The $P_{protein}$ values are all above 0.300 for both backbone structural descriptors.

Table 7. Evaluation of the goodness of fit (model 2) for amino acid type-specific distribution functions $f_a^{(k)}(x)$: Secondary structures and 5 conformational states. For dataset description, table headings, amino acid type occurrences N_a and calculations of degrees of freedom df (note the change in degrees of freedom as a result of the linear fit), see legends of Table 3 and 4.

Amino acid type	DSSP				5 conformational states			
	df	χ^2	*P*	$P_{protein}$	*df*	χ^2	*P*	$P_{protein}$
ALA	239	273.83	6.03E-02	0.340	437	506.27	1.22E-02	0.408
ARG	227	171.05	9.98E-01	0.493	413	487.22	6.85E-03	0.402
ASN	239	290.19	1.31E-02	0.319	437	487.17	4.86E-02	0.429
ASP	237	242.52	3.89E-01	0.382	433	438.23	4.21E-01	0.481
CYS	175	210.40	3.49E-02	0.330	309	312.94	4.27E-01	0.510
GLN	231	229.00	5.25E-01	0.394	421	455.65	1.18E-01	0.447
GLU	237	213.45	8.62E-01	0.429	433	360.97	9.95E-01	0.581
GLY	239	282.89	2.70E-02	0.328	437	425.68	6.42E-01	0.501
HIS	197	233.30	3.92E-02	0.333	353	422.53	6.44E-03	0.408
ILE	235	266.23	7.90E-02	0.345	429	462.73	1.26E-01	0.448
LEU	237	256.34	1.85E-01	0.362	433	423.22	6.23E-01	0.500
LYS	235	245.60	3.04E-01	0.374	429	470.71	8.04E-02	0.439
MET	201	226.71	1.03E-01	0.350	361	333.96	8.43E-01	0.543
PHE	231	233.73	4.37E-01	0.387	421	653.42	2.54E-12	0.257
PRO	235	284.12	1.56E-02	0.321	305	367.11	8.46E-03	0.568
SER	237	304.34	2.05E-03	0.299	433	518.4	2.95E-03	0.391
THR	233	234.07	4.68E-01	0.389	302	364.4	7.99E-03	0.568
TRP	197	199.73	4.32E-01	0.390	353	385.5	1.13E-01	0.457
TYR	223	232.14	3.23E-01	0.377	405	390.88	6.84E-01	0.511
VAL	239	262.58	1.41E-01	0.356	437	415.3	7.65E-01	0.514

The propensity model is also much better applicable to the distance functions than model 1. Except for the lysine-phenylalanine distance function with ≥1 residue sequence separation, all $P_{protein}$ values are larger than 0.05 (data not shown). The statistical computations for the evaluation of goodness of fit in equation (13) with regard to pairwise residue-residue distance functions are summarized at the right side of Table 5. Even in this stronger test, there are cases when the goodness of fit cannot be rejected. The number of pair functions with a significance $P > 0.05$ involving a specific amino acid type is presented. Many pair functions (up to ~50%) are well fitted by equation (13) for generally rare amino acid types such as cysteine or tryptophane but not only (for example, arginine, glutamine, isoleucine or valine). From this data, sequence separation ≥1 appears to produce the largest number of good fits but the preference seems not dramatic. Maybe, secondary structures have an influence here and propensities are relatively well suited to describe amino acid type preferences in terms of local backbone-conformational descriptors.

To summarize, the computation show that amino acid type-dependent and protein-independent propensities are a much better approximation to the physical reality in protein structures than model 1. At least at the level of an average protein, the weakest test applied here, a good fit in equations (12) or (13) can typically not be rejected with statistical criteria. Apparently, propensities calculated for individual proteins coincide well from a statistical point of view with propensities calculated over a large learning set, although several cases of clear disagreement can be found. Hence, the propensities $\Pi_a^{(\Sigma)}(x)$ calculated from a structure set can be transferred to a new protein only to a certain extent. A significant, non-statistical error remains that depends partially on the environment variable selected (accessibility and backbone conformation are better fitted than residue-residue distances). This limited transferability of propensities introduces noise into any type of a propensity-based structure prediction algorithm. Calculating the propensities from a larger learning set does not reduce this systematic, non-statistical error. Many observations that we have collected in section (2.2.) are essentially connected with this limitedness of model 2.

5. Statistical Models 3 and 4: Removal of Inhomogeneity Caused by Amino Acid Composition

5.1. Restriction of Amino Acid Composition Variability by the Assumption of Protein-Independent Propensities

It can be demonstrated easily that the existence of protein-independent propensities for one-dimensional environment parameters such as solvent accessibility of backbone conformation descriptors restricts the variability of amino acid composition of proteins. If $v_a^{(k)}$ denotes the fractional occurrence of amino acid type $\boldsymbol{a}$ in the sequence of protein $\boldsymbol{k}$, then $g^{(k)}(x)$ can be represented as

$$g^{(k)}(x) = \sum_a v_a^{(k)} f_a^{(k)} = \sum_a v_a^{(k)} \Pi_a(x) g^{(k)}(x) \qquad (15)$$

relying on equations (7) and (12). Thus, we obtain

$$\sum_a \nu_a^{(k)} \Pi_a(x) = 1 \tag{16}$$

from (15) in addition to

$$\sum_a \nu_a^{(k)} = 1, \tag{17}$$

the normalization condition of fractional occurrences. Equation (16) is rather a system of equations (one for every x with $g^{(k)}(x) \neq 0$). Thus, it is not excluded that there are more linearly independent equations than amino acid types! Similarly, restrictions in the composition of amino acid residue pairs are the result of protein-independent propensities for residue-residue distance functions. One might dream up that only such sequence-structure matches do exist that (16) is automatically fulfilled. The very moderate significances in the goodness-of-fit calculations for model 2 imply that this is not the case.

This difficulty can be avoided with amino acid composition-specific propensities. We generalize equation (12) by including a factor that takes into account the specifics of protein k

$$f_a^{(k)}(x) = C^{(k)}(x) \cdot \Pi_a(x) \cdot g^{(k)}(x). \tag{18}$$

Whereas $g^{(k)}(x)$ describes the availability of spatial sites with the property x and $\Pi_a(x)$ evaluates the ability of amino acid type a to occupy the site x, the term $C^{(k)}(x)$ is dependent on the amino acid composition of protein k (we sum up over all amino acid types b)

$$C^{(k)}(x) = \left[\sum_b \nu_b^{(k)} \Pi_b(x) \right]^{-1}. \tag{19}$$

The product $C^{(k)}(x) \cdot \Pi_a(x)$ can be considered as a protein- and amino acid type-specific propensity. We will refer to the approach leading to equations

(18) (and (19)) as model 3. A calculation of the simultaneous goodness-of-fit test for all amino acid types for this model with regard to secondary structure descriptor DSSP revealed an improvement in the significance from ~10^{-11} (see Table 2) to ~10^{-5}; still a low value <0.001. For all but five amino acid types (cysteine, glycine, histidine, proline and serine; by the way, also in model 2), we found significances above 0.05 in the amino acid type-specific test (data not shown).

We have tested an even more complicated model 4 with

$$f_a^{(k)}(x) = C^{(k)}(x) \cdot D_a^{(k)} \cdot \Pi_a(x) \cdot g^{(k)}(x) \tag{20}$$

with the equation system (21) and (22)

$$D_a^{(k)} = \sum_y g^{(k)}(y) C^{(k)}(y) \Pi_a(y) \tag{21}$$

$$C^{(k)}(x) = \sum_b v_a^{(k)} \Pi_a(x) D_a^{(k)} \tag{22}$$

where we sum up over all residue environment states ***y*** in equation (20). The derivation of (21) and (22) is analogous to that of (16) if the relationship $\sum_x f_a^{(k)}(x) = 1$ is taken into account. It should be noted that the function $g^{(k)}(x)$ is known prior to prediction since this is the definition of a structural template. In the case of DSSP, we find the significance being ~0.001 within one order of magnitude for the simultaneous test for all amino acid types (good but not excitingly high) and ***P***-values above 0.05 for all but two (alanine and histidine) amino acid types.

6. Concluding Remarks

Obviously, there are other systematic factors that cause inhomogeneity in the statistical tests. Sequence length is another important attribute of proteins in this respect. For example, the probability of a residue to be buried is certainly much larger in a protein with longer sequence than in a small globule. It can be demonstrated (data not shown) that a restriction of the learning set on proteins with similar length and/or amino acid composition improves the ***P***-values in statistical tests (8) and (12).

Generally, the so-called knowledge-based functions for the evaluation of a match between an amino acid and a structural site have problems with highly strained residue conformations. If there are a few sites in protein ***k*** where amino acid residues ***a*** have to occupy highly non-optimal environments, then the protein-specific function $f_a^{(k)}$ is strongly different from $f_a^{(\Sigma)}$ and the extrapolation error to the new protein structure will be especially dramatic. We have seen that a single conformationally strained phenylalanine in 2sn3 or a few buried lysines or accessible leucines (*i.e.*, outliers in the distributions) can drastically reduce the ***P***-values computed in statistical tests for model 2. Apparently, the superposition of decomposed single residue-single site matches does not work in these cases and the sequence context becomes more important.

In the middle of the nineties of the last century, the scientific journals in the protein structure field were full of communications, which considered methodical variants and applications of so-called threading techniques that relied on equation (2). For the time before December 1994 (Eisenhaber *et al.*, 1995b), we have counted more than 300 papers that could be related to that area.

Less than 10 years later, the same threading methods have a mere homeopathic existence in the field. Although they have never experienced a public, generally accepted critical analysis of methodical progress and conceptual restrictions except for comparative structure prediction tests, the more pragmatically oriented majority of the scientific community has moved away from them since they did not prove their advantage for structural and functional annotation of non-characterized proteins, the major task in the genome-sequencing era. Much more simple sequence-sequence and sequence-profile comparison techniques with the additional benefit of statistical evaluation of false positive match have occupied the scene and annotation transfer from characterized proteins has become an essential method for tertiary structure prediction.

With the concepts (1) of transferability (the evaluation of the correctness of approximation of $f_a^{(k)}(x)$ for the unknown protein ***k*** with functions calculated from a structural set Σ, especially with $f_a^{(\Sigma)}(x)$; see this text) and (2) of discriminative power (the evaluation of differences between $f_a^{(\Sigma)}(x)$ and $g^{(\Sigma)}(x)$, test whether the function $f_a^{(\Sigma)}(x)$ is indeed amino acid type-specific; see Sunyaev *et al.* (1998), it is possible to quantify to which extent knowledge-based functions are indeed useful for predictions. We have seen here that the condition (1) is not at all fulfilled for the statistical model 1 whereas it is not strictly true for model 2. It is also known that even a combination of all generally applied residue environment descriptors does have enough discriminative power to distinguish more than two thirds of all amino acid types from an average residue. If knowledge-based functions cannot be transferred from structural sets to new proteins without considerable, non-statistical error or if the functions for different amino acid types differ from each other only by statistical noise, there is little reason to blame other details of a prediction technique for a possible failure in structure prediction.

The techniques described in this work can be applied not only for the evaluation of classical threading potentials but also for any propensity-based function. It is of major importance to trace systematic sources of inhomogeneity such as residue composition, protein length, etc. and to modify the statistical model or to consider only subgroups of the learning set of proteins, which are homogeneous with respect to this factor.

Acknowledgement

This work received support from the Russian Fund of Fundamental Research (grant 94-07-13186) and from the Boehringer-Ingelheim Foundation (program for visiting scientists). Special thanks are given to the members of the seminar of computer genetics of the EIMB RAS in Moscow, especially to M.S. Gelfand, V.Yu. Makeev, A.A. Mironov, and M.A. Roytberg for their many questions and suggestions with respect to major statements of this paper. F.E. acknowledges financial support from Boehringer Ingelheim International. We are grateful to Sebastian Maurer-Stroh for critically reading the manuscript.

References

Bork, P., and Gibson, T.J. 1996. Applying motif and profile searches. Meth. Enzymol. 266: 162-184

Bork, P., and Koonin, E.V. 1998. Predicting function from protein sequence: Where are the bottlenecks? Nat. Genet. 13: 313-318

Bowie, J.U., Lüthy, R., and Eisenberg, D. 1991. A method to identify protein sequences that fold into a known three-dimensional stricture. Science 253: 164-170

Breiman, L. 1973. Statistics. With a view toward applications. Houghton Mifflin Company, Boston.

Cid, H., Bunster, M., Canales, M., and Gazitua, F. 1992. Hydrophobicity and structural classes in proteins. Protein Engineering 5: 373-375

Colloc'h, N., Etchebest, C., Thoreau, E., Henrissat, B., and Mornon, J.-P. 1993. Comparison of three algorithms for the assignment of secondary structure in proteins: The advantage of a consensus assignment. Protein Engineering 6: 377-382

Crippen, G.M. 1996. Failures of inverse folding and threading with gapped alignment. Proteins: Structure, Functions, Genetics. 26: 167-171

Crippen, G.M., and Viswanadhan, V.N. 1985. Sidechain and backbone potential function for conformational analysis of proteins. Int. J. Peptide Prot.Res. 25: 487-509

Eisenhaber, F., and Argos, P. 1993. Improved strategy in analytic surface calculation for molecular systems: handling of singularities and computational efficiency. J. Comput. Chem. 14: 1272-1280

Eisenhaber, F., Lijnzaad, P., Argos, P., Sander, C., and Scharf, M. 1995a. The double cubic lattice method: Efficient approaches to numerical integration of surface area and volume and for generating dot surfaces of molecular assemblies. J. Comput. Chem. 16: 273-284

Eisenhaber, F., Persson, B., and Argos, P. 1995b. Protein structure prediction: recognition of primary, secondary, and tertiary structural features from amino acid sequence. CRC Crit. Rev. Biochem. Mol. Biol. 30: 1-94

Finkelstein, A.V., Badretdinov, A.Ya., and Gutin, A.M. 1995a. Why do protein architectures have Boltzmann-like statistics ? Proteins: Structure, Functions, Genetics. 23: 142-150

Finkelstein, A.V., Gutin, A.M., and Badretdinov, A.Ya. 1995b. Perfect temperature for protein structure prediction and folding. Proteins: Structure, Functions, Genetics 23: 151-162

Fischer, D., and Eisenberg, D. 1997. Assigning folds to the proteins encoded by the genome of *Mycoplasma genitalium*. Proc. Natl. Acad. Sci. USA 94: 11929-11934

Furuichi, E., and Koehl, P. 1998. Influence of protein structure databases on the predictive power of statistical pair potentials. Proteins: Structure, Functions, Genetics. 31: 139-149

Garnier, J., Osguthorpe, D.J., and Robson, B. 1978. Analysis of the accuracy and implications of simple methods for predicting the secondary structure of globular proteins. J. Mol. Biol. 120: 97-120

Gibrat, J.-F., Garnier, J., and Robson, B. 1987. Further developments of protein secondary structure prediction using information theory. New parameters and consideration of residue pairs. J. Mol. Biol. 198: 425-443

Godzik, A., Kolinski, A., and Skolnick, J. 1995. Are proteins ideal mixtures of amino acids? Analysis of energy parameter sets. Protein Sci. 4: 2107-2117

Huynen, M., Doerks, T., Eisenhaber, F., Orengo, C., Sunyaev, S.R., Yuan, Y., and Bork, P. 1998. Homology-based fold predictions for *Mycoplasma genitalium* proteins. J. Mol. Biol. 280: 323-326

Jaroszewski, L., Rychlewski, L., Zhang, B., and Godzik, A. 1998. Fold prediction by a hierarchy of sequence, threading and modeling methods. Protein Sci. 7: 1431-1440

Jernigan, R.L., and Bahar, I. 1996. Structure-derived potentials and protein simulations. Curr. Opin. Struct. Biol. 6: 195-209

Jones, D.T., and Thornton, J.M. 1996. Potential energy functions for threading. Curr. Opin. Struct. Biol. 6: 210-216

Juffer, A., Eisenhaber, F., Hubbard, S.J., Walther, D., and Argos, P. 1995. Comparison of atomic solvation parameter sets: applicability and limitations in protein folding and binding. Protein Sci. 4: 2499-2509

Kabsch, W., and Sander, C. 1983. Dictionary of protein secondary structures: Pattern recognition of hydrogen-bonded and geometrical features. Biopolymers 22: 2577-2637

Kocher, J.-P.A., Rooman, M.J., and Wodak, S.J. 1994. Factors influencing the ability of knowledge-based potentials to identify native sequence-structure matches. J. Mol. Biol. 235: 1598-1613

Koppensteiner, W.A., and Sippl, M.J. 1998. Knowledge-based potentials - back to the roots. Biochemistry (Moscow), Engl.transl. 63: 293-299

Kyngäs, J., and Valjakka, J. 1998. Unreliability of the Chou-Fasman parameters in predicting protein secondary structure. Protein Engineering 11: 345-348

Lee, B., and Richards, F.M. 1971. The interpretation of protein structures: estimation of static accessibility. J. Mol. Biol. 55: 379-400

Lemer, C.M.R., Rooman, M.J., and Wodak, S.J. 1995. Protein structure prediction by Threading methods: Evaluation of current techniques. Proteins: Structure, Functions, Genetics 23: 337-355

Matsuo, Y., and Nishikawa, K. 1994. Protein database search and structure prediction by 3D-1D compatibility method. Protein Engineering 7: 1163-1163

Miller, S., Janin, J., Lesk, A.M., and Chothia, C. 1987. Interior and surface of monomeric proteins. J. Mol. Biol. 196: 641-656

Miyazawa, S., and Jernigan, R.L. 1985. Estimation of interresidue contact energies from protein crystal structures: Quasi-chemical approximation. Macromolecules 18: 534-552

Nishikawa, K., and Matsuo, Y. 1993. Development of pseudoenergy potentials for assessing protein 3-D - 1-D compatibility and detecting weak homologies. Protein Engineering 6: 811-820

Ota, M. and Nishikawa, K. 1995. Performance of a 3-D-1-D homology search. Protein Engineering 8: 950-951

Park, J., Karplus, K., Barret, C., Hughey, R., Haussler, D., Hubbard, R.E., and Chothia, C. 1998. Sequence comparisons using multiple sequences detect three times as many remote homologues as pairwise methods. J. Mol. Biol. 284: 1201-1210

Press, W.H., Teukolsky, S.A., Vetterling, W.T., and Flannery, B.P. 1992. Numerical recipes in C. The art of scientific computing. Cambridge University Press, Cambridge.

Rooman, M.J. and Wodak, S.J. 1995. Are database-derived potentials valid for scoring both forward and inverted protein folding? Protein Engineering 8: 849-858

Sippl, M.J. 1990. Calculation of conformational ensambles from potentials of mean force. An approach to the knowledge-based prediction of local structures in globular proteins. J. Mol. Biol. 213: 859-883

Sippl, M.J. 1994. Boltzmann's principle, knowledge-based mean-field potentials and protein folding. An approach to the computational determination of protein structures. Journal of Computer-Aided Molecular Design 7: 473-501

Smith, T.F., Lo Conte, L., Bienkowska, J., Gaitatzes, C., Rogers, G.R., Jr., and Lathrop, R. 1998. Current limitations to protein threading approaches. J. Comp. Biol. 4: 217-225

Sunyaev, S.R., Eisenhaber, F., Argos, P., Kuznetsov, E.N., and Tumanyan, V.G. 1998. Are knowledge-based potentials derived from protein structures sets discriminative with respect to amino acid types? Proteins: Structure, Functions, Genetics 31: 225-246

Tanaka, S., and Scheraga, H.A. 1976. Medium and long-range interaction parameters between amino acids for predicting three-dimensional structure of proteins. Macromolecules 9: 945-950

Thomas, P.D., and Dill, K.A. 1996. Statistical potentials extracted from protein structures: How accurate are they? J. Mol. Biol. 257: 457-469

Appendix: Statistical Criteria for a k-Sample Homogeneity Test

Let us consider $\boldsymbol{k}$ sets of observations $\{\boldsymbol{x}\}_1$, $\{\boldsymbol{x}\}_2$,..., $\{\boldsymbol{x}\}_k$ of a residue environment parameter $\boldsymbol{x}$ comprising $\boldsymbol{n}_1, \boldsymbol{n}_2, ..., \boldsymbol{n}_k$ observations respectively. $\boldsymbol{N}$ is the total number of observations. For example, $\boldsymbol{k}$ may denote the protein index and each of the $\boldsymbol{k}$ sample sets may contain all observations of the backbone conformational state for one of the protein structures in the learning set for one amino acid type.

For the samples $\{x\}_1$, $\{x\}_2$,..., $\{x\}_k$ with respective empirical distribution functions $\boldsymbol{F}^{(1)*}(x)$, $\boldsymbol{F}^{(2)*}(x)$,..., $\boldsymbol{F}^{(k)*}(x)$, the homogeneity hypothesis $\boldsymbol{H}_0$ is the condition

$$\boldsymbol{H}_0\text{:}\ \ \boldsymbol{F}^{(1)}(\boldsymbol{x}) = \boldsymbol{F}^{(2)}(\boldsymbol{x}) = \ldots = \boldsymbol{F}^{(k)}(\boldsymbol{x})\ \cdot$$

If the hypothesis $\boldsymbol{H}_0$ can be accepted, a statistical distribution law $\boldsymbol{F}(x)$ common for all $\boldsymbol{k}$ sample sets can be suggested. The $\boldsymbol{k}$-sample χ^2-homogeneity test and the rank test of Kruskal and Wallis can be used to examine $\boldsymbol{H}_0$ (see a textbook on statistics; *e.g.*, Breiman 1973).

The χ^2-criterion tests the average deviation amongst the $\boldsymbol{k}$ distributions. The full range of $\boldsymbol{x}$ values is subdivided into $\boldsymbol{I}$ bins. The number of observations from the j-th sample in the i-th bin is denoted as $\boldsymbol{n}_{ij}$. With $\boldsymbol{n}_j$, we indicate the total number of observations of the j-th sample. The overall frequency of $\boldsymbol{x}$ values in the i-th bin for all samples is $\boldsymbol{p}_i$. If the homogeneity hypothesis is true, then the statistic χ^2

$$\chi^2 = \sum_j \sum_i \frac{(\boldsymbol{n}_{ij} - \boldsymbol{n}_j \boldsymbol{p}_i)^2}{\boldsymbol{n}_j \boldsymbol{p}_i} \qquad \text{(A1)}$$

is χ^2-distributed with $(\boldsymbol{k}\text{-}1)(\boldsymbol{I}\text{-}1)$ degrees of freedom.

The Kruskal-Wallis test is sensitive to median shifts among distributions. This test is especially powerful if all $\boldsymbol{k}$ distribution density functions have a similar form. The Kruskal-Wallis test is based on the use of the ranks of observed values instead of the values themselves. The initial data from all $\boldsymbol{k}$ samples have to be sorted. The rank of the i-th observation in the j-th sample is denoted as $\boldsymbol{R}_{ij}$. The Kruskal-Wallis statistic $\boldsymbol{K}$ is

$$K = \left(\frac{12}{N(N+1)} \sum_{j=1}^{k} \frac{R_{.j}^2}{n_j} \right) - 3(N+1), \tag{A2}$$

where N is the total number of observations in all samples and

$$R_{.j} = \sum_{i=1}^{n_j} R_{ij} \ ,$$

where n_j is the number of values in the j-th sample. The statistic K is asymptotically distributed as χ^2 with k-1 degrees of freedom if the homogeneity hypothesis is true. The Kruskal-Wallis criterion is strictly applicable only if all observations are different to allow an unambiguous determination of ranks. If not too many observations are equal, they can be given equal ranks and the Kruskal-Wallis test remains appropriate.

Each statistical criterion examines only some properties of the distribution functions $F^{(1)}(x)$, $F^{(2)}(x)$,..., $F^{(k)}(x)$ such as average deviation or shift. Therefore, a positive decision about the hypothesis H_0 can be taken only if several criteria independently support this conclusion.

In traditional test situations, the data in the value samples are used to compute a statistic (χ^2 or K) with a subsequent search for the corresponding P-value (significance) in statistical tables. In our case, the number of degrees of freedom and N are very large. They are beyond the tabulated limits of the statistical distributions. We applied computerized subroutines to calculate the P-values (with the help of numerical recipes from Press *et al.* (1992).

From: *Frontiers in Computational Genomics*

4

Protein Structure Modeling in Functional Genomics

Jürgen Kopp, Manuel C. Peitsch, Torsten Schwede

Abstract

Functional genomics can be greatly assisted by three-dimensional structures, giving insights to the molecular function of proteins. Although structural genomics projects are aiming to provide protein structures on a large scale, in the foreseeable future this can be achieved only for a small fraction of all protein sequences. Comparative modeling techniques are used to complement experimental methods by building models for protein sequences which are related to known structures. According to their quality, applications of three-dimensional models range from low-resolution functional assignments to molecular simulations at atomic level of detail.

1. Introduction

Three-dimensional protein structures are key to a comprehensive understanding of protein function and represent valuable information when trying to assign a function to every gene and gene product in a genome. For the first time mankind looked at its own blueprint in 2001 (Lander *et al.*, 2001; Venter *et al.*, 2001). While the number of completely sequenced genomes increases rapidly, the massive amount of data requires accurate analysis and gene-by-gene annotation to unravel the secrets of a genome

(Ashburner *et al.*, 2000). Highly sophisticated annotation pipelines have been developed by many bioinformatics scientists for similarity and pattern-based functional annotations of conceptual transcripts. These methods allow for a fairly shallow level of functional annotation, which in the vast majority of the cases will need experimental confirmation. In the last few years experimental techniques to analyze the transcriptome and the proteome improved substantially (Vukmirovic and Tilghman, 2000). However, functional information is missing for most of the genes or predicted open reading frames and assigning function to all genes in a given genome is the grand challenge of biology for the years to come. Three-dimensional protein structures supplied by an efficient combination of experimental structure determination and comparative modeling will open wide areas of structural space to answer functional questions. This chapter outlines how modeling of protein structures can provide valuable information for functional genomics.

2. Experimental Structure Determination

Perhaps the most widely known example of functional insights into biology through structure determination may not be from a protein but from DNA. The correct interpretation of the X-ray diffraction patterns recorded by Rosalind Franklin as a double helical structure by Watson and Crick directly led to the greatest biological insight of the last century: the modest statement that it did not escape their notice that the structure "immediately suggests a possible copying mechanism for the genetic material" (Watson and Crick, 1953) was the starting point for the genetic revolution. In most current research projects the elucidation of a protein structure comes near the end of the biological discovery process. After the biochemical functions and roles of a protein have been well characterized by molecular biology and genetic approaches, the experimental three-dimensional structure determination is the 'ultimate' step for understanding these molecular functions on an atomic base.

Today, several well established experimental techniques for macromolecular structure determination are available and used routinely: Starting with the pioneering work of Perutz and Kendrew, X-ray –crystallography evolved as the main technique to elucidate whole protein structures. This method can provide high-resolution structures of large proteins or protein complexes at an atomic level of detail. Prerequisite is the formation of highly ordered single crystals of the protein. The process of protein crystal formation is still not completely understood, and is mainly based on a systematic trial-and-error approach to screen experimental conditions (Abola *et al.*, 2000). Crystallization therefore is one of the bottlenecks in protein crystallography. Nuclear magnetic resonance (NMR) is a complementary experimental technique that can be applied to soluble proteins with a maximum molecular weight of about 30 kDa (Montelione *et al.*, 2000). The accuracy of NMR protein structures is usually described as comparable to medium resolution X-ray structures, although a direct comparison of accuracy of the two methods is difficult (Lee and Kollman, 2001).

Despite of the tremendous progress seen in recent years for soluble globular proteins, the experimental structure determination of membrane proteins remains a challenging endeavor. In such cases where the membrane proteins are forming 2-dimensional crystals, electron microscopy and electron crystallography have proven to be an interesting alternative (Baumeister and Steven, 2000; Glaeser, 1999). In combination with computational molecular modeling these techniques can provide insights into molecular mechanisms, which were formerly only achievable by X-ray studies (de Groot *et al.*, 2000; de Groot *et al.*, 2001; de Groot and Grubmüller, 2001).

High-Throughput Structure Determination

Major improvements of a variety of biochemical and physical techniques during the last few years have allowed for a quantum leap in the rate at which structures can be determined experimentally (Burley, 2000). High-level protein expression systems provide the amount of protein necessary for crystallization trials, which are carried out and analyzed automatically. High-field NMR spectrometers, cryogenic handling of crystals and the availability of new powerful tunable synchrotron sources (for multi wavelength anomalous diffraction; MAD) allow for an efficient collection of high quality data. Massively improved and automated programs for data processing and structure refinement accompany this progress. It is now therefore possible to realize a high-throughput structural genomics approach. Several projects are initiated to set up large-scale structural studies of proteins (Brenner, 2001). Although it is expected that most of the projects will be based on X-ray crystallography, a recent report describes a structural genomics approach using NMR with a starting base of 500 small proteins (Yee *et al.*, 2002).

Strategies in Structural Genomics – Experiment and Modeling

Despite of technical improvements in experimental structure determination, it will not be possible to solve a structure for every single gene product having reasonable biochemical and technical resources. In fact, for the foreseeable future the number of experimentally determined protein structures will remain about two orders of magnitude smaller than the number of known protein sequences (Sanchez *et al.*, 2000).

Therefore, computational methods, such as comparative structure modeling and fold recognition, have gained much interest to complement the experimental approaches. Considering a combination of high-throughput structure determination and comparative modeling, several different strategies have been evaluated to maximize the return of structural information, while minimizing the total cost of the project (Vitkup *et al.*, 2001). The selection of the targets for structure determination depends on the accuracy of the modeling methods. Assuming that comparative models based on a template structure with more than 30% sequence identity should be reasonably reliable, and that only 90% of the structural space will be covered in this approach, it is estimated that about 16,000 carefully selected, independent structure determinations are necessary to cover the structural diversity of proteins. Presuming a global coordination of the target selection, Vitkup *et al.* expect that this task can be accomplished within a decade.

3. Computational Methods for Structure Prediction

Ab initio Structure Prediction

From Anfinsen's pioneering folding experiments on ribonuclease A (Anfinsen, 1973) and numerous similar studies (Fersht and Daggett, 2002), it is known for the vast majority of globular proteins that the amino acid sequence is the only determinant for folding to its native structure. Current concepts of protein folding assume energy funnels that allow a defined protein chain to take multiple pathways to its native fold, which is at the global free energy minimum (Dill and Chan, 1997).

Since obviously the protein's sequence contains the necessary information to find its native structure, several ideas were explored in the last years to predict three-dimensional protein structures merely using the amino acid sequence as primary source of information (Bonneau and Baker, 2001).

Recently, IBM addressed this problem and announced the start of a five-year project called BLUE GENE. The goal is to build a massively parallel computer that will allow computationally intensive modeling of bio-molecular phenomena such as protein folding (Allen *et al.*, 2001).

Currently, the most successful *ab initio* protein structure prediction method is the "ROSETTA" approach – named after the stone that allowed the deciphering of Egyptian hieroglyphics. In a blind prediction experiment (CASP4), Baker and co-workers' (Bonneau *et al.*, 2001) prediction method performed significantly better than other methods (Lesk *et al.*, 2001). For 17 of the 21 domains attempted with fewer than 300 residues, fragments longer than 50 residues were predicted with RMSDs of < 6.5 Å. However, in consideration of the extremely low overall accuracy and the limitation to small fragments, *ab initio* structure prediction remains a goal for the future, rather than a present reality (Lazaridis and Karplus, 2000). It is conceivable that *ab initio* prediction will increase our understanding of the principles of protein structures, but regarding the fast progress in structural genomics, it is not unlikely that *ab initio* methods will reach maturity at a time when they are no longer needed.

Comparative Modeling

Despite of some progress in *ab initio* structure prediction, protein structure homology modeling (or comparative modeling) remains the most accurate of all current structure prediction methods (Tramontano *et al.*, 2001). Comparative modeling is based on the observation that small changes in a protein sequence normally result in small changes in its three-dimensional structure (Chothia and Lesk, 1986). Comparative modeling aims to predict the structure of a given protein sequence based on the known three-dimensional structure of one or more homologous proteins. This can be considered as the equivalent of a "structural extrapolation". Since three-dimensional structures of proteins within a family are more conserved than their sequences (Lesk and Chothia, 1980), we can assume that two proteins are structurally similar, when a similarity can be detected on the sequence level. Approximately 700,000 protein sequences are currently stored in sequence databases (Bairoch and Apweiler, 2000) compared to about 18,000 experimentally determined protein structures in the PDB database (Berman *et al.*, 2000). In any given genome, 20% to 50% of all sequences (open reading frames) show significant homology to at least one known protein structure, and are thus amenable to comparative homology modeling (Fischer and Eisenberg, 1997; Huynen *et al.*, 1998; Jones, 1999; Rychlewski *et al.*, 1998; Sanchez and Sali, 1998). Ongoing structural genomics projects are aiming to provide at least one template structure for most structural folds in the near future (Sanchez *et al.*, 2000). It is anticipated that comparative modeling will be applicable to most protein sequences in less than 10 years (Marti-Renom *et al.*, 2000; Sanchez and Sali, 1997a).

All homology modeling methods are based on the following four steps: template selection (fold assignment), alignment of target and template, model building and evaluation (assessment). In case the obtained model is not satisfying, these steps can be repeated or modified until satisfactory results are obtained. A schematic flow chart of the comparative modeling procedure is shown in Figure 1. The following paragraph will briefly summarize some of the techniques used for homology modeling; a more exhaustive review is given in Ref. (Marti-Renom *et al.*, 2000).

Template Selection

The first step in comparative modeling is to select the best template structures from protein structure databases such as PDB (Berman *et al.*, 2000), SCOP

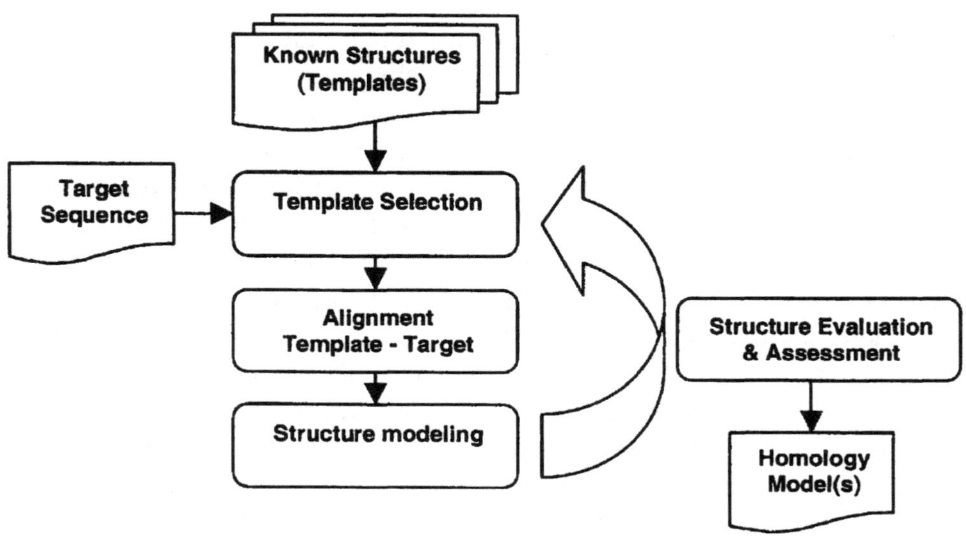

Figure 1. Flowchart of homology modeling. All homology-modeling methods consist of the steps template selection, target-template alignment, model building and model evaluation. These steps can be repeated iteratively if modeling errors are discovered in the evaluation step.

(Lo Conte *et al.*, 2002) or CATH (Pearl *et al.*, 2000). Using the target sequence as a query for searching, several programs are available to scan the sequences of these structure databases. Pair wise comparison methods, such as BLAST (Altschul *et al.*, 1990) or FASTA (Pearson, 1990) are widely used to identify closely related protein sequences. Multiple sequence comparison methods can enhance the sensitivity of the search. A frequently used program for this task is PSI-BLAST (Altschul *et al.*, 1997). For a given starting query, the program will collect a set of homologous sequences from the database, generate a weighted multiple sequence alignment of these sequences, derive a position specific scoring matrix from this alignment and use this matrix to search for further homologues. This process can be repeated iteratively, until a suitable template structure is identified or no more homologous sequences are found. Several related methods have been developed to identify remote sequence-structure relationships when the sequence identity between the target and the template drops below 25%, *e.g.* using profiles for targets and templates (Bates *et al.*, 2001) or Hidden Markov Models (Karplus *et al.*, 1999; Karplus and Hu, 2001).

In cases where no related sequences are available, and profile methods can therefore not increase the sensitivity, threading (or 3D-template matching) methods are useful. A pair wise comparison of a sequence and a protein structure is performed to determine if the sequence may be compatible with a fold by using a structure dependent scoring function to optimize the alignment. By threading the target sequence through a library of three-dimensional protein folds, possible templates are identified.

Although higher overall sequence similarity between the target and the template is a good criterion to select suitable template structures, there are other factors that have to be taken into account. It is important to consider the accuracy of different template structures, *e.g.* indicated by resolution and free R-factor for crystal structures, and the number of restraints per residues for NMR structures. In general, templates with a high resolution and low free R-factor are preferred. Even if the overall quality of a template structure is good, local inaccuracy might occur in parts of the protein with high mobility, often indicated by high B-factors. Several tools that allow evaluating the quality of protein structures are summarized in Table 2. On the other hand, the experimental conditions of the template structures have to be taken into consideration when choosing a template for model building. The choice of the template will be influenced by structural differences that play a biological role, such as the presence or absence of cofactors, ligands or substrates, the protein's oligomerization state, or different domain conformations.

Alignment of Target Sequence and Template

The alignment of target sequence and template structure needs particular attention since currently no comparative modeling method will recover from an incorrect alignment. Several specialized programs for multiple sequence alignments are available, *e.g.* CLUSTALW is widely used for this purpose (Thompson *et al.*, 1994). The weighted combination of different alignment methods (Notredame *et al.*, 1998; Notredame *et al.*, 2000) can significantly improve the alignment accuracy. Closely related proteins, *i.e.* over 40% sequence identity, are usually aligned correctly. When sequence identity drops below 30% in the so-called "twilight zone" (Rost, 1999), the alignment becomes non-trivial and a significant percentage of residues are likely to be misaligned (Johnson and Overington, 1993; Thompson *et al.*, 1999). Especially for low-homology sequence pairs local similarity may result in misleading alignments (Saqi *et al.*, 1998).

In any case, it is highly beneficial to include structural information in the alignment. For an automated procedure, multiple structural alignments of superposed template structures provide a good starting alignment, to which closely related sequences are added. Similarly a profile of sequences for the target protein is generated. Aligning the two profiles then generates the final alignment (Barton and Sternberg, 1987; Taylor *et al.*, 1994). In cases of low sequence-identity between target and template, it is mandatory to check and edit the alignment manually by inspecting the template structure to avoid the placement of insertions or deletions in structurally unfavorable regions such as secondary structure elements or tightly packed core regions. The evaluation of the quality of a three-dimensional model is more meaningful than the evaluation of alignment quality. Therefore it is advisable in ambiguous cases to build models for all alternative alignments and choose the best resulting three-dimensional model (Sanchez and Sali, 1997b).

Model Building

Several methods have been developed to build a three-dimensional model for a target protein sequence based on a given template structure and the alignment between target and template. The two most widely used techniques are "modeling by rigid body assembly", as implemented *e.g.* in programs such as COMPOSER (Sutcliffe *et al.*, 1987) or SWISS-MODEL (Guex and Peitsch, 1997), and "modeling by satisfaction of spatial restraints" used *e.g.* by MODELLER (Sali and Blundell, 1993) or CPHmodels (Lund *et al.*, 1997).

Modeling by rigid body assembly – This method is based on the observation that for a set of related protein structures, it is possible to identify a conserved core, where most residues can be superposed with small deviations, and structurally variable loop regions. As first step in model building, the selected template structures are superposed and the backbone atoms of the structurally conserved core regions are used to construct the framework of the model. For modeling loop regions no structural information can be derived from the template structures. These are filled in by scanning a database of known protein structures to identify fragments that are compatible with the anchor residues. When no compatible fragments can be identified in the library, other methods *e.g. ab initio* loop prediction can be applied. As next step, the resulting backbone model is completed by adding side chains, taking into account the corresponding side chains in the template structures, their backbone dependent conformational preferences, and their local packing environment. Finally, a short energy minimization or molecular dynamics step is used to release conformational strain and to improve the stereochemistry of the model.

Modeling by satisfaction of spatial restraints – This group of methods is based on the assumption that corresponding distances and angles between aligned residues in the template structure should be similar in the target. Therefore the first step in modeling consists in deriving a large number of constraints from a set of superposed template structures and assigning them to the target sequence based on the sequence alignment. Additional restraints from molecular mechanics force fields are normally added to enforce proper stereochemistry of bond length, angles, dihedrals and non-bonded atomic contacts. A model is then calculated by minimizing the violations of all these restraints (Sali and Blundell, 1993). The strength of "modeling by satisfaction of spatial restraints" is that information from different sources, such as NMR, cross-linking experiments or fluorescence spectroscopy can be combined with homology-derived restraints.

Modeling of loop regions – While the structural core of a given protein fold family normally is structurally well conserved, the insertion and deletion of residues is causing structural variability in loop regions connecting secondary structure elements. Loop regions may play an important role by contributing to specific interactions, such as *e.g.* active sites, recognition sites of receptors, or antigen-binding sites in immunoglobulins. Therefore, loop modeling will need close attention when building models to study protein interaction at the protein surface. Loop modeling has to predict the correct conformation of a short fragment, taking into account the fragment's sequence, the geometry of the anchor residues and the structure of the surrounding protein residues. Many different *ab initio* and database search methods have been developed to find compatible loop conformations. Unfortunately, current loop building algorithms are not reliable enough to construct loops longer than 5 residues. For an exhaustive list of techniques, see (Fiser *et al.*, 2000). Recently some progress has been made in modeling loops up to 8 residues with an acceptable accuracy by introducing a method accounting for the shape of the potential energy curve in the evaluation of the conformational free energy (Xiang *et al.*, 2002).

Errors in Protein Models

As for any predictive method, the protein model quality determines its usability and adequate level of interpretation. Therefore it is essential to evaluate the reliability of protein modeling methods. Comparing protein models with their experimental structures can identify errors in homology models. The biannual meetings on Critical Assessment of Techniques for Protein Structure Prediction (CASP) provide a forum for protein structure prediction and evaluation methods. The results of the last meeting held in December 2001 are summarized in a special issue of Proteins (Moult *et al.*, 2001). Sequences of unknown structures are presented as modeling targets, and the models are submitted to the organizers before the 3D structures of the prediction targets are determined by X-ray crystallography or NMR methods. The CASP meeting has been extended to include World Wide Web based completely automated protein structure modeling servers (Fischer *et al.*, 2001). Several methods for weekly fully automated assessment of prediction servers, *e.g.* EVA (Eyrich *et al.*, 2001; Rost and Eyrich, 2001), LiveBench (Bujnicki *et al.*, 2001), have been established to extend the evaluation to a statistically significant number of targets.

Errors in comparative models can be divided in five categories (Peitsch *et al.*, 2000; Sanchez and Sali, 1997b; Tramontano *et al.*, 2001):

- Template selection: In case of low homology between template and target (*i.e.* less than 20% sequence identity) incorrect target structures might be selected. It is not easy to distinguish models based on wrong template structures from those based on wrong alignments, since in both cases evaluation methods identify low model quality.
- Inaccurate alignments are the largest source of errors in comparative modeling, predominantly when the target-template sequence identity is lower than 30%. Often regions with misaligned residues can be identified

in the three-dimensional model, and then used to iteratively modify the alignment to improve model quality.

- Loop rebuilding: Regions in a model where no structurally equivalent residues are present in the target structure (insertions / deletions) can only be modeled successfully if the region is relatively short (less than 5-8 residues). This is further limited by the accuracy of the modeled environment, *i.e.* the conformation of the anchor residues and the neighboring residues.
- Distortions and subunits displacements: Different protein structures may exhibit local structural changes due to sequence divergence, crystal-packing constraints, different experimental conditions, and domain motions. Therefore, it is possible that even for correct alignments the model deviates significantly from the experimental structure.
- Side chain packing: The packing of side chains in the core of a protein may change with increasing sequence divergence due to its local packing. The correct conformation of side chains is critical in active site and contact regions such as ligand binding sites.

Evaluation of Protein Models

Since the "true" structure of the protein is not known, evaluation of a protein model aims to estimate the quality of the predicted model coordinates. Several evaluation programs and servers for different aspects have been developed (see Table 2).

Model evaluation starts with assessing fold selection and overall alignment. Correctly selected templates should have significant sequence similarity to the target, and functionally or structurally relevant residues should be well aligned. Models based on wrong templates or largely incorrect alignments can be identified by energy based scoring functions (Melo and Feytmans, 1998; Sippl, 1993). Sequence identity above 30% is a good indicator for the accuracy of protein models. Below 30 % the target-template sequence identity is not reliable as a measurement for model accuracy, since alignment errors become frequent and current modeling algorithms cannot recover from this type of error (Sali *et al.*, 1995; Sanchez and Sali, 1997b).

Statistical approaches identify errors in protein models by comparing them with distributions of spatial features, compiled from high-resolution X-ray structures. Additionally, deviations in stereochemistry, such as bond lengths, bond angles, peptide and aromatics side chain planarity, or clashes between

Table 1. World Wide Web Servers and Software Tools for homology modeling. This list is supposed to provide a starting point and is by far not exhaustive.

Protein Sequence Databases	
SwissProt /TrEMBL	www.expasy.org/sprot/ (Bairoch and Apweiler, 2000)
GenBank	www.ncbi.nih.gov/Genbank/ (Benson *et al.*, 2002)
Structure Databases	
PDB	www.pdb.org (Berman *et al.*, 2000)
CATH	www.biochem.ucl.ac.uk/bsm/cath/ (Orengo *et al.*, 1999)
SCOP	scop.mrc-lmb.cam.ac.uk/scop/ (Lo Conte *et al.*, 2002)
Homology Modeling Software	
COMPOSER	www-cryst.bioc.cam.ac.uk (Sutcliffe *et al.*, 1987)
MODELLER	guitar.rockefeller.edu/modeller/ (Sali and Blundell, 1993)
WhatIf	www.cmbi.kun.nl/whatif/ (Vriend, 1990)
InsightII, Quanta	www.accelrys.com (Accelrys Inc, St. Diego, CA, USA)
SYBYL	www.tripos.com (Tripos Inc. St.Luis, MO, USA)
Homology Modeling Server	
SWISS-MODEL	www.expasy.org/swissmod/ (Guex and Peitsch, 1997)
CPHmodels	www.cbs.dtu.dk/services/CPHmodels/ (Lund *et al.*, 1997)
3D-Jigsaw	www.bmm.icnet.uk/~3djigsaw/ (Bates *et al.*, 2001)
SDSC1	cl.sdsc.edu/hm.html

atoms may indicate regions of low quality. Methods taking into account several of these aspects to assess three-dimensional protein structures are *e.g.* WhatCheck (Hooft *et al.*, 1996), ProCheck (Laskowski *et al.*, 1993), ANOLEA (Melo and Feytmans, 1998), and ProsaII (Sippl, 1993) (Table 2).

Table 2. Tools for protein structure evaluation

Software Packages	
WhatCheck	www.cmbi.kun.nl/swift/whatcheck/ (Hooft *et al.*, 1996)
ProCheck	www.biochem.ucl.ac.uk/~roman/procheck/procheck.html (Laskowski *et al.*, 1993; Morris *et al.*, 1992)
ANOLEA	www.fundp.ac.be/pub/ANOLEA.html (Melo and Feytmans, 1998)
ProsaII	www.came.sbg.ac.at/Services/prosa.html (Sippl, 1993)
PROVE	www.ucmb.ulb.ac.be/UCMB/PROVE/ (Pontius *et al.*, 1996)
Evaluation Servers	
EVA	maple.bioc.columbia.edu/eva/ (Eyrich *et al.*, 2001)
LifeBench	bioinfo.pl/LiveBench/ (Bujnicki *et al.*, 2001)
Critical Assessment of Techniques for Protein Structure Prediction	
CASP / CAFASP	predictioncenter.llnl.gov/ (Moult *et al.*, 2001)

Applications of Comparative Modeling

Protein structure homology modeling has been shown to be an efficient way to obtain structural information for proteins where no experimental structures are available. Applications are increasingly widespread, including *e.g.* design of protein mutants, structural interpretation of inherited human diseases (Erlandsen and Stevens, 1999; Peitsch and Tschopp, 1995; Wattenhofer *et al.*, 2002) or single nucleotide polymorphisms (Wang and Moult, 2001), interpretation of site-directed mutagenesis results (Srinivasan *et al.*, 1999), rationalizing membrane targeting from calculated electrostatic properties (Murray and Honig, 2002), confirming functional assignments based on low sequence homology (Duret *et al.*, 1998) or providing the building blocks in the assembly of large protein complexes (Spahn *et al.*, 2001). A more complete overview of homology modeling applications is provided in Ref. (Johnson *et al.*, 1994).

The accuracy of the protein models determines, which kind of problems can be addressed (Figure 2): For models based on very low sequence homology (less than 30% identical residues between template and target), the model can be used to confirm the correct fold selection. In cases where all members

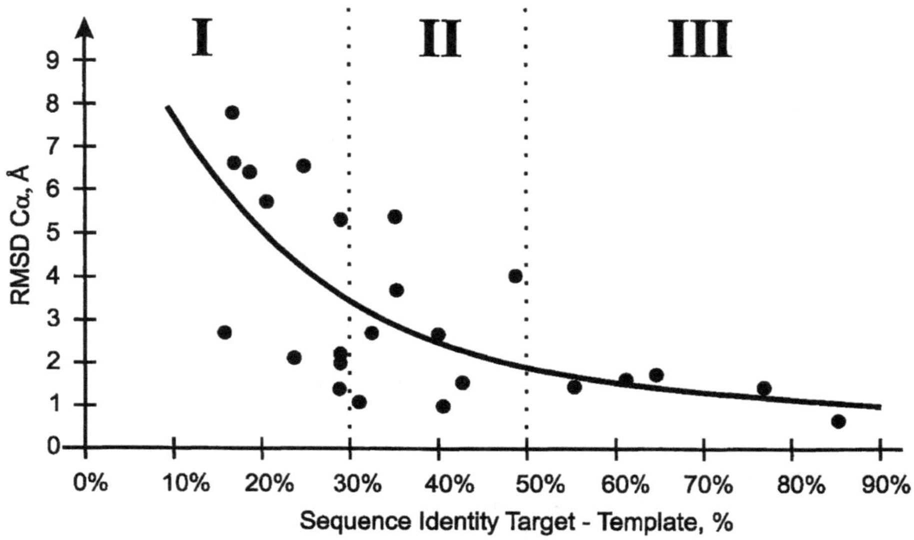

Figure 2. Applications of homology modeling depend on the expected model accuracy. The expected accuracy is illustrated as Cα-RMSD of protein models to experimental control structures in recent CASP experiments (illustration modified from Vitkup *et al.*, 2001). As expected, sequence identity between target and template can be used as a good indicator. Models based on low sequence homology (I) may serve for annotation by fold recognition; often active sites can be identified. Medium quality models based on templates sharing 30-50% sequence identity (II) generally have active sites and functional residues built correctly, and may therefore be used to support site directed mutagenesis experiments. The quality of models >50% sequence identity (III) can be comparable to medium resolution NMR, or low resolution X-ray structures, and are therefore suitable for numerous applications, such as structure based drug design or docking experiments.

of a given fold family are known to belong to a certain functional class, the fold assignment will give a first hypothesis about the function.

Models based on alignments sharing 30%-50% identical residues are expected to have more than 80% of their Cα-positions modeled within 3.5 Å of their correct positions. Frequently, the active site residues and ligand binding sites are better conserved than the rest of the fold and therefore modeled more accurately (Sanchez and Sali, 1998). These models often allow locating the active sites and thereby confirm a functional assignment, or even predict specific properties of binding sites (Xu *et al.*, 1996).

Models based on template structures sharing more than 50% sequence identity have been shown to reach an average accuracy comparable to low resolution X-ray structures (3 Å resolution), or medium resolution NMR structures (10 distance restraints per residue) (Sanchez and Sali, 1997b). Besides of the already mentioned applications, these high quality models can be used for applications such as docking and structure based drug design (Ring *et al.*, 1993; Schafferhans and Klebe, 2001), or detailed molecular simulations (Sotriffer *et al.*, 1999).

Genome Scale Homology Modeling

Whole genome sequencing programs and structural genomics projects are producing a plethora of data that can no longer be handled in a case-by-case approach. This motivates the development of fully automated rapid and accurate expert systems for comparative modeling that can be applied to whole genomes. Automation makes modeling available also to non-expert users, allowing them to use three-dimensional models to plan their experiments more efficiently. Besides of the practical aspect, automation also stimulates the development of better algorithms. Automatic modeling methods are more objective by removing human bias, and – since it is increasingly difficult for modeling experts to keep overview of the fast growing amount of information – will provide an indispensable tool for experts and non-experts alike.

The SWISS-MODEL server has pioneered automated modeling (Peitsch, 1995) by providing access to a fully automated pipeline via email, and later the World Wide Web. Other web based automated services are available in the meantime (Table 1). Several large scale modeling experiments for complete genomes have been undertaken. Starting 1995 Peitsch and Guex initiated large-scale automated protein modeling by submitting every full genome transcript to the SWISS-MODEL server, and by 1996 (Peitsch, 1996; Peitsch and Guex, 1997) six available fully sequenced genomes had been subjected to structural modeling. An average of 15% of all open reading frames were amenable to modeling. Three years later, in the 3D-Crunch experiment Peitsch *et al.* submitted all sequences registered in the SwissProt and the trEMBL databases to the SWISS-MODEL pipeline. The resulting models were made available in the SWISS-MODEL repository (Peitsch *et al.*, 1997). A second large scale modeling pipeline, ModPipe, was established by Sali *et al.* to model proteins from *Saccaromyces cerevisiae* and four other genomes (Sanchez and Sali, 1998; 1999). These studies not only made substantial numbers of three-dimensional models publicly available, but aimed at the same time to evaluate the reliability of the modeling methods.

4. Relating Structure to Function

The function of proteins can be divided into three categories: The influence a protein has on the properties of an organism as a whole (phenotypic function), the role of a protein in the context of a cell system (cellular function) and the properties related directly to the interactions with other molecules (molecular function) (Bork *et al.*, 1998). Molecular function can be described on different levels of detail, *e.g.* whether the protein is an enzyme, which type of reaction it catalyzes and which exact specificity or stereochemistry is observed. The biochemical activity and specificity of proteins are determined by their three-dimensional structures, which are much better conserved than the amino acid sequences. Typical examples for this observation are hemoglobin and myoglobin: While on the sequence level they seem not to be very similar, their close relationship is evident when one compares the three-dimensional structures.

Therefore three-dimensional structures can provide valuable information about the molecular function of an uncharacterized protein. During the last few years the first cases of proteins with unknown functions solved by structural genomics projects were published, *e.g.* (Chance *et al.*, 2002; Yee *et al.*, 2002). Two systematic approaches aim to elucidate the three-dimensional structures for functionally uncharacterized proteins in the organisms *Haemophilus influenzae* and *Methanococcus jannaschii* (Eisenstein *et al.*, 2000; Zarembinski *et al.*, 1998). Although few in number, the results are encouraging.

A four-step approach was suggested to analyze and interpret functional features of protein structures obtained from structural genomics: evaluating information from fold, active site recognition, ligand and cofactor binding and macromolecular interactions with DNA, RNA or other proteins (Eisenstein *et al.*, 2000).

Functional Information from the Protein Fold

The relationship between fold and function of proteins has been investigated extensively (Hegyi and Gerstein, 1999; Thornton *et al.*, 1999) and it is obvious from this analyses that a close association of fold and function in a one-to-one sense does not exist. Hegyi and Gerstein identified the five most versatile folds (TIM-barrel, alpha-beta hydrolase, Rossmann, P-loop containing NTP hydrolase, ferredoxin fold), which accommodate from six to as many as 16 functions. Furthermore, the two most versatile enzymatic functions (hydrolases and o-glycosyl-glucosidases) are associated with seven folds each. These observations demonstrate that the predictive power of a functional assignment through the fold is rather limited. Thornton *et al.* draw the conclusion that one has to look at the details of a molecule's active site, to assign a specific function (Thornton *et al.*, 2000).

However, fold recognition can provide useful functional hypotheses. The structure of protein MJ0226 from *Methanococcus jannaschii* (Hwang *et al.*, 1999) revealed a new protein fold family. Nevertheless it was possible by comparison with all other known structures in the databases to identify for a part of the protein a structural similarity with several nucleotide-binding proteins. This observation lead to the hypothesis that MJ0226 might interact with nucleotides, and subsequent biochemical essays confirmed its enzymatic activity as NTPase. Interestingly it has been observed that MJ0226 will effectively hydrolyze non-standard nucleotides but not the standard ones. These studies implicate that its cellular function may be to prevent mutations by protecting DNA from the incorporation of modified purine bases, such as dXTP or dITP.

In the case of the functionally uncharacterized *ycaC* gene product (YcaC) from *E. coli*, its structure was found to have a fold similar to that of N-carbamoylsarcosine amidohydrolase (CSHase). Although the sequence similarity was lower than 20% (Colovos *et al.*, 1998), in the superposition of the two structures the active-site nucleophilic cysteine and a non-proline cis-peptide bond of CSHase aligned spatially with corresponding residues of YcaC. This implicates that these two proteins catalyze similar hydrolytic reactions by a common mechanism.

It is well known that the number of different protein folds in nature is limited, and that a significant fraction is already known (Chothia, 1992; Govindarajan *et al.*, 1999). *In silico* fold recognition aims to predict which of the already known three-dimensional protein structures might be compatible with a given amino acid sequence. Fold recognition can be considered as the first step in homology modeling, and consequently both methods have converged in recent years (Bates *et al.*, 2001). Most current algorithms for fold recognition are using combinations of different sequence, structure, and annotation based scoring methods; well known examples are *e.g.* 3D-PSSM (Kelley *et al.*, 2000) or GenTHREADER (Jones, 1999). Using GenTHREADER Jones identified for 46 % of the proteins from predicted coding regions in the *Mycoplasma genitalium* genome significant relationships to protein folds of known structures. Although these fold relationships may not give a detailed functional picture, they are a starting point for structural modeling, and the fold assignment can provide good working hypotheses to be tested experimentally.

Using Active Site Topology

Enzymatic function is built by the local arrangement of amino acid residues composing an active site with precisely defined geometry. Among homologous enzymes with similar function, the particular spatial arrangement of functionally important residues is often conserved. The amino acids contributing to these three-dimensional functional motifs may play a catalytic role, or contribute to the specific structural arrangement determining substrate and ligand specificity. Kasuya and Thornton analyzed the three-dimensional conformation of the 20 most abundant PROSITE patterns, and extracted a representative three-dimensional structural fragment for each of the patterns (Kasuya and Thornton, 1999). The observed three-dimensional motifs were characteristic for a specific enzymatic function. Allowing some spatial flexibility in the three-dimensional pattern matching, a similar approach was used to generate a library of active site descriptors called "fuzzy functional forms" (Fetrow and Skolnick, 1998). These libraries of three-dimensional patterns can be used to search protein structures for the occurrence of known functional motifs. This approach is not limited to experimental structures. On the contrary its strength becomes apparent when applied to protein structure models of low quality, *e.g.* "fuzzy functional forms" derived from glutaredoxin/thioredoxin disulfide oxidoreductases were used to search models of putative proteins from the yeast genome generated by threading. Sixteen of the models were found to have an active site similar to disulfide oxidoreductase, although the sequence similarity between these proteins is described as "virtually non-existent" (Fetrow and Skolnick, 1998).

Ligand and Cofactor Binding

Often proteins require ligands or cofactors for their function. In newly determined protein structures bound cofactors or ligands often help to locate the active site, and can give a good indication for the protein function or mechanism. For example, Kim and coworkers (Zarembinski *et al.*, 1998) unexpectedly found an ATP molecule in the crystal structure of protein MJ0577 from *Methanococcus jannaschii*. The protein required these ATP molecules from its recombinant expression host *E. coli.* The initial hypothesis, that MJ0577 might be an ATPase, has been shown to be correct by biochemical experiments. Although the biochemical function could be assigned correctly, the specific biological role of this ATPase activity has to be determined by biochemical experiments. Again we see that the structure based functional genomics approach can lead to hypotheses that can be tested by simple biochemical experiments (Shapiro and Harris, 2000).

Structural Modeling and Functional Assignment

Functional assignment based on sequence information alone is prone to result in a number of wrong assignments for low sequence similarity. Taking into account information from three-dimensional protein structure modeling provides a way to increase sensitivity without compromising specificity at the same time. Since conserved functional features of a protein are more easily recognized in a three-dimensional structure than in its underlying protein sequence, protein models are a powerful means to confirm sequence based functional hypotheses.

Duret *at al.* used this approach to identify insulin-like proteins in the genome of *C. elegans* (Duret *et al.*, 1998). Although an insulin-receptor like gene in *C. elegans* had been reported and shown to play a role in metabolic regulation (Kimura *et al.*, 1997), the ligand(s) to activate this receptor were not apparent. Starting from an alignment of known insulin-like peptides from other organisms, profile methods were applied to identify possible related proteins in the *C. elegans* genome. Although ten proteins could be identified, these proteins shared some features very unusual for insulin-like proteins: The equivalent of a C-propeptide, which is usually found between the B and A chains, is absent. Additionally in two sub-forms, several basic residues responsible for post-translational processing were missing, indicating that the proteins might act as single chains rather than two separate peptides.

Using comparative protein structure modeling it was possible to demonstrate that all ten proteins would be compatible with an insulin-like three-dimensional structure. Especially the typical disulfide pattern observed between the A and B chains in this protein family was conserved in all ten *C. elegans* insulin like proteins, with additional disulfide bonds predicted to be formed in one sub-family. Remarkably, residues observed in correlated mutations were located close in space, contributing to a stabilization of the hydrophobic core. In this example, comparative modeling was used to demonstrate that despite of only remote sequence similarity, and several atypical sequence features, the analyzed proteins could adopt an insulin-like structure and play a similar functional role.

In a more general perspective, this approach opens the possibility to verify genome annotations based on very low sequence similarities by building the three-dimensional models for all gene products. Independent of the overall accuracy of the model, it should be possible to test for the presence or absence of functionally relevant arrangements of residues, and putative active sites, thereby confirming the functional assignments. These specific functional assignments can be transferred to other members of the same family related in primary sequence, where the functional relevant residues are conserved. This approach will propagate the functional assignment correctly through the databases, making it accessible to other problems as well.

5. Structure-Based Simulation of Molecular Function

Once an atomic model of a protein structure is available, this enables a more detailed simulation of the molecular events responsible for a protein's function. A tremendous progress in these molecular simulations has been observed in recent years, due to increased availability of computational resources, as well as improvements of the underlying algorithms. Many aspects of molecular function have been targeted by molecular modeling techniques, such as ligand binding (Sotriffer *et al.*, 1999), enzymatic catalysis (Warshel and Florian, 1998), electrostatic properties (Honig and Nicholls, 1995), ion channel conductivity (Berneche and Roux, 2001). Most of these techniques are computationally expensive, and are therefore not suitable for routine large-scale applications in functional genomics. A more exhaustive review of the progress in molecular modeling is given in Ref. (Wang *et al.*, 2001). Exemplarily we would like to present two recent impressive studies that illustrate how simulations based on molecular models could explain protein functions that were not directly explicable by experimental methods.

Electrostatic Control of Membrane Targeting

The electrostatic properties of a protein molecule play an important role in many functional aspects (Honig and Nicholls, 1995). Accurate calculations of the molecular electrostatic potential of proteins can be achieved based on three-dimensional protein structures by numerical approximation of the

Poisson-Boltzmann equation, taking into account different electrochemical properties (dielectric constants, ionic strength) of the solvent and the protein (Rocchia *et al.*, 2001). Many proteins involved in signal transduction and vesicle trafficking contain C2 domains whose membrane association is often regulated by calcium. Calcium binding can drive the association of some C2 domains to negatively charged membranes and others to neutral, zwitterionic membranes. The structures of over ten C2 domains have been determined to date. All have a quite similar fold, which consists of an eight-stranded anti-parallel β-sandwich that serves as a scaffold for variable surface loops. Many C2 domains bind calcium ions through a cluster of aspartic acid residues located in the loop region at one end of the domain. Calcium-independent C2 domains lack one or more of the calcium-coordinating residues. Honig *et al.* used finite-difference Poisson-Boltzmann calculations to describe the electrostatic interactions between C2 domains of known structures and phospholipid membranes (Murray and Honig, 2002). Non-specific electrostatic interactions were identified to play a major role in membrane targeting. Consequently, this opens the possibility to gain insight into the function of new C2 domains: Based on the known C2 domain structure homology models can be constructed for new C2-sequences. By analyzing their electrostatic properties it should be possible to predict the membrane association directly in silico.

Water Permeation across Biological Membranes by Aquaporin

Osmotic water regulation is essential for all life forms. Aquaporins are integral membrane proteins that facilitate highly efficient and specific the passive permeation of water across biological membranes with an activation energy nearly as low as the one associated with the self-diffusion rate in bulk water (Zeidel *et al.*, 1992). The interesting question is: How is this extremely high water conductance achieved, maintaining strict selectivity? How are protons excluded, even though they are known to be conducted well by hydrogen bonded water chains?

This question has recently been addressed using molecular dynamics techniques. The three-dimensional structure of human aquaporin has been determined independently by Engel *et al.* using electron crystallography techniques (de Groot *et al.*, 2001), and Sui *et al.* using X-ray crystallography (Sui *et al.*, 2001). The structural models of aquaporin show a membrane pore formed by six trans-membrane helices in the shape of an hourglass; the pore center is formed by two highly conserved asparagine-proline-alanine (NPA) motifs.

Since kinetic properties and selectivity of water conductance can not be explained based on a static model, Grubmüller *et al.* (de Groot and Grubmüller, 2001) and Schulten *et al.* (Tajkhorshid *et al.*, 2002) used extensive molecular dynamics simulations to analyze the flow and the orientation of the water molecules through the channel pore. As a result of these simulations, both groups showed that continuous hydrogen-bonded water chains, a prerequisite for proton conductance, are not observed. Instead, the interaction of the water molecules with the protein induces a strong translational and orientational control of the water molecules, involving a dipole inversion in the pore region near the NPA motifs. Without going into the experimental details, we would like to emphasize that these findings based on molecular modeling simulations would not have been accessible to experimental observation alone. Moreover, it allowed simulating the effects of mutations introduced into the channel to increase water permeability (Tajkhorshid *et al.*, 2002).

Which SNPs are Harmful?

The genetic variation between individual humans is caused mainly by single nucleotide polymorphisms (SNPs). The effects of genetic variations range from neutral mutations over increased susceptibly for complex diseases, individual variation in drug response, to rare single-allele mendelian inherited diseases (Brookes, 1999; Fredman *et al.*, 2002). A large body of knowledge about disease related genetic variations have been accumulated in specific databases *e.g.* OMIM (McKusick, 2000), or PAHdb (Scriver *et al.*, 2000), including non-synonymous mutations in coding regions that are likely to affect protein function. It has been estimated that any two individuals differ by about 100,000 amino acid altering substitutions.

Many diseases have been associated with alterations of certain genes. Changes in coding regions of proteins may affect protein function by altering protein structure. Although the underlying mutations are known on the amino acid sequence level, the molecular and functional changes can only be fully understood in the light of the three-dimensional structure of the protein. Well-known examples of proteins involved in monogenic inherited human diseases are phenylalanine hydroxylase (phenylketonuria) (Erlandsen and Stevens, 1999), or p53 (cancer) (Cho *et al.*, 1994). Recent studies analyzing the effects on SNPs involved in human inherited diseases impressively showed the importance of detailed structural analysis (Sunyaev *et al.*, 2001; Wang and Moult, 2001) of the genetic data.

6. Conclusion and Outlook

A paradigm shift is currently observed in many areas of life science research: Starting from a situation where no structural information was available for most proteins of biological interest, the number of available structures is growing rapidly due to recent developments in experimental structure determination techniques. Structural genomics projects, aiming to determine structures for at least one example of each protein family, will generate the foundation to reliably build comparative models for many globular soluble protein domains. Within the next decade, structural information will be available for the majority of all non-membrane proteins.

Integrating the extensive knowledge of biological function with the growing wealth of structural information will synergistically lead to a more detailed understanding of many processes of life. The availability of protein structures at the very beginning of research projects will change our way to explore biological phenomena, and will encourage completely new approaches; *e.g.* knowing the binding site geometry of a whole family of proteins might allow to improve the specificity of a drug molecule *in silico*, and thereby minimize adverse side effects.

We are now at the beginning of this exciting process, and the first emerging results are very encouraging. Protein structures reflect the molecular mechanisms of life, and we expect life science research will benefit in many aspects from protein structure information in the near future.

References

Abola, E., Kuhn, P., Earnest, T., and Stevens, R.C. 2000. Automation of X-ray crystallography. Nat. Struct. Biol. 7 Suppl: 973-977.

Allen, F., Almasi, G., Andreoni, W., and Zhou, R. 2001. Blue Gene: a vision for protein science using petaflop supercomputer. IBM Systems Journal. 40: 310-327.

Altschul, S.F., Gish, W., Miller, W., Myers, E.W., and Lipman, D.J. 1990. Basic local alignment search tool. J. Mol. Biol. 215: 403-410.

Altschul, S.F., Madden, T.L., Schaffer, A.A., Zhang, J., Zhang, Z., Miller, W., and Lipman, D.J. 1997. Gapped BLAST and PSI-BLAST: a new generation of protein database search programs. Nucleic Acids Res. 25: 3389-3402.

Anfinsen, C.B. 1973. Principles that govern the folding of protein chains.

Science. 181: 223-230.

Ashburner, M., Ball, C.A., Blake, J.A., Botstein, D., Butler, H., Cherry, J.M., Davis, A.P., Dolinski, K., Dwight, S.S., Eppig, J.T., Harris, M.A., Hill, D.P., Issel-Tarver, L., Kasarskis, A., Lewis, S., Matese, J.C., Richardson, J.E., Ringwald, M., Rubin, G.M., and Sherlock, G. 2000. Gene ontology: tool for the unification of biology. The gene ontology consortium. Nat. Genet. 25: 25-29.

Bairoch, A., and Apweiler, R. 2000. The SWISS-PROT protein sequence database and its supplement TrEMBL in 2000. Nucleic Acids Res. 28: 45-48.

Barton, G.J., and Sternberg, M.J. 1987. A strategy for the rapid multiple alignment of protein sequences. Confidence levels from tertiary structure comparisons. J. Mol. Biol. 198: 327-337.

Bates, P.A., Kelley, L.A., MacCallum, R.M., and Sternberg, M.J. 2001. Enhancement of protein modeling by human intervention in applying the automatic programs 3D-JIGSAW and 3D-PSSM. Proteins. Suppl. 5: 39-46.

Baumeister, W., and Steven, A.C. 2000. Macromolecular electron microscopy in the era of structural genomics. Trends Biochem. Sci. 25: 624-631.

Benson, D.-A., Karsch-Mizrachi, I., Lipman, D.-J., Ostell, J., Rapp, B.-A., and Wheeler, D.-L. 2002. GenBank. Nucleic Acids Res. 30: 17-20.

Berman, H.M., Westbrook, J., Feng, Z., Gilliland, G., Bhat, T.N., Weissig, H., Shindyalov, I.N., and Bourne, P.E. 2000. The protein data bank. Nucleic Acids Res. 28: 235-242.

Berneche, S., and Roux, B. 2001. Energetics of ion conduction through the K+ channel. Nature. 414: 73-77.

Bonneau, R., and Baker, D. 2001. *Ab initio* protein structure prediction: Progress and prospects. Ann Rev. Biophys. Biomol. Struct. 30: 173-189.

Bonneau, R., Tsai, J., Ruczinski, I., Chivian, D., Rohl, C., Strauss, C.E., and Baker, D. 2001. Rosetta in CASP4: Progress in *ab initio* protein structure prediction. Proteins. Suppl. 5: 119-126.

Bork, P., Dandekar, T., Diaz-Lazcoz, Y., Eisenhaber, F., Huynen, M., and Yuan, Y. 1998. Predicting function: from genes to genomes and back. J. Mol. Biol. 283: 707-725.

Brenner, S.E. 2001. A tour of structural genomics. Nat. Rev. Genet. 2: 801-809.

Brookes, A.J. 1999. The essence of SNPs. Gene. 234: 177-186.

Bujnicki, J.M., Elofsson, A., Fischer, D., and Rychlewski, L. 2001. LiveBench-1: continuous benchmarking of protein structure prediction servers. Protein Sci. 10: 352-361.

Burley, S.K. 2000. An overview of structural genomics. Nat. Struct. Biol. 7

Suppl: 932-934.

Chance, M.R., Bresnick, A.R., Burley, S.K., Jiang, J.S., Lima, C.D., Sali, A., Almo, S.C., Bonanno, J.B., Buglino, J.A., Boulton, S., Chen, H., Eswar, N., He, G., Huang, R., Ilyin, V., McMahan, L., Pieper, U., Ray, S., Vidal, M., and Wang, L.K. 2002. Structural genomics: a pipeline for providing structures for the biologist. Protein Sci. 11: 723-738.

Cho, Y., Gorina, S., Jeffrey, P.D., and Pavletich, N.P. 1994. Crystal structure of a p53 tumor suppressor-DNA complex: understanding tumorigenic mutations. Science. 265: 346-355.

Chothia, C., and Lesk, A.M. 1986. The relation between the divergence of sequence and structure in proteins. EMBO J. 5: 823-826.

Chothia, C. 1992. Proteins. One thousand families for the molecular biologist. Nature. 357: 543-544.

Colovos, C., Cascio, D., and Yeates, T.O. 1998. The 1.8 A crystal structure of the ycaC gene product from *Escherichia coli* reveals an octameric hydrolase of unknown specificity. Structure. 6: 1329-1337.

de Groot, B.L., Heymann, J.B., Engel, A., Mitsuoka, K., Fujiyoshi, Y., and Grubmuller, H. 2000. The fold of human aquaporin 1. J. Mol. Biol. 300: 987-994.

de Groot, B.L., Engel, A., and Grubmuller, H. 2001. A refined structure of human aquaporin-1. FEBS Lett. 504: 206-211.

de Groot, B.L., and Grubmüller, H. 2001. Water permeation across biological membranes: mechanism and dynamics of aquaporin-1 and GlpF. Science. 294: 2353-2357.

Dill, K.A., and Chan, H.S. 1997. From Levinthal to pathways to funnels. Nat. Struct. Biol. 4: 10-19.

Duret, L., Guex, N., Peitsch, M.C., and Bairoch, A. 1998. New insulin-like proteins with atypical disulfide bond pattern characterized in *Caenorhabditis elegans* by comparative sequence analysis and homology modeling. Genome Res. 8: 348-353.

Eisenstein, E., Gilliland, G.L., Herzberg, O., Moult, J., Orban, J., Poljak, R.J., Banerjei, L., Richardson, D., and Howard, A.J. 2000. Biological function made crystal clear - annotation of hypothetical proteins via structural genomics. Curr. Opin. Biotechnol. 11: 25-30.

Erlandsen, H., and Stevens, R.C. 1999. The structural basis of phenylketonuria. Mol. Genet. Metab. 68: 103-125.

Eyrich, V.A., Marti Renom, M.A., Przybylski, D., Madhusudhan, M.S., Fiser, A., Pazos, F., Valencia, A., Sali, A., and Rost, B. 2001. EVA: continuous automatic evaluation of protein structure prediction servers. Bioinformatics. 17: 1242-1243.

Fersht, A.R., and Daggett, V. 2002. Protein folding and unfolding at atomic resolution. Cell. 108: 573-582.

Fetrow, J.S., and Skolnick, J. 1998. Method for prediction of protein function from sequence using the sequence-to-structure-to-function paradigm with application to glutaredoxins/thioredoxins and T1 ribonucleases. J. Mol. Biol. 281: 949-968.

Fischer, D., and Eisenberg, D. 1997. Assigning folds to the proteins encoded by the genome of Mycoplasma genitalium. Proc. Natl. Acad. Sci. USA. 94: 11929-11934.

Fischer, D., Elofsson, A., Rychlewski, L., Pazos, F., Valencia, A., Rost, B., Ortiz, A., and Dunbrack Jr., R. 2001. CAFASP2: The second critical assessment of fully automated structure prediction methods. Proteins. Suppl. 5: 171-183.

Fiser, A., Do, R., and Sali, A. 2000. Modeling of loops in protein structures. Protein Sci. 9: 1753-1773.

Fredman, D., Siegfried, M., Yuan, Y.P., Bork, P., Lehvaslaiho, H., and Brookes, A.J. 2002. HGVbase: a human sequence variation database emphasizing data quality and a broad spectrum of data sources. Nucleic Acids Res. 30: 387-391.

Glaeser, R.M. 1999. Review: electron crystallography: present excitement, a nod to the past, anticipating the future. J. Struct. Biol. 128: 3-14.

Govindarajan, S., Recabarren, R., and Goldstein, R.A. 1999. Estimating the total number of protein folds. Proteins. 35: 408-414.

Guex, N., and Peitsch, M.C. 1997. SWISS-MODEL and the Swiss-PdbViewer: an environment for comparative protein modeling. Electrophoresis. 18: 2714-2723.

Hegyi, H., and Gerstein, M. 1999. The relationship between protein structure and function: a comprehensive survey with application to the yeast genome. J. Mol. Biol. 288: 147-164.

Honig, B., and Nicholls, A. 1995. Classical electrostatics in biology and chemistry. Science. 268: 1144-1149.

Hooft, R.W., Vriend, G., Sander, C., and Abola, E.E. 1996. Errors in protein structures. Nature. 381: 272.

Huynen, M., Doerks, T., Eisenhaber, F., Orengo, C., Sunyaev, S., Yuan, Y., and Bork, P. 1998. Homology-based fold predictions for *Mycoplasma genitalium* proteins. J. Mol. Biol. 280: 323-326.

Hwang, K.Y., Chung, J.H., Kim, S.H., Han, Y.S., and Cho, Y. 1999. Structure-based identification of a novel NTPase from *Methanococcus jannaschii.* Nat. Struct. Biol. 6: 691-696.

Johnson, M.S., and Overington, J.P. 1993. A structural basis for sequence comparisons. An evaluation of scoring methodologies. J. Mol. Biol. 233: 716-738.

Johnson, M.S., Srinivasan, N., Sowdhamini, R., and Blundell, T.L. 1994. Knowledge-based protein modeling. Crit Rev. Biochem. Mol. Biol. 29: 1-68.

Jones, D.T. 1999. GenTHREADER: an efficient and reliable protein fold recognition method for genomic sequences. J. Mol. Biol. 287: 797-815.

Karplus, K., Barrett, C., Cline, M., Diekhans, M., Grate, L., and Hughey, R. 1999. Predicting protein structure using only sequence information. Proteins. Suppl. 3: 121-125.

Karplus, K., and Hu, B. 2001. Evaluation of protein multiple alignments by SAM-T99 using the BAliBASE multiple alignment test set. Bioinformatics. 17: 713-720.

Kasuya, A., and Thornton, J.M. 1999. Three-dimensional structure analysis of PROSITE patterns. J. Mol. Biol. 286: 1673-1691.

Kelley, L.A., MacCallum, R.M., and Sternberg, M.J. 2000. Enhanced genome annotation using structural profiles in the program 3D-PSSM. J. Mol. Biol. 299: 499-520.

Kimura, K.D., Tissenbaum, H.A., Liu, Y., and Ruvkun, G. 1997. *daf-2*, an insulin receptor-like gene that regulates longevity and diapause in *Caenorhabditis elegans*. Science. 277: 942-946.

Lander, E.S., Linton, L.M., Birren, B., Nusbaum, C., Zody, M.C., Baldwin, J., Devon, K., Dewar, K., Doyle, M., FitzHugh, W., Funke, R., Gage, D., Harris, K., Heaford, A., Howland, J., Kann, L., Lehoczky, J., LeVine, R., McEwan, P., McKernan, K., Meldrim, J., Mesirov, J.P., Miranda, C., Morris, W., Naylor, J., Raymond, C., Rosetti, M., Santos, R., Sheridan, A., Sougnez, C., Stange-Thomann, N., Stojanovic, N., Subramanian, A., Wyman, D., Rogers, J., Sulston, J., Ainscough, R., Beck, S., Bentley, D., Burton, J., Clee, C., Carter, N., Coulson, A., Deadman, R., Deloukas, P., Dunham, A., Dunham, I., Durbin, R., French, L., Grafham, D., Gregory, S., Hubbard, T., Humphray, S., Hunt, A., Jones, M., Lloyd, C., McMurray, A., Matthews, L., Mercer, S., Milne, S., Mullikin, J.C., Mungall, A., Plumb, R., Ross, M., Shownkeen, R., Sims, S., Waterston, R.H., Wilson, R.K., Hillier, L.W., McPherson, J.D., Marra, M.A., Mardis, E.R., Fulton, L.A., Chinwalla, A.T., Pepin, K.H., Gish, W.R., Chissoe, S.L., Wendl, M.C., Delehaunty, K.D., Miner, T.L., Delehaunty, A., Kramer, J.B., Cook, L.L., Fulton, R.S., Johnson, D.L., Minx, P.J., Clifton, S.W., Hawkins, T., Branscomb, E., Predki, P., Richardson, P., Wenning, S., Slezak, T., Doggett, N., Cheng, J.F., Olsen, A., Lucas, S., Elkin, C., Uberbacher, E., Frazier, M., *et al.* 2001. Initial sequencing and analysis of the human genome. Nature. 409: 860-921.

Laskowski, R.A., MacArthur, M.W., Moss, D.S., and Thornton, J.M. 1993. PROCHECK: a program to check the stereochemical quality of protein structures. J. Appl. Cryst. 26: 283-291.

Lazaridis, T., and Karplus, M. 2000. Effective energy functions for protein structure prediction. Curr. Opin. Struct. Biol. 10: 139-145.

Lee, M.R., and Kollman, P.A. 2001. Free-energy calculations highlight differences in accuracy between X-ray and NMR structures and add value to protein structure prediction. Structure. 9: 905-916.

Lesk, A.M., and Chothia, C. 1980. How different amino acid sequences determine similar protein structures: the structure and evolutionary dynamics of the globins. J. Mol. Biol. 136: 225-270.

Lesk, A.M., LoConte, L., and Hubbard, T.J.P. 2001. Assessment of Novel Fold Targets in CASP4: Predictions of three-dimensional structures, secondary structures, and interresidue contacts. Proteins. Suppl. 5: 98-118.

Lo Conte, L., Brenner, S.E., Hubbard, T.J.P., Chothia, C., and Murzin, A.G. 2002. SCOP database in 2002: refinements accommodate structural genomics. Nucleic Acids Res. 30: 264-267.

Lund, O., Frimand, K., Gorodkin, J., Bohr, H., Bohr, J., Hansen, J., and Brunak, S. 1997. Protein distance constraints predicted by neural networks and probability density functions. Protein Eng. 10: 1241-1248.

Marti-Renom, M.A., Stuart, A.C., Fiser, A., Sanchez, R., Melo, F., and Sali, A. 2000. Comparative protein structure modeling of genes and genomes. Annu. Rev. Biophys. Biomol. Struct. 29: 291-325.

McKusick, V.A. 2000. McKusick-Nathans Institute for Genetic Medicine, Johns Hopkins University (Baltimore, MD) and National Center for Biotechnology Information, National Library of Medicine (Bethesda, MD) Online Mendelian Inheritance in Man, OMIM. World Wide Web URL: http://www.ncbi.nlm.nih.gov/omim/.

Melo, F., and Feytmans, E. 1998. Assessing protein structures with a non-local atomic interaction energy. J. Mol. Biol. 277: 1141-1152.

Montelione, G.T., Zheng, D., Huang, Y.J., Gunsalus, K.C., and Szyperski, T. 2000. Protein NMR spectroscopy in structural genomics. Nat. Struct. Biol. 7 Suppl: 982-985.

Morris, A.L., MacArthur, M.W., Hutchinson, *E.G.*, and Thornton, J.M. 1992. Stereochemical quality of protein structure coordinates. Proteins. 12: 345-364.

Moult, J., Fidelis, K., Zemla, A., and Hubbard, T. 2001. Critical assessment of methods of protein structure prediction (CASP): Round IV. Proteins. Suppl. 5: 2-7.

Murray, D., and Honig, B. 2002. Electrostatic control of the membrane targeting of C2 domains. Mol. Cell. 9: 145-154.

Notredame, C., Holm, L., and Higgins, D.G. 1998. COFFEE: an objective function for multiple sequence alignments. Bioinformatics. 14: 407-422.

Notredame, C., Higgins, D.G., and Heringa, J. 2000. T-Coffee: A novel method for fast and accurate multiple sequence alignment. J. Mol. Biol. 302: 205-217.

Orengo, C.A., Pearl, F.M.G., Bray, J.E., Todd, A.E., Martin, A.C., Lo, C.L., and Thornton, J.M. 1999. The CATH Database provides insights into protein structure/function relationships. Nucleic Acids Res. 27: 275-279.

Pearl, F.M., Lee, D., Bray, J.E., Sillitoe, I., Todd, A.E., Harrison, A.P., Thornton, J.M., and Orengo, C.A. 2000. Assigning genomic sequences to CATH. Nucleic Acids Res. 28: 277-282.

Pearson, W.R. 1990. Rapid and sensitive sequence comparison with FASTP and FASTA. Methods Enzymol. 183: 63-98.

Peitsch, M.C. 1995. Protein modelling by E-Mail. Bio/Technology. 13: 658-660.

Peitsch, M.C., and Tschopp, J. 1995. Comparative molecular modelling of the Fas-ligand and other members of the TNF family. Mol. Immunol. 32: 761-772.

Peitsch, M.C. 1996. ProMod and Swiss-Model: Internet-based tools for automated comparative protein modelling. Biochem. Soc. Trans. 24: 274-279.

Peitsch, M.C., and Guex, N. 1997. Large-scale comparative protein modelling. In: Proteome research: new frontiers in functional genomics. M.R. Wilkins, K.L. Williams, R.O. Appel, and D.F. Hochstrasser, editors. Springer p. 177-186.

Peitsch, M.C., Wilkins, M.R., Tonella, L., Sanchez, J.C., Appel, R.D., and Hochstrasser, D.F. 1997. Large-scale protein modelling and integration with the SWISS-PROT and SWISS-2DPAGE databases: the example of *Escherichia coli.* Electrophoresis. 18: 498-501.

Peitsch, M.C., Schwede, T., and Guex, N. 2000. Automated protein modelling—the proteome in 3D. Pharmacogenomics. 1: 257-266.

Pontius, J., Richelle, J., and Wodak, S.J. 1996. Deviations from standard atomic volumes as a quality measure for protein crystal structures. J. Mol. Biol. 264: 121-136.

Ring, C.S., Sun, E., McKerrow, J.H., Lee, G.K., Rosenthal, P.J., Kuntz, I.D., and Cohen, F.E. 1993. Structure-based inhibitor design by using protein models for the development of antiparasitic agents. Proc. Natl. Acad. Sci. USA. 90: 3583-3587.

Rocchia, W., Alexov, E., and Honig, B. 2001. Extending the applicability of the nonlinear poisson-boltzmann equation: multiple dielectric constants and multivalent ions. J. Phys. Chem. B. 105: 6507-6514.

Rost, B. 1999. Twilight zone of protein sequence alignments. Protein Eng. 12: 85-94.

Rost, B., and Eyrich, V. 2001. EVA: Large-scale analysis of secondary structure prediction. Proteins. Suppl. 5: 192-199.

Rychlewski, L., Zhang, B., and Godzik, A. 1998. Fold and function predictions for *Mycoplasma genitalium* proteins. Fold Des. 3: 229-238.

Sali, A., and Blundell, T.L. 1993. Comparative protein modelling by satisfaction of spatial restraints. J. Mol. Biol. 234: 779-815.

Sali, A., Potterton, L., Yuan, F., van Vlijmen, H., and Karplus, M. 1995. Evaluation of comparative protein modeling by MODELLER. Proteins. 23: 318-326.

Sanchez, R., and Sali, A. 1997a. Advances in comparative protein-structure modelling. Curr. Opin. Struct. Biol. 7: 206-214.

Sanchez, R., and Sali, A. 1997b. Evaluation of comparative protein structure modeling by MODELLER-3. Proteins. Suppl. 1: 50-58.

Sanchez, R., and Sali, A. 1998. Large-scale protein structure modeling of the *Saccharomyces cerevisiae* genome. Proc. Natl. Acad. Sci. USA. 95: 13597-13602.

Sanchez, R., and Sali, A. 1999. ModBase: a database of comparative protein structure models. Bioinformatics. 15: 1060-1061.

Sanchez, R., Pieper, U., Melo, F., Eswar, N., Marti-Renom, M.A., Madhusudhan, M.S., Mirkovic, N., and Sali, A. 2000. Protein structure modeling for structural genomics. Nat. Struct. Biol. 7 Suppl: 986-90.

Saqi, M.A., Russell, R.B., and Sternberg, M.J. 1998. Misleading local sequence alignments: implications for comparative protein modelling. Protein Eng. 11: 627-630.

Schafferhans, A., and Klebe, G. 2001. Docking ligands onto binding site representations derived from proteins built by homology modelling. J. Mol. Biol. 307: 407-427.

Scriver, C.R., Waters, P.J., Sarkissian, C., Ryan, S., Prevost, L., Cote, D., Novak, J., Teebi, S., and Nowacki, P.M. 2000. PAHdb: a locus-specific knowledgebase. Hum. Mutat. 15: 99-104.

Shapiro, L., and Harris, T. 2000. Finding function through structural genomics. Curr. Opin. Biotechnol. 11: 31-35.

Sippl, M.J. 1993. Recognition of errors in three-dimensional structures of proteins. Proteins. 17: 355-362.

Sotriffer, C.A., Flader, W., Cooper, A., Rode, B.M., Linthicum, D.S., Liedl, K.R., and Varga, J.M. 1999. Ligand binding by antibody IgE Lb4: assessment of binding site preferences using microcalorimetry, docking, and free energy simulations. Biophys. J. 76: 2966-2977.

Spahn, C.M., Beckmann, R., Eswar, N., Penczek, P.A., Sali, A., Blobel, G., and Frank, J. 2001. Structure of the 80S ribosome from *Saccharomyces cerevisiae*—tRNA-ribosome and subunit-subunit interactions. Cell. 107: 373-386.

Srinivasan, N., Antonelli, M., Jacob, G., Korn, I., Romero, F., Jedlicki, A., Dhanaraj, V., Sayed, M.F., Blundell, T.L., Allende, C.C., and Allende, J.E. 1999. Structural interpretation of site-directed mutagenesis and specificity of the catalytic subunit of protein kinase CK2 using comparative modelling. Protein Eng. 12: 119-127.

Sui, H., Han, B.G., Lee, J.K., Walian, P., and Jap, B.K. 2001. Structural basis of water-specific transport through the AQP1 water channel. Nature. 414: 872-878.

Sunyaev, S., Ramensky, V., Koch, I., Lathe, W., Kondrashov, A.S., and Bork, P. 2001. Prediction of deleterious human alleles. Hum. Mol. Genet. 10: 591-597.

Sutcliffe, M.J., Haneef, I., Carney, D., and Blundell, T.L. 1987. Knowledge based modelling of homologous proteins, Part I: Three-dimensional frameworks derived from the simultaneous superposition of multiple structures. Protein Eng. 1: 377-384.

Tajkhorshid, E., Nollert, P., Jensen, M.O., Miercke, L.J.W., O'Connell, J., Stroud, R.M., and Schulten, K. 2002. Control of the selectivity of the aquaporin water channel family by global orientational tuning. Science. 296: 525-530.

Taylor, W.R., Flores, T.P., and Orengo, C.A. 1994. Multiple protein structure alignment. Protein Sci. 3: 1858-1870.

Thompson, J.D., Higgins, D.G., and Gibson, T.J. 1994. CLUSTAL W: improving the sensitivity of progressive multiple sequence alignment through sequence weighting, position-specific gap penalties and weight matrix choice. Nucleic Acids Res. 22: 4673-4680.

Thompson, J.D., Plewniak, F., and Poch, O. 1999. BAliBASE: a benchmark alignment database for the evaluation of multiple alignment programs. Bioinformatics. 15: 87-88.

Thornton, J.M., Orengo, C.A., Todd, A.E., and Pearl, F.M. 1999. Protein folds, functions and evolution. J. Mol. Biol. 293: 333-342.

Thornton, J.M., Todd, A.E., Milburn, D., Borkakoti, N., and Orengo, C.A. 2000. From structure to function: approaches and limitations. Nat. Struct. Biol. 7 Suppl: 991-994.

Tramontano, A., Leplae, R., and Morea, V. 2001. Analysis and assessment of comparative modeling predictions in CASP4. Proteins. Suppl. 5: 22-38.

Venter, J.C., Adams, M.D., Myers, E.W., Li, P.W., Mural, R.J., Sutton, G.G., Smith, H.O., Yandell, M., Evans, C.A., Holt, R.A., Gocayne, J.D., Amanatides, P., Ballew, R.M., Huson, D.H., Wortman, J.R., Zhang, Q., Kodira, C.D., Zheng, X.H., Chen, L., Skupski, M., Subramanian, G., Thomas, P.D., Zhang, J., Gabor-Miklos, G.L., Nelson, C., Broder, S., Clark, A.G., Nadeau, J., McKusick, V.A., Zinder, N., Levine, A.J., Roberts, R.J.,

Simon, M., Slayman, C., Hunkapiller, M., Bolanos, R., Delcher, A., Dew, I., Fasulo, D., Flanigan, M., Florea, L., Halpern, A., Hannenhalli, S., Kravitz, S., Levy, S., Mobarry, C., Reinert, K., Remington, K., Abu-Threideh, J., Beasley, E., Biddick, K., Bonazzi, V., Brandon, R., Cargill, M., Chandramouliswaran, I., Charlab, R., Chaturvedi, K., Deng, Z., Di-Francesco, V., Dunn, P., Eilbeck, K., Evangelista, C., Gabrielian, A.E., Gan, W., Ge, W., Gong, F., Gu, Z., Guan, P., Heiman, T.J., Higgins, M.E., Ji, R.R., Ke, Z., Ketchum, K.A., Lai, Z., Lei, Y., Li, Z., Li, J., Liang, Y., Lin, X., Lu, F., Merkulov, G.V., Milshina, N., Moore, H.M., Naik, A.K., Narayan, V.A., Neelam, B., Nusskern, D., Rusch, D.B., Salzberg, S., Shao, W., Shue, B., Sun, J., Wang, Z., Wang, A., Wang, X., Wang, J., Wei, M., Wides, R., Xiao, C., Yan, C., *et al.* 2001. The sequence of the human genome. Science. 291: 1304-1351.

Vitkup, D., Melamud, E., Moult, J., and Sander, C. 2001. Completeness in structural genomics. Nat. Struct. Biol. 8: 559-566.

Vriend, G. 1990. WHAT IF: a molecular modeling and drug design program. J. Mol. Graph. 8: 52-56.

Vukmirovic, O.G., and Tilghman, S.M. 2000. Exploring genome space. Nature. 405: 820-822.

Wang, W., Donini, O., Reyes, C.M., and Kollman, P.A. 2001. Biomolecular simulations: recent developments in force fields, simulations of enzyme catalysis, protein-ligand, protein-protein, and protein-nucleic acid noncovalent interactions. Annu. Rev. Biophys. Biomol. Struct. 30: 211-243.

Wang, Z., and Moult, J. 2001. SNPs, protein structure, and disease. Hum Mutat. 17: 263-270.

Warshel, A., and Florian, J. 1998. Computer simulations of enzyme catalysis: finding out what has been optimized by evolution. Proc. Natl. Acad. Sci. USA. 95: 5950-5955.

Watson, J.D., and Crick, F.H.C. 1953. Molecular structure of nucleic acids. Nature. 171: 737-738.

Wattenhofer, M., Di Iorio, M.V., Rabionet, R., Dougherty, L., Pampanos, A., Schwede, T., Montserrat-Sentis, B., Arbones, M., Iliades, T., Pasquadibisceglie, A., D'Amelio, M., Alwan, S., Rossier, C., Dahl, H.H.M., Petersen, M.B., Estivill, X., Gasparini, P., Scott, H.S., and Antonarakis, S.E. 2002. Mutations in the TMPRSS3 gene are a rare cause of childhood non-syndromic deafness in Caucasian patients. J. Mol. Med. 80: 124-131.

Xiang, Z., Soto, C.S., and Honig, B. 2002. Evaluating conformational free energies: The colony energy and its application to the problem of loop prediction. Proc. Natl. Acad. Sci. USA. 99: 7432-7437.

Xu, L.Z., Sanchez, R., Sali, A., and Heintz, N. 1996. Ligand specificity of brain lipid-binding protein. J. Biol. Chem. 271: 24711-24719.

Yee, A., Chang, X., Pineda Lucena, A., Wu, B., Semesi, A., Le, B., Ramelot, T., Lee, G.M., Bhattacharyya, S., Gutierrez, P., Denisov, A., Lee, C.H., Cort, J.R., Kozlov, G., Liao, J., Finak, G., Chen, L., Wishart, D., Lee, W., McIntosh, L.P., Gehring, K., Kennedy, M.A., Edwards, A.M., and Arrowsmith, C.H. 2002. An NMR approach to structural proteomics. Proc. Natl. Acad. Sci. USA. 99: 1825-1830.

Zarembinski, T.I., Hung, L.W., Mueller-Dieckmann, H.J., Kim, K.K., Yokota, H., Kim, R., and Kim, S.H. 1998. Structure-based assignment of the biochemical function of a hypothetical protein: a test case of structural genomics. Proc. Natl. Acad. Sci. USA. 95: 15189-15193.

Zeidel, M.L., Ambudkar, S.V., Smith, B.L., and Agre, P. 1992. Reconstitution of functional water channels in liposomes containing purified red cell CHIP28 protein. Biochemistry. 31: 7436-7440

From: *Frontiers in Computational Genomics*

5

Identifying Domains, Repeats and Motifs from Protein Sequences

Chris P. Ponting and Alex Bateman

Abstract

The newly-sequenced genomes provide previously undreamt-of opportunities to generate experimentally-testable hypotheses. These hypotheses arise from the identification of gene regions that are homologous, having arisen from a common ancestor, that often possess similarities in function. These homologues encode families of protein domains, repeats or motifs, that often have been assembled by evolution into multiple distinct combinations, or architectures. This article suggests strategies for the analysis of protein sequences in terms of their domain, repeat and motif architectures with an emphasis on statistical evaluation of sequence similarities and the prediction of function from sequence.

1. Introduction

A genome may be thought of as a melting pot of genes, descended from ancestral sequences that were divided, fissured and fused, or that were stolen wholesale or piecemeal from well-acquainted genomes (International Human Genome Sequencing Consortium, 2001). The modern generation of genes is missing countless past variations that failed the survival tests of resourcefulness. That all this has been surmised from a million or more known protein sequences might be considered extraordinary. Yet our abilities as scientists to imbue sequence with information is attributable in part to the repetitions in genomes, and therefore, to their abilities throughout history, to duplicate genes. Duplication and mutation generates potential for new genetic, molecular and organismal function from the old, pre-existing gene set.

Gene, and indeed genome, duplications of yesteryear have left their marks throughout modern genomes. Approximately two percent of the gene set of the unicellular yeast *Saccharomyces cerevisiae* encode protein kinases that perform radically differing roles (Hunter and Plowman, 1997). One percent of the plant *Arabidopsis thaliana*'s gene complement represents helical repeats of ill-determined function (Aubourg *et al.*, 2000). These figures for coding repeats pale in comparison to mobile DNA elements that make up approximately 45% of the human genome (International Human Genome Sequencing Consortium, 2001).

The ubiquity and proliferation of gene duplications has been harnessed by bioinformaticians dedicated to the categorisation of all gene products by homology. Many ancestral gene duplications are readily apparent from comparison of the modern protein sequences since their similarities are far greater than would be expected from comparison of pairs of random, or evolutionarily unrelated, sequences. Such gene pairs are 'homologous', meaning that they are descended from a common ancestral gene. Gene homologues that have arisen from gene duplication *within* a genome are termed 'paralogues', whereas others that arose due to the creation of new species are 'orthologues'.

2. The Interpretation of Multiple Sequence Alignments

Well constructed multiple alignments maximise the conservation of amino acid identity and physicochemical characteristics among homologues' sequences. Such alignments are articulate representations of structural and functional predictions that arise directly from inferences of homology. Aligning particular amino acids from different sequences implies that they occupy a common spatial location in an optimal superposition of homologues' three dimensional structures.

Alignments based on the superposition of known protein structures are considered to be the 'gold standard' against which other alignments should be compared. Alignments do not simply tell us which residues have similar spatial positions in homologous proteins, they can also tell us which residues are functionally or structurally important. The genes that code for proteins have mutated over time giving rise to variations in the protein sequence. Mutations that alter amino acid residues at sites important for the function or stability of the protein are less likely to be conserved than those elsewhere. This leads to the observation that residues in multiple alignment columns that are highly conserved have structural and/or functional roles. The types of amino acids in a conserved column can also allow for more specific predictions of the role of each residue. Alignment columns that contain conserved hydrophobic residues are likely to form a hydrophobic core that contributes to the stability of the domain. Columns that have a strongly conserved polar residue may contribute to an active site.

One conceptualisation of an alignment is a 'folding' of the aligned sequences along the winding path taken by the consensus tertiary structure of the homologues. Multiple alignment columns can be thought of as being projected onto the consensus backbone structure. Each side chain of the consensus structure thus represents not one but all of the residues in an alignment column. Although this conception is an idealisation, since extremely rarely are tertiary structures of most homologues known, it has several benefits. Firstly, it demonstrates that conserved blocks without gaps in structurally-based alignments almost always represent secondary structural elements, β-strands or helices. Secondly, the residues that intervene between conserved alignment blocks often form irregular and ill-conserved structures that link between secondary structures, usually at the solvent-accessible periphery of the molecule. Finally, it shows that conservation of amino acid similarity is greatest within structural domains (see below) and is least in the polypeptides that lie outside of the domain limits.

These conclusions are drawn from extensive comparisons of homologues' structures. In nearly all cases (for exceptions see Russell and Ponting, 1998), the structures of homologues are recognisably similar and superimposable. However, it is not the case that the sequences of *all* homologues are recognisably similar (Park *et al.*, 1998). Frequently, two proteins are only deemed to be homologous upon comparison of their tertiary structures and consideration of common functional groups or mechanisms, rather than by simple comparison of their sequences. This implies that intervening between homologues whose sequences are highly similar (a common example is of mouse and human orthologues) and homologues whose sequences that are not discernibly similar (for example, many sequences whose last common ancestor existed one or two billion years ago), there are intermediate sequences whose similarities are subtle, yet sufficient to warrant assignment of homology.

3. The Significance of Sequence Similarities

This begs an important question: how can similarities be assessed reliably for their significance? It is clear that as the volume of sequence data pouring into databases increases, so does the likelihood that sequence comparisons associated with a particular alignment score are found *simply by chance*. This means that a robust measure of sequence similarity significance must take account of the degree of sequence similarities (the alignment score) and the size of the database that was searched. Fortunately, the statistics for scores of local similarities (those that do not necessarily involve the whole of the query sequence) provide such a robust measure (Karlin and Altschul, 1990).

The algorithm of choice in searching for significant local similarities is BLAST (Altschul *et al.*, 1997; http://www.ncbi.nlm.nih.gov/BLAST/). BLAST incorporates Karlin-Altschul statistics of local alignment scores (see http://www.ncbi.nlm.nih.gov/BLAST/tutorial/Altschul-1.html) and estimates the significance of sequence similarities in the form of *Expect* (or E-) values. The value of E is calculated from an alignment score x, and represents the number of alignments expected to score at least x simply by chance in a random database of a size equivalent to the one searched. An E-value of 0.01 suggests there is a 1% chance that you would have observed an alignment with this score or higher by chance alone.

Although in theory E-values lower than 1 represent likely significant relationships, in practice thresholds of 0.01 and 0.1 are more appropriate for BLAST and FASTA (Pearson and Lipman, 1988) searches, respectively. It is emphasised however, that the use of an E-value threshold may not be appropriate for sequences that are compositionally biased. For such cases, it is recommended either that sequences are pre-filtered using the SEG algorithm (Wootton and Federhen, 1993) or that a PSI-BLAST server that takes account of compositional variation (see http://www.ncbi.nlm.nih.gov/blast/) is used. Whatever the E-values of sequence similarities, it is recommended that results of database searches are treated with an appropriate degree of scepticism. True homologues need to be identified both with significant E-values and with alignments that demonstrate patterns of amino acid conservation that are consistent with those of other homologues.

4. Building Multiple Alignments

The patterns of amino acid conservation among putative homologues are best observed by constructing a multiple alignment. This is a two-step procedure (Bork and Gibson, 1996; Ponting and Birney, 2000). A preliminary alignment is first constructed using approaches such as Clustal-W (Thompson *et al.*, 1994) or PSI-BLAST (Altschul *et al.*, 1997), and is then refined by manual editing. Unfortunately unless all the proteins in the alignment are quite similar manual intervention to produce optimal alignment is unavoidable. Guidelines for hand-editing alignments are given elsewhere (Bork and Gibson, 1996; Ponting and Birney, 2000), but the most salient points can be summarised thus: minimise gaps ("insertion/deletion positions") to regions between (predicted) secondary structures, and maximise conservation of core hydrophobic residues within secondary structures.

Aligned residues represent correspondences not only in spatial locations but also in functionality. An amino acid known in one homologue, if aligned with an identical amino acid from a second homologue, implies that these amino acids perform identical roles for both homologues. Perfect conservation of two or more polar residues (aspartic and glutamic acids, asparagine, arginine, cysteine, glutamine, histidine, lysine, serine and threonine) usually identifies these as active site residues of enzymes, or metal-binding sites, or, in the case of cysteine, as possible contributors to disulphide bridges. The absence of absolutely conserved amino acids in alignments suggests that (at least one) of the homologues may not possess enzymatic activity.

5. Domains, Repeats and Motifs

The similarities between protein sequences can be classified into three different types: domains, repeats and motifs. For each type the optimal methods for detection and boundary definition differ. Consequently, we propose three analysis protocols here that are tailored for each of the three types.

There are, however, some analytical steps in these protocols that are held in common. The first step in each protocol is a BLAST search with your sequence against current databases in order to detect significant similarities to previously characterised proteins. For well studied proteins a large amount of information can be gleaned quickly from annotation in protein sequence databases such as Swiss-Prot (Bairoch and Apweiler, 2000), and from the primary literature. Sequence databases contain information and links concerning protein function, enzymatic activity, subcellular location, mutations and variations and much more. Clearly this gold-mine should be consulted before further time-consuming analysis is embarked upon. However, even if your protein is related to a well characterised protein you should carry out further analyses to find other more subtle, yet perhaps equally informative, similarities. New similarities can become evident due to improving sensitivity of new search methods or from sequences newly deposited in databases. To aid the process of identifying homologies within protein sequences several web-based annotation resources exist that also save on time and effort. These protein family-based resources provide tools to search your sequence for similarity to all known families, domains and motifs. These databases are summarised in Table 1.

The domain composition of your protein will often give strong clues about the cellular localisation of the protein. Proteins containing signalling domains such as the protein serine/threonine-specific kinase , for example, are likely to be intracellular. Other domains, such as those that are disulphide-rich, are only found in secreted proteins. However care should be exercised as many domains can be found in several cellular locations, such as immunoglobulin (Ig) domains that occur primarily in secreted proteins, but are also found in intracellular muscle proteins.

To complement the analysis of domains using domain family databases, other sequence-based signals and features can also be predicted. Signal peptides can be detected with software such as SignalP (see URL http://www.cbs.dtu.dk/services/SignalP/); if present this gives strong evidence that the protein is secreted. Transmembrane spanning α-helices can be predicted with high accuracy. Tools such as Tmpred (URL: http://www.ch.embnet.org/software/TMPRED_form.html) and TMhmm (URL: http://www.cbs.dtu.dk/services/TMHMM-1.0/) can be used. Predictions of transmembrane regions provide a prediction of localisation that can be combined with other information such as the domain organisation. Although prediction of individual helices is often good, prediction of overall topology is poor and should be treated with caution. In addition, some proteins such as porins contain transmembrane spanning β-strands that at present are unable to be detected reliably.

Table 1. Tools to search sequences for similarity to all known families, domains and motifs.

Database	No. Families	Annotation	Speciality	Notes
Pfam	2,697 (version 6.0)	Yes		Uses HMMer profile HMMs. High coverage: 2/3 of proteins have a match to a Pfam domain http://www.sanger.ac.uk/Software/Pfam/
SMART	517 (Version 3.2)	Yes	Signaling, extracellular and nuclear domains	Uses HMMer profile HMMs. Many alignments contain more divergent sequences than other databases. Will also predict TM, low complexity and signal peptides http://smart.embl-heidelberg.de/
Prosite patterns	1048 (Release 16.22)	Yes		Very fast to search but prone to false positives http://www.expasy.ch/prosite/
Prosite Profiles	303	Yes	Signaling domains	Uses the pftools software. http://www.isrec.isb-sib.ch/software/PFSCAN_form.html
PRINTS	1460 (release 29)	Yes	G protein coupled receptors and ion channels	Uses FingerPRINTScan software http://www.bioinf.man.ac.uk/dbbrowser/PRINTS/
DOMO	8,877	Links only		Automatically generated clustering of Swiss-Prot and PIR. http://www.infobiogen.fr/services/domo/
Prodom	51,303 (Version 2000.1)	Links only		Automatically generated from PSI-BLAST, essentially complete coverage. http://www.toulouse.inra.fr/prodom.html
Interpro	3,204 (Release 2.0)	Yes		Derived from Pfam, Prosite, Prints and PRODOM
BLOCKS	998 (version 12)	No		Derived from Interpro release 1.0 http://www.blocks.fhcrc.org/

6. Domains

Protein domains are defined as discrete structural units of proteins that can be found in different contexts. Protein domains are the common currency of protein function, with individual domains often conveying independent functions within a protein. Many domains conserve a function even in different protein contexts. Understanding the domain organisation of a protein, therefore, often aids the understanding of protein function. Domains can vary in size from 25 to 500 amino acid residues (see Figure 1). All domains have a stabilising core. In the case of domains containing fewer than 50 amino acids the core is often provided by a zinc ion chelated by cysteine and/or histidine residues (as in zinc fingers), or in the domains in secreted proteins by disulphide bridges. This results in alignments of small domains often being characterised by columns of conserved cysteine and possibly histidine residues. For larger domains, the stable core is most

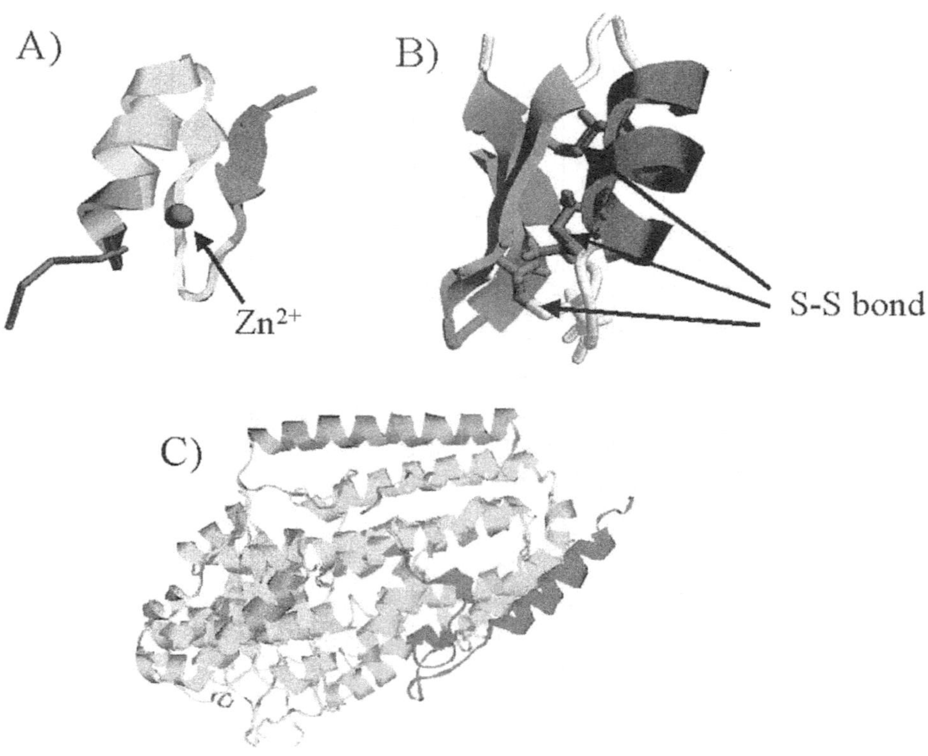

Figure 1. Protein domains, large and small. All structure figures have been produced using Rasmol (Sayle and Milner-White, 1995). A) The 31 residue long classical C2H2 zinc finger domain. The zinc ion is shown by a dark sphere. B) 37 amino acid long snake toxin domain that is stabilised by three disulphide bridges. C) This α-helical domain from lipoxygenase is over 500 amino acids in length. ***Colour plate at back of book.***

frequently provided by hydrophobic residues. Although most domains contain fewer than 250 amino acids some, such as the lipoxygenase domain (Figure 1C) contain up to 500 amino acids.

Determining Domain Boundaries

The most reliable way to determine the boundaries of a domain is to use a known structure. Compact regions of the structure can readily be identified by eye or using computer algorithms (see the links in http://www.bmm.icnet.uk/people/rob/CCP11BBS/foldrec.html). However, for the large majority of domains a known structure of a homologue is not yet available. For these cases it is often possible to define accurate domain boundaries from sequence alone. Methods that predict these boundaries have been developed and are used as the basis for automatic domain definition, such as in the Prodom (Corpet *et al.*, 2000) and DOMO (Gracy and Argos, 1998) databases. These methods rely on an initial sequence comparison step, followed by a second parsing step that attempts to find the maximal extent of domains based on a number of rules. In the simplest terms these rules are:

(i) No domain can extend beyond the N- or C- terminus of any sequence.
(ii) No two domains can overlap in a sequence.

For a given set of sequences containing homologous domains these rules define the maximal amount by which a domain can be extended in the N- and C-terminal directions, (Figure 2).

Once alerted to the presence of a domain in a sequence, its identity domain should be ascertained. Many populous domain families are detectable using SMART, Pfam and PROSITE profiles web-based resources (Table 1). If the domain of interest appears not to belong to any of these typical families then the following protocol for its analysis is suggested:

(i) Find a list of potential relatives using PSI-BLAST with a threshold of 0.002. The longer this list of homologues the more likely one is to correctly delineate the domain boundaries.
(ii) Determine the boundaries and extent of the domain as described above.
(iii) Align all domain homologues, including sequences that flank the domains' N- and C- termini using ClustalW. If no sequence conservation is discernible in the termini extensions, the alignment may be trimmed back.

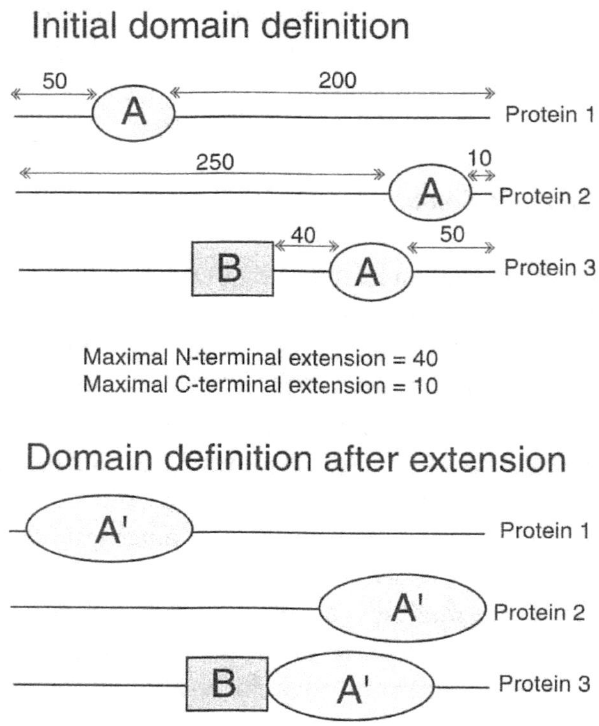

Figure 2. Definition of domains from sequence alone. Sequence comparison shows a region of similarity (a domain), called A, shared between three proteins. A second domain shown is called B. The extended region of similarity is called A' and is the maximal extent of the predicted domain. ***Colour plate at back of book.***

(iv) Undertake a search of protein sequence databases using a HMM constructed from the multiple alignment. Construct and calibrate the HMM ('hmmbuild' and 'hmmcalibrate' in the HMMER2 package; http://hmmer.wustl.edu/) using default parameters. This is subsequently compared with database sequences and generates global alignments ('hmmsearch' in HMMER2). Accept domain sequences as representing homologues if their protein *E*-values are less than 0.1 in a search of highly non-redundant databases such as nrdb90 (Holm and Sander, 1998), and add these sequences to the previously-derived alignment. Also add other sequences that can be shown using PSI-BLAST or other methods to be significantly similar to members in the alignment.

(v) Repeat step (iii) and (iv) iteratively until no further candidate homologues are detected.

An Example of Using Domains to Interpret Biology: The ACT Domain

The discovery of homologous domains in different proteins can be used to interpret and predict biological function. The delineation of the ACT domain (Aravind and Koonin 1999), for example, allows us to make many new biological inferences about proteins in which it is found. The ACT domain has a ferredoxin-like fold, found in the well-studied enzyme, phosphoglycerate dehydrogenase (PGD). It is known that the binding of serine to this domain causes allosteric inhibition of the PGD enzyme (Grant *et al*, 1996).

By sequence analysis, ACT domains are predicted in a wide range of metabolic enzymes that are regulated by amino acids; for these, at least 15 different enzyme activities are known, which are regulated by over 10 different amino acid types (see Table 2 and Figure 3). From an alignment of representative ACT domains (Figure 4) we see that conservation is greatest within the secondary structure elements. Furthermore, from our knowledge of PGD we can infer that ACT domains form homodimers, and from the threonine dehydratase structure we can infer that ACT domains can form tandem domain dimers. So it seems likely that the functional unit for ACT domains is a dimeric form (Figure 5). Indeed, this is consistent with the highest level of sequence conservation being present within this interface region. Furthermore from this structure and the location of its serine-binding site at the dimer interface, we predict that other amino acids will bind this same interface region in other ACT domain dimers.

Based on experimental evidence for the kinetics of PGD we can infer that the regulation of all these enzymes is likely due to an allosteric mechanism mediated by the ACT domain. From structural models based on the PGD crystal structure we expect the regulating amino acid to bind the ACT domain resulting in small conformational changes that cause activation or inhibition by an allosteric mechanism. By examining these models the amino acids that are involved in non-specific amino acid-binding and in determining specificity may be predicted. At present for some of the ACT domain-containing proteins it is not known if any amino acid regulates the enzyme. Based on the presence of an ACT domain and functional knowledge, however, we predict the following three possible regulations:

(i) PII uridylyl transferase is inhibited by glutamine, this protein is in a cascade of events that regulates glutamine-synthetase.
(ii) L-serine dehydratase is activated by serine.

Table 2. Metabolic enzymes containing ACT domains.

Protein	Example	EC number	Pathway	Inhibited/activated by
Homoserine dehydrogenase	DHOM_MYCTU	1.1.1.3	Third step in conversion of l-aspartate to homoserine	Inhibited by threonine
Malate oxidoreductase	MAO1_BACSU	1.1.1.38		?
D-3-phosphoglycerate dehydrogenase	SERA_ECOLI	1.1.1.95	First committed step in the "phosphorylated" pathway of l-serine biosynthesis	Inhibited by serine
Phenylalanine-4-hydroxylase	PH4H_HUMAN	1.14.16.1	Rate-limiting step in phenylalanine catabolism	Activated by phenylalanine
Tyrosine-3monooxygenase	TY3H_DROME	1.14.16.2	Rate-limiting step in catecholamine biosynthesis	Activated by phosphorylation
Tryptophan-5-monooxygenase	TR5H_HUMAN	1.14.16.4	Rate-limiting enzyme in the biosynthesis of serotonin	Inhibited by L-Dopa
Aspartokinase	AK_YEAST	2.7.2.4	First step in the common biosynthetic pathway leading from asp to diaminopimelate and lysine, to methionine, isoleucine and threonine.	Regulated by lysine.
GTP pyrophosphokinase	RELA_ECOLI	2.7.6.5	First step in the metabolism of ppGpp	?
PII uridylyl transferase	GLND_ECOLI	2.7.7.59	Involved in stringent response	*Predict inhibition by glutamine*
Phosphoserine phosphatase,	SERB_ARCFU	3.1.3.3	Serine biosynthesis	Inhibited by serine
Formyltetrahydrofolate deformylase	PURU_ECOLI	3.5.1.10	*de novo* purine biosynthesis	Activated by methionine and inhibited by glycine
Acetolactate synthase small regulatory subunit	ILVH_ECOLI	4.1.3.18	First step in valine and isoleucine biosynthesis.	Inhibited by valine
L-serine dehydratase beta chain	SDHB_PEPAS	4.2.1.13	Gluconeogenesis from serine	*Predict activation by serine*
Threonine dehydratase,	THD1_BACSU	4.2.1.16	First step of isoleucine biosynthesis	Isoleucine inhibits, valine activates
Prephenate dehydrogenase,	PHEA_BACSU	4.2.1.51	L-phenylalanine biosynthesis	*Predict inhibition by phenylalanine*
TyrR	TYRR_ECOLI	Non enzyme	Regulates many other genes involved in aromatic amino acid biosynthesis	Regulated by Phenylalanine, Trptophan and Tyrosine

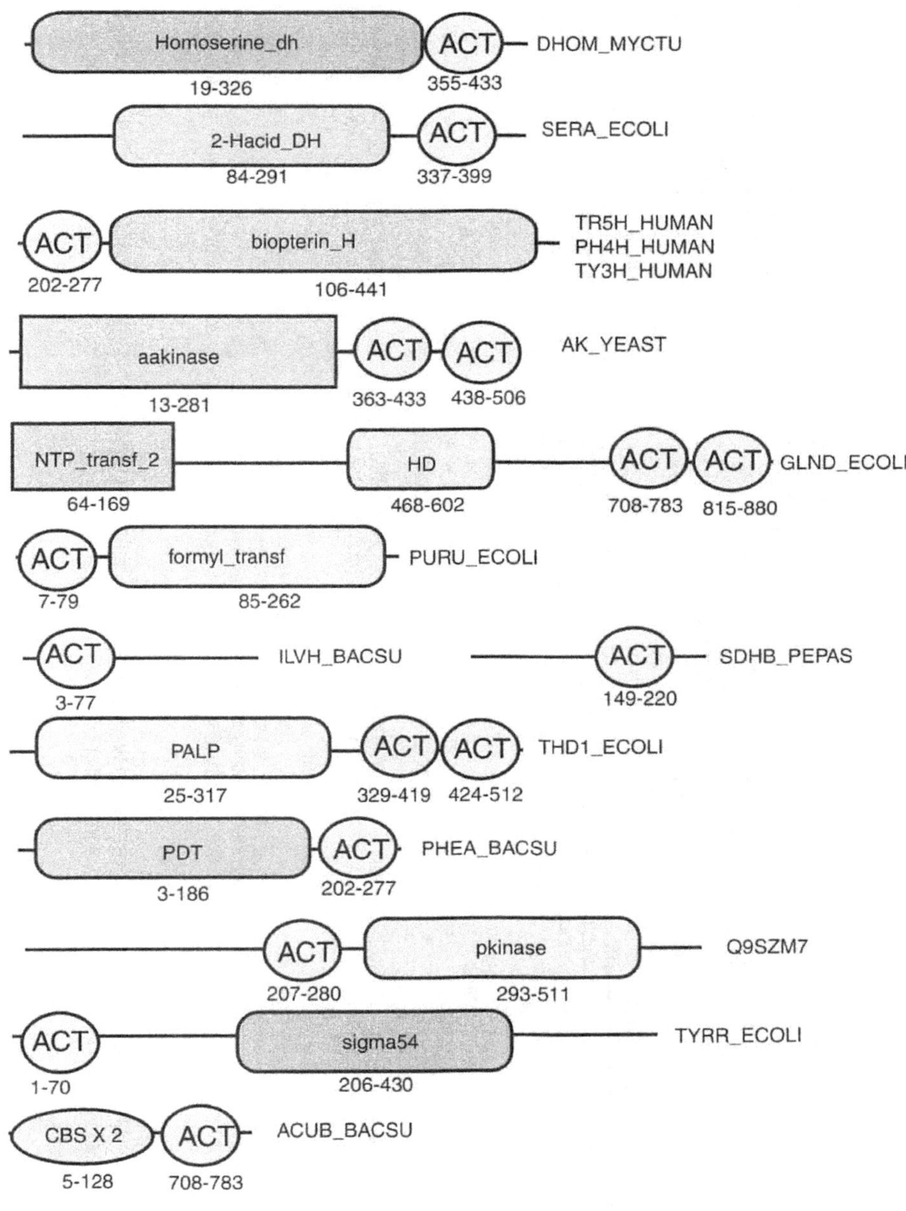

Figure 3. Domain organisations of ACT domain-containing proteins. Boxes and circles represent domains; within each domain is its Pfam identifier and start and end points of the domain are given below. The Swiss-Prot sequence identifier is given for a representative of each protein. ***Colour plate at back of book.***

```
ACUB_BACSU     138 SQIEIKVNDI...TKSLAEISSLCQDLQVKILSVLVYPHD....DPGVKVLVFRVKT...MNPLPFLQALQRNGH..HVVWPSEQRD 212
AK1H_ECOLI     316 AMFSVSGPGMKGMVGMAARVFAAMSRARISVVLITQSSSE........YSISFCVP......QSDCVRAERAM....LEEFYLELKE 384
AK1H_ECOLI     397 AIISVVGDGLRTLRGISAKFFAALARANINIVAIAQGSSE........RSISVVVNN...DDATTGVRVTHQM....LFNTDQVIEV 468
AK_YEAST       363 FVINIHSNKKTLSHGFLAQIFTILDKYKLVVDLISTSEVH..........VSMALPI...PDADSLKSLRQAEE...KLRILGSVDI 433
AK_YEAST       438 SIVSLVGKHMKQYIGIAGTMFTTLAEEGINIEMISQGANE........INISCVIN......ESDSIKALQCI....HAKLLSERTN 506
DHOM_MYCTU     355 YYVSMNVADK...PGVLSAVAAEFAKREVSIAEVRQEGVVDEGGRRVGARIVVVTHL...ATDAALSETVDALDD..LDVVQGVSSV 433
GCVR_ECOLI       8 LVITALGADR...PGIVNTITRHVSSCGCNIEDSRLAMLG......EEFTFIMLLSGS.WNAITLIESTLPLKGA..ELDLLIVMKR 82
GCVR_ECOLI      95 VWVQVDVADS...PHLIERFTALFDAHHMNIAELVSRTQPAENERAAQLHIQITAHSPASADAANIEQAFKALCT..ELNAQGSINV 176
GLND_ECOLI     708 TEIFIWSPDR...PYLFAAVCAELDRRNLSVHDAQIFTTR....DGMAMDTFIVLE....PDGNPLSADRHEVIRFGLEQVLTQSSW 783
GLND_ECOLI     815 SFLELIALDQ...PGLLARVGKIFADLGISLHGARITTIG.....ERVEDLFIIAT...........ADRRALNN..ELQQEVHQRL 880
ILVH_BACSU       3 RIITLTVVNR...SGVLNRITGLFTKRHYNIESITVGHTE....TAGVSRITFVVHV...EGENDVEQLTKQLNK..QIDVLKVTDI 77
MAO1_BACSU       8 RTLMIETPSV...PGNLGRVATAIGLLGGDIGEVETVKVG...PNYTMRNITVQVE.....NEEQLQEVIAAVQA..LGEGIRLHTV 81
PH4H_HUMAN      35 ISLIFSLKEE...VGALAKVLRLFEENDVNLTHIESRPSR.....LKKDEYEFFTHLD.KRSLPALTNIIKILRH..DIGATVHELS 110
PHEA_BACSU     202 TLMVMLPQDD..QSGALHRVLSAFSWRNLNLSKIESRPTK.....TGLGHYFFIIDIEKAFDDVLIPGAMQEL....EALGCKVRLL 277
PHEB_BACSU      69 ITLFFHLEDR...SGALSQLLQAVADSGSNVLSIHQTIPL.....QGRANVTLSIST..SAMEEDIHTLMNKLR...KFDFVEKVEI 142
PURU_ECOLI       7 KVLRTICPDQ...KGLIARITNICYKHELNIVQNNEFVDH......RTGRFFMRTEL...EGIFNDSTLLADLDS..ALPEGSVREL 79
RELA_ECOLI     667 LVVRVVANDR...SGLLRDITTILANEKVNVLGVASRSDT....KQQLATIDMTIEI...YNLQVLGRVLGKLNQ..VPDVIDARRL 741
SDHB_PEPAS     149 STILLEYPEQ...RGVISYVSSLLTGSEYNIESLNTKK......NKLTNIVTLTVE....IDKPLTESLKSAILG..VERFTTAKYV 220
SERA_ECOLI     337 RRLMHIHENR...PGVLTALNKIFAEQGVNIAAQYLQTSA....QMGYVVIDIEAD......EDVAEKALQAMKA..IPGTIRARLL 408
SERA_ECOLI SS      .EEEEEEE.......HHHHHHHHH....EEEEEEEEE......EEEEEEEEE........HHHHHHHHHHHH......EEEEEE.
SERB_ARCFU      47 YVVTILGKDR...VGIVRDITRAFLDFGINIERTSLTARE......ELISIEFLVD....LGQRDAAEVRKRLRR..EAERLGLDIV 118
SERX_YEAST     398 VRVLYIHQNV...PGVLKTVNDILS..NHNIEKQFSDSNG........EIAYLMADIS.SVDQSDIKDIYEQLNQ..TSAKISIRLL 468
SERY_YEAST     398 VRVLYIHRNV...PGVLKTVNDILS..DHNIEKQFSDSHG........EIAYLMADIS.SVNQSEIKDIYEKLNQ..TSAKVSIRLL 468
SPOT_ECOLI     627 TEIKVEMFNH...QGALANLTAAINTTTSNIQSLNTEEKD.....GRVYSAFIRLT......ARDRVHLANIMRK..IRVMPDVIKV 697
TR5H_HUMAN      18 ATLIFSLKNE...VGGLIKALKIFQEKHVNLLHIESRKSK.....RRNSEFEIFVDC..DTNREQLNDIFHLLKS..HTNVLSVTPP 92
TYRA_BACSU     299 YDLYVDVPDH...PGVISEITAILAAERISITNIRIIETR....EDINGILRISFQ.....SDDDRKRAEQCIEA..RAEYETFYAD 371
TYRR_ECOLI       1 MRLEVFCEDR...LGLTRELLDLLVLRGIDLRGIEIDPIG....RIYLNFAELEFE........SFSSLMAEIRR..IAGVTDVRTV 70
```

Figure 4. An alignment of representative ACT domains produced with the Belvu program (Erik Sonnhammer, unpublished). The Swiss-Prot identifier for each sequence is given. The known secondary structure taken from DSSP of PDB entry 1psd is shown for SERA_ECOLI. ***Colour plate at back of book.***

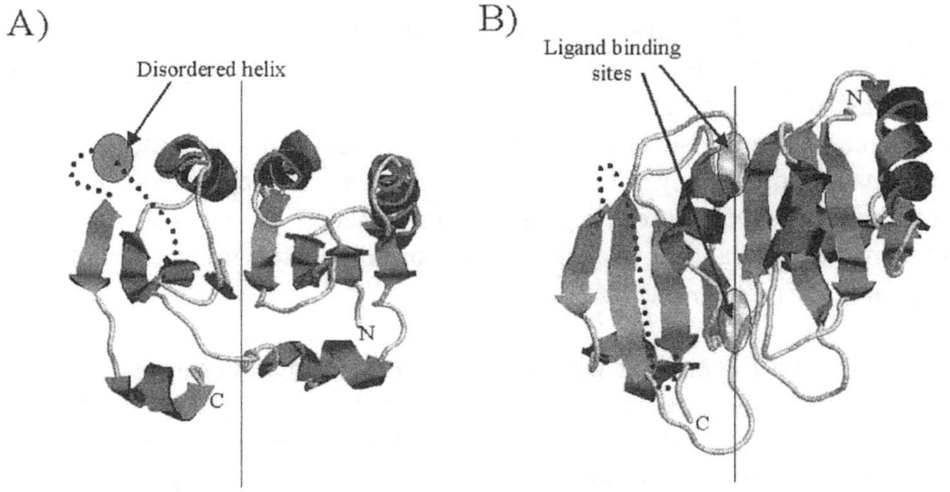

Figure 5. The orientation of ACT domain dimers. The structures show the pseudo dimeric arrangement from the C-terminal ACT domains of threonine dehydratase (pdb entry 1thd). The site of ligand binding inferred from phosphoglycerate dehydrogenase is shown in part B. Part of this structure is disordered, this region is shown by a dotted line. ***Colour plate at back of book.***

(iii)Prephenate dehydrogenase is inhibited by phenylalanine.

The ACT domain-containing enzyme phenylalanine-4-hydroxylase is mutated in individuals with phenylketonuria (PKU). This is one of the most common of metabolic disorders, affecting about 1 in 10,000 births. A number of mutations that cause PKU fall within the ACT domain of this enzyme whose activity is proposed to be regulated by phenylalanine. So it has been possible on the basis of sequence similarity alone to unify the regulation of a diverse range of enzymes.

7. Repeats

Significantly similar sequences that are never found singly in proteins are termed repeats. Repeats come in three main flavours: those that occur in closed, compact structures such as β-propellers (Murzin, 1992), those that form semi-closed structures such as proteins containing leucine-rich repeats (Kobe and Deisenhofer, 1995), and others that form extended, superhelical structures (Groves and Barford, 1999; Andrade *et al.*, 2000). Many of the challenges in detecting repeats mirror those of domains. Additional problems, however, arise from interpretation of database searches seeded with a query sequence that contains multiple repeats. A repeat-containing query sequence may be aligned accurately against a database sequence containing homologous repeats, in many different ways. Such large numbers of legitimate pairwise alignments in search outputs are confusing. It is important, however, to emphasise that multiple instances of distinct alignments should be considered a strong indicator of the *presence* of repeats in the query sequence. In other words, repeats are indicated if a region of the query sequence significantly matches two or more distinct regions of a database sequence, or *vice versa*.

It is important to consider whether these multiple matches are significant. Unfortunately, most database search algorithms, including PSI-BLAST and HMMer2, generate statistics that are relevant only to optimal ("top-scoring") alignments rather than sub-optimal ("repeat") alignments. What is required is an estimate of the probability that two regions of a single protein sequence align purely by chance. Fortunately, this calculation is provided by PROSPERO (Mott and Tribe, 1999). A similar estimate, although for ungapped alignments, can be determined using MACAW (Schuler *et al.*, 1991) and a searchspace equal to the square of the protein length. An additional method, REPRO (Heringa and Argos, 1993), is able to detect repeats but does not provide statistical significance estimates.

Once the presence of repeats has been detected, the identity of the repeats should be ascertained. Many typical repeats' families are detectable using SMART, Pfam and PROSITE profiles. In addition, REP may be used. This is a profile-based method for detecting eleven high-abundance repeat types which also calculates sub-optimal alignment statistical significance (probability) values (Andrade *et al.*, 2000). If the repeats of interest appear not to conform to these typical families then the following protocol is suggested:

(i) Determine the boundaries and extent of the repeat unit. Repeats may be visualised using Dotter (Sonnhammer and Durbin, 1995) or else extracted from PROSPERO (Mott and Tribe, 1999) output. Assess the start and end of the most N- and C-termini of the protein or of other constituent domains/repeats. Note that most repeats occur in tandem arrays and in integral numbers, and they are only rarely interrupted by long insertions between repeats.
(ii) Align those sequences found by PROSPERO with significance in a multiple alignment. Add other sequences that can be shown using PSI-BLAST to be significantly similar to the repeats found by PROSPERO. Ensure that no repeats overlap. Use the same criteria in aligning repeats'

sequences as for aligning those of domains.

(iii) Undertake a database search with a HMM constructed from the multiple alignment. Construct and calibrate the HMM ('hmmbuild' and 'hmmcalibrate' in the HMMER2 package; http://hmmer.wustl.edu/) using default parameters. This is subsequently compared with database sequences and generates global alignments ('hmmsearch' in HMMER2). The HMMER2 package includes a heuristic procedure that is highly efficient at detecting repeats. This procedure discriminates *against* single alignments with marginal scores, and *for* several alignments with marginal scores. Accept repeat sequences as representing homologues if their protein E-values are less than 0.1 in a search of highly non-redundant databases such as nrdb90 (Holm and Sander, 1998), and add these sequences to the previously-derived alignment.

(iv) Repeat step (iii) iteratively until no further candidate homologues are detected.

(A)

```
> 1 CG12188   vs   CG12188 E = 5.810687x10^-08

   22 PAIHEQCKQLERAKTSDLLKAKIQQRPNREDLERLHILEEDE    63  CG12188
      |:: |: : |::|: :| | ::|| ||    :| : :||  ::
   68 PSLAEKQRMLKKARLADQLNSQIQHRPGPLELIKKNILHTEK   109  CG12188
```

(B)

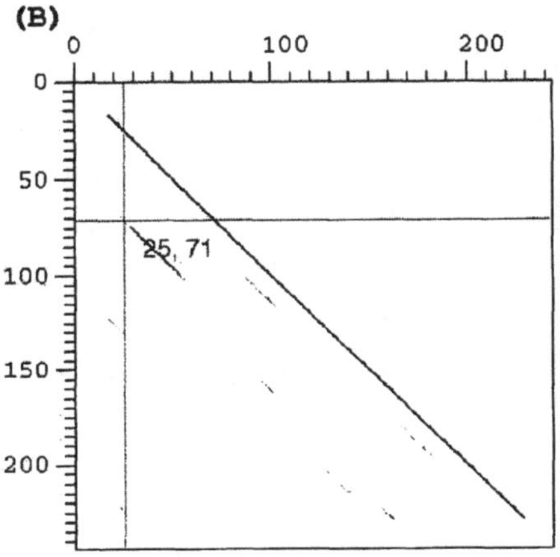

(C)

```
CG10860/54-79          ESLSLKLQLRPDKQDLINRNILHQVN
CG10860/92-117         ARLIRRLSMRPTAEELVERNILKTQS
CG10860/130-155        SYLLRKLSFRPTVEELKEKKIIRFND
DJ257A7.2/43-68        DFFRRRLSQRPTAEELEQRNILKPRN
DJ257A7.2/81-10        RRLTRKLSQRPTVEELRERKILIRFS
KIAA0680/60-85         AVLERKISTRQSREELIRRGVLKELP
KIAA0680/477-502       DTLAIKLGNRPSKKELEDKNILQRTS
KIAA0680/515-540       TKLVRRLSQRPTTEELEQRNILKQKN
KIAA0680/553-578       RRLSRKLSLRPTVAELQARRILRFNE
DJ1042K10.5/27-52      AVLQLKLQQRRTREELVSQGIMPPLK
DJ1042K10.5/71-96      DYLKRKIRSRPERSELVRMHILEETS
DJ1042K10.5/115-140    DDLNEKIAQRPGPMELVEKNILPVES
CG12188/38-63          DLLKAKIQQRPNREDLERLHILEEDE
CG12188/84-109         DQLNSQIQHRPGPLELIKKNILHTEK
F26H9.2/391-416        IKLARKLSERPTPEELEDRNILRRRG
F26H9.2/428-453        KLLLRKLSFRPTIKELKEQQIIQFAD
Consensus/80%          ..L.b+lpbRPs.pEL.pppILp..p
```

Figure 6. Analysis of RPEL protein sequence repeats. A) The result of the best scoring self-self match of *Drosophila* protein CG12188 using prospero. B) A dot plot of the first 250 amino acids of CG12188 against itself. The strong diagonal line shows the 100% identical self match. Other off-diagonal lines represent potential repeats. C) An alignment of RPEL repeats from a variety of proteins. The sequence identifier is followed by the start and end points of the repeat. An alignment of RPEL repeats can be found in the Pfam database with accession number PF02755. ***Colour plate at back of book.***

An Example of Detecting Repeats: CG12188

To illustrate this protocol, we will describe the detection of tandem repeats in the *Drosophila melanogaster* hypothetical protein CG12188. Self-comparison of the CG12188 sequence that had been pre-filtered using SEG strongly indicated a pair of tandem repeats near the N-terminus (Figure 6). Three iterations of PSI-BLAST using the CG12188 sequence (amino acids 1-140; *E*-value inclusion threshold of 0.01) as query, revealed similar repeats in mammalian and invertebrate proteins (Figure 6). A HMM calculated from a multiple alignment of these repeats (which we shall call "RPEL" repeats, after four highly-conserved amino acids) was compared with nrdb90 (Holm and Sander, 1998) using HMMER2. This resulted in the identification of three further repeats in these proteins (Figure 6). A further HMMER2 search failed to reveal additional repeats. The function of RPEL repeats remains unknown, but RPEL may be DNA-binding repeats given the observation of a SAP motif in CG12188 (Aravind and Koonin, 2000).

8. Motifs

The motif differs from repeats since it often occurs singly, and it differs from domains in that it doesn't possess a distinct compact structure. Motifs can either represent a small region of a domain, such as an active or binding site, or else may be a region of a protein sequence outside of domain bounds that has no discernible tertiary or secondary structure in the absence of its binding partner. Examples of the latter include the DNA-binding AT-hook (Aravind and Landsman, 1998) and the Cdc42/Rac-interaction binding (CRIB) motif (*e.g.* Mott *et al.*, 1999); each of these motif types is unlikely to possess a secondary structure, when in isolation. A more rare, third type of sequence motif is one that is observed in different folds. Examples of these include the DNA-binding helix-hairpin-helix motif (Doherty *et al.*, 1996) and sialidase-like Asp-boxes (Copley *et al.*, 2001).

Each of these three motif types has properties that are of interest for reasons that relate to function, structure or evolution. The active- or binding-site motif often specifies function and thus is important in defining the repertoire of functions possessed within a homologous family. The motif that lacks defined structure in isolation is, for obvious reasons, under-represented in structure databases, and thus may be intractable to structural genomics projects without reference to experimental information. Finally, the motif that occurs in different non-homologous contexts might have arisen via convergent evolution.

An Example of Detecting Motifs: CG17290

A PSI-BLAST search of current databases using *Drosophila* CG11131 sequence as query revealed significant similarities to several other fly hypothetical proteins. Upon closer inspection of the PSI-BLAST alignments it was apparent that two short motifs, 'DGY' and 'YLP' are repeated in these homologues with varying frequencies and patterns (Figure 7). The nrdb90 database was iteratively searched with HMMs derived from multiple

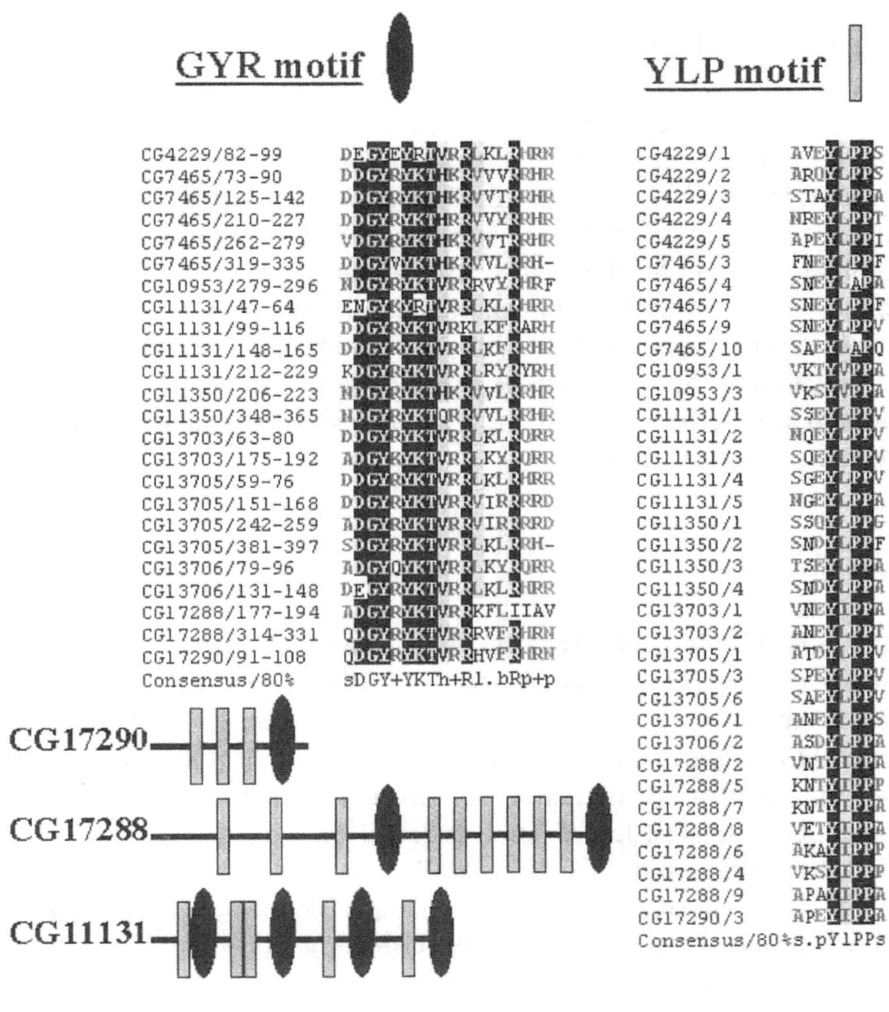

Figure 7. Alignment of GYR and YLP motifs in *Drosophila* proteins. The motifs are shown graphically for three of these proteins. Alignments of GYR and YLP motifs can be found in the Pfam database with accession numbers PF02756 and PF02757 respectively. ***Colour plate at back of book.***

alignments of DGY motifs; the initial alignment contained 8 such motifs extracted from PSI-BLAST alignments. Simultaneously, this ungapped alignment was used to search nrdb90 using MoST. Both of these approaches identified 29 copies of the DGY motifs in 9 *Drosophila* hypothetical proteins. YLP motifs were identified in these 9 protein sequences by comparing them with a HMM derived from 8 such motifs, and a reduced *E*-value acceptance threshold due to their extremely short length (Figure 7). The molecular functions of these motifs are unclear, but absolute conservation of tyrosine residues in each might indicate that they are substrates for tyrosine kinases.

9. Conclusions and Perspectives

Genomes and sequence databases contain a wealth of information that awaits evaluation and appreciation. Sequence analysis enables access to this information for all biologists and, when applied appropriately, generates new hypotheses that are amenable to experimental investigation. No single protocol, appropriate for the analysis of *all* sequences, exists. Rather each protein sequence requires individual attention to infer as much as possible about its structure and function without recourse to over-prediction. As regions of proteins however, can be categorised into three types, domains, repeats and motifs, there are only a few strategies for their analysis that each can be reused to great effect.

References

Altschul S.F., Madden, T.L., Schaffer, A.A., Zhang, J., Zhang, Z., Miller, W. and Lipman, D.J. 1997. Gapped BLAST and PSI-BLAST: a new generation of protein database search programs. Nucl. Acids Res. 25: 3389-3402.

Andrade, M.A., Ponting, C.P., Gibson, T.J., and Bork, P. 2000. Homology-based method for identification of protein repeats using statistical significance estimates. J. Mol. Biol. 298: 521-537.

Aravind, L., and Koonin, E.V. 2000. SAP - a putative DNA-binding motif involved in chromosomal organization. Trends Biochem. Sci. 25: 112-114.

Aravind, L., and Koonin, E.V. 1999. Gleaning non-trivial structural, functional and evolutionary information about proteins by iterative database searches. J. Mol. Biol. 287: 1023-1040.

Aravind, L., and Landsman, D. 1998. AT-hook motifs identified in a wide variety of DNA-binding proteins. Nucl. Acids Res. 26: 4413-4421.

Aubourg, S., Boudet, N., Kreis, M., and Lecharny, A. 2000. In *Arabidopsis thaliana*, 1% of the genome codes for a novel protein family unique to plants. Plant Mol. Biol. 42: 603-613.

Bairoch A, Apweiler R. 2000. The SWISS-PROT protein sequence database and its supplement TrEMBL in 2000. Nucleic Acids Res. 28: 45-48.

Bork, P., and Gibson, T.J. 1996. Applying motif and profile searches. Methods Enzymol. 266: 162-184.

Copley, R.R., Russell, R.B., and Ponting, C.P. 2001. Sialidase-like Asp-boxes: Sequence-similar structures within different protein folds. Protein Sci. 10: 285-292.

Corpet, F., Servant, F., Gouzy, J., and Jahn, D. 2000. ProDom and ProDom-CG: tools for protein domain analysis and whole genome comparisons. Nucl. Acids Res. 28: 267-269.

Doherty, A.J., Serpell, L.C., and Ponting, C.P. 1996. The helix-hairpin-helix DNA-binding motif: a structural basis for non-sequence-specific recognition of DNA. Nucl. Acids Res. 24: 2488-2497.

Gracy, J., and Argos, P. 1998. Automated protein sequence database classification. II. Delineation Of domain boundaries from sequence similarities. Bioinformatics 14: 174-187.

Grant, G.A., Schuller D.J and Banaszak L.J. 1996. A model for the regulation of D-3-phosphoglycerate dehydrogenase, a Vmax-type allosteric enzyme. Protein Sci. 5: 34-41.

Groves, M.R., and Barford, D. 1999. Topological characteristics of helical repeat proteins. Curr. Opin. Struct. Biol. 9: 383-389.

Heringa, J., and Argos, P.1993. A method to recognize distant repeats in protein sequences. Proteins. 17: 391-411.

Holm, L., and Sander, C. 1998. Removing near-neighbour redundancy from large protein sequence collections. Bioinformatics. 14: 423-9.

Hunter, T., and Plowman, G.D. 1997. The protein kinases of budding yeast: six score and more. Trends Biochem. Sci. 22: 18-22.

International Human Genome Sequencing Consortium. 2001. Initial sequencing and analysis of the human genome. Nature. 409: 860-921

Karlin, S., and Altschul, S.F. 1990. Methods for assessing the statistical significance of molecular sequence features by using general scoring schemes. Proc. Natl. Acad. Sci. USA. 87: 2264-2268.

Kobe, B., and Deisenhofer, J. 1995. Proteins with leucine-rich repeats. Curr. Opin. Struct. Biol. 5: 409-416.

Mott, H.,R, Owen, D., Nietlispach, D., Lowe, P.N., Manser, E., Lim, L., and Laue, E.D. 1999. Structure of the small G protein Cdc42 bound to the GTPase-binding domain of ACK. Nature 399: 384-388.

Mott, R., and Tribe, R. 1999. Approximate statistics of gapped alignments. J. Comput. Biol. 6: 91-112.

Murzin, A.G. 1992. Structural principles for the propeller assembly of beta-sheets: the preference for seven-fold symmetry. Proteins. 14: 191-201.

Park, J, Karplus, K., Barrett, C., Hughey, R., Haussler, D., Hubbard, T., and Chothia, C. 1998. Sequence comparisons using multiple sequences detect three times as many remote homologues as pairwise methods. J. Mol. Biol. 284: 1201-1210.

Pearson, W.R., and Lipman, D.J. 1998. Improved tools for biological sequence comparison. Proc. Natl. Acad. Sci. USA. 85: 2444-2448

Ponting, C.P., and Birney, E. 2000. Identification of domains from protein sequences. In: Protein Structure Prediction, D. Webster, Ed., Humana Press, USA.

Russell, R.B., and Ponting, C.P. 1998. Fold irregularities that hinder sequence analysis. Curr. Opin. Struct. Biol. 8: 364-371.

Sayle R.A., and Milner-White E.J. 1995. RASMOL: biomolecular graphics for all.

Trends Biochem. Sci. 20;374.

Schuler, G.D., Altschul, S.F., and Lipman, D.J. 1991. A workbench for multiple alignment construction and analysis. Proteins. 9: 180-190.

Sonnhammer, E.L., and Durbin, R. 1995. A dot-matrix program with dynamic threshold control suited for genomic DNA and protein sequence analysis. Gene. 167: GC1-10.

Thompson, J.D., Higgins, D.G., and Gibson, T.J. 1994. CLUSTAL W: improving the sensitivity of progressive multiple sequence alignment through sequence weighting, position-specific gap penalties and weight matrix choice. Nucl. Acids Res. 22: 4673-4680.

Wootton, J.C. and Federhen, S. 1993. Statistics of local complexity in amino acid sequences and sequence databases. Comput. Chem. 17: 149-163.

From: *Frontiers in Computational Genomics*

6

Exploiting the Variations in the Genomic Associations of Genes to Predict Pathways and Reconstruct their Evolution

Martijn A. Huynen and Berend Snel

Abstract

Various methods have been proposed to predict functional interactions between proteins using the genomic association of their genes. Here we exploit the variation in this genomic association to increase the resolution of these methods. The metabolic distance between enzymes, as defined by the number of substrates that separate them in a pathway, is used to quantify the directness of functional interaction. We show a positive correlation between the strength of the genomic association and the directness of the functional interaction: *i.e.* the stronger the genomic association is, the shorter the average metabolic distance between the enzymes. Subsequently we map the variation and reconstruct the evolution of a single pathway in detail: iron-sulfur cluster assembly in Proteobacteria and mitochondria. By deconstructing the evolution of this pathway to the level of gene-gain and gene-loss events we identify (sub)modules within it. Proteins that are within the same sub-module are expected to have a more direct functional interaction than ones that are in different sub-modules.

1. Introduction

The sequencing of complete genomes from cellular species has led to an explosive increase in the knowledge about the phylogenetic distribution of (predicted) pathways (*e.g.* Huynen *et al.,* 1999; Wolf *et al.,* 1999; Boucher and Doolittle, 2000; Hynds *et al.,* 2000; Lange *et al.,* 2001), to new methods for the prediction of the function of proteins and their involvement in pathways (Eisenberg *et al.,* 1999; Huynen *et al.,* 2000), and to an increase in the rate of experimental function determination of enzymes (*e.g.* Luttgen *et al.,* 2000; Stec *et al.,* 2000; Verhees *et al.,* 2001). One of the most interesting aspects of the genome sequences is the variation in the predicted pathways between the species. Specifically the many different, incomplete appearances of what are considered standard pathways, force us to reconsider our naïve, species-independent pathway concept. In terms of the prediction of pathways, they undermine the applicability of using the evolutionarily conserved genomic association of genes to predict the involvement of their proteins in the same pathway: *i.e.* if the functional interactions between proteins are not conserved in evolution, how do we expect to predict them based on the conserved association of their genes? We will, in this chapter outline two approaches to actually exploit the variation in pathways and in the genomic association of genes. The first one exploits the abundance of genomic data to quantify the correlation between on the one hand the genomic association of genes and on the other hand the functional interaction of the encoded proteins. Such an approach increases the specificity at which we can make predictions about functional interactions, *e.g.* rather than predicting that two proteins are involved in the same pathway, we can, at a certain level of genomic association, with high reliability actually predict them to be involved in the same biochemical reaction. The second approach is qualitative: we unravel the evolution of a single pathway, iron-sulfur cluster assembly in Proteobacteria and mitochondria, to the level of the gain and loss of genes. Aside from revealing a fascinating history in itself, this allows us to predict sub-modules within the pathway, and thus to be more specific in our predictions about the functions and functional interactions of its proteins.

2. Results

Quantifying the Correlation Between Genomic Association and Functional Interaction

Various types of genomic contexts have been proposed to identify functional properties of genes: 1) the fusion of genes (Marcotte *et al.,* 1999; Enright *et al.*, 1999), 2) the conserved co-occurrence of genes as neighbors or in potential operons (Overbeek *et al.*, 1998; Dandekar *et al.*, 1998; Overbeek *et al.*, 1999) 3) the co-occurrence of genes among genomes (Huynen and Bork, 1998; Pellegrini *et al.*, 1999; Tatusov *et al.*, 2001) and 4) the sharing of regulatory elements (Gelfand *et al.*, 2000; McGuire *et al.*, 2000). The first three types of genomic contexts are variations on the theme that the genomic association of genes correlates with functional association of the encoded proteins. An extension of the co-occurrence idea is to use also the evolutionary distances between proteins or the similarity of their evolutionary trees (Goh *et al.*, 2000; Pazos and Valencia, 2001). Interestingly there have been indications of a correlation between the strength of the genomic association and the directness of the functional interaction: gene fusion and the conservation of gene-order are more likely to reflect a physical interaction between their proteins than the mere co-occurrence of genes in the same operon or genome (Huynen *et al.*, 2000), and the more often genes occur together in potential operons, the more likely they are to physically interact (Mushegian and Koonin 1996; Huynen *et al.*, 2000). Combining various types of genomic association decreases the fraction of false positives (Eisenberg *et al.*, 1999). Furthermore, when there is prior knowledge about a protein's likely involvement in a process, either based on experimental data, or based on the conservation of gene order, the coordinated loss or gain of genes can more specifically pinpoint in which sub-process the protein plays a role (Aravind *et al.*, 2000, Ettema *et al.*, 2001).

In this type of research, where one tries to find genomic correlates of functional interactions and pathways, determining what exactly constitutes a functional interaction between proteins is generally the most time-consuming and subjective aspect, requiring either painstaking manual examination of the experimental data, or arguably crude text mining techniques (Eisenberg *et al.*, 1999). Furthermore, the functional interaction is generally only scored qualitatively: proteins are either considered to functionally interact or not. We here focus on a specific class of proteins, enzymes, which allows us to use their metabolic distance to obtain a more quantitative expression for the functional interaction between proteins. The metabolic distance is here defined as the number of substrates that separates two enzymes on the metabolic map (Ettema *et al.*, 2001). Using the COG database for orthology relations (Tatusov *et al.*, 2001) and the KEGG database for metabolic distances (Kanehisa and Goto, 2000), we show that all types of genomic association show a correlation with the average metabolic distance between the encoded enzymes (Figures 1a, 1b, 1c). The stronger the genomic association between two genes, the closer their two enzymes are in metabolic space. The strength of the relation between the genomic and metabolic

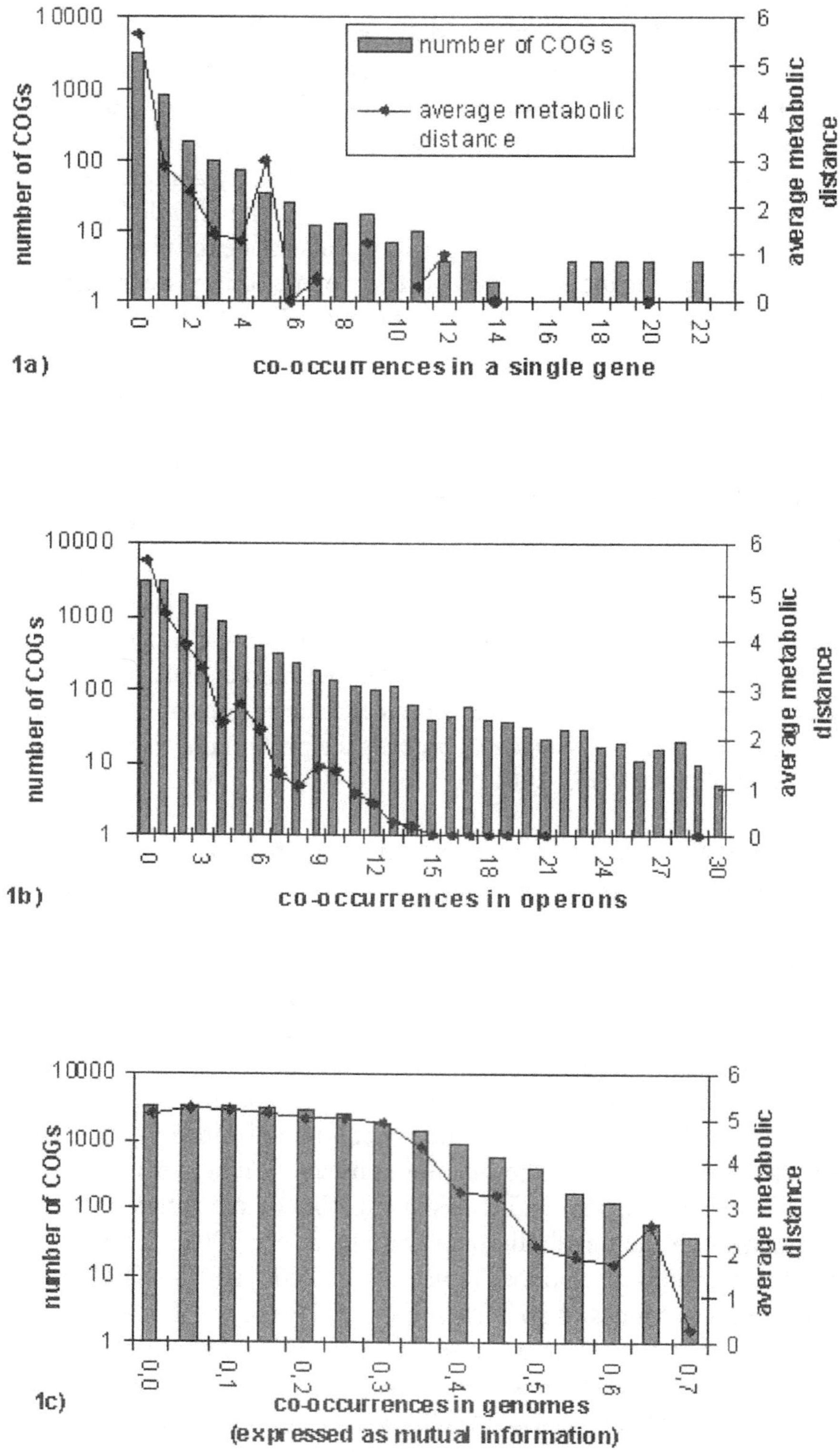
10000
1000
100
10
1
number of COGs
number of COGs
average metabolic distance
6
5
4
3
2
1
0
average metabolic distance
0
2
4
6
8
10
12
14
16
18
20
22
1a)
co-occurrences in a single gene
10000
1000
100
10
1
number of COGs
6
5
4
3
2
1
0
average metabolic distance
0
3
6
9
12
15
18
21
24
27
30
1b)
co-occurrences in operons
10000
1000
100
10
1
number of COGs
6
5
4
3
2
1
0
average metabolic distance
0,0
0,1
0,2
0,3
0,4
0,5
0,6
0,7
1c)
co-occurrences in genomes
(expressed as mutual information)

associations is illustrated by the fact that when two enzymes occur 15 or more times together in an operon, in all cases observed they are subunits of an enzyme complex performing the same metabolic step. These quantitative trends indicate the value of having many genomes: the more widespread a specific type of genomic association, the more likely it is to reflect a "universal" type of functional association in which the proteins completely depend upon each other. Furthermore they allow us to increase the resolution at which one can predict functional interactions: variations in the genomic association of a set of genes indicate the presence of subsets and allow us to identify them.

Correlating Co-Expression with Gene-Order Conservation in Eukaryotes

Comparing the three types of signals we find a trend that has been observed before (Huynen *et al.*, 2000): in terms of the number of confidently predicted interactions the conservation of gene order is the most powerful technique, at least for the presently sequenced genomes, which are strongly dominated by the prokaryotes. Although there are anecdotal examples in eukaryotes of conservation of gene clusters that are involved in similar functions, like the Homeodomain cluster, in large scale, quantitative analyses they do not play a significant role (Huynen *et al.*, 2001a). The main basis for gene-order

Figure 1. Correlation between the strength of the functional interaction of enzymes and that of the genomic association of their genes. The directness of the functional interaction is calculated by taking the average metabolic distance between enzymes that have a specific genomic association. Metabolic distances are from KEGG (Kanehisa and Goto, 2000), they are defined as the number of substrates that separate two enzymes, for enzymes that are involved in the same reaction the distance is zero. Orthologous groups of sequences were obtained from the COG-database (Tatusov *et al.*, 2001). COGs containing multiple Enzyme Codes were discarded. 1a) Gene fusion data are from the COG database. 1b) Gene-order conservation was quantified with STRING (Snel *et al.*, 2000), using the orthology definitions from COG. 1c) Co-occurrence of genes is expressed as mutual information (Kullback, 1959; Huynen *et al.*, 2001), which is a mathematical expression for the amount of information that the presence of one gene gives about the presence of another one. As high mutual information scores both result from complementary distributions as from identical distributions, giving the information that the other gene is absent or present respectively, the complementary distributions were discarded. In all the three types of genomic context the strength of the genomic association correlates with the directness of the functional association. The strictly modular behavior of some pathways, in which the enzymes of a pathway are either all absent or all present in a genome, results in higher than zero metabolic distances for the highest scores of genomic co-occurrence. An example of such a pathway is histidine biosynthesis, of which all enzymes are either present or absent. This cause the high average metabolic distance at the mutual information score of 0.65 (mutual information scores were binned).

conservation in prokaryotes is that the genes that are involved tend to be in operons, with divergently transcribed genes as a distant second (Tamames *et al.,* 1997; Huynen and Snel, 2000). Divergently transcribed, co-regulated genes have also been observed in *Saccharomyces cerevisiae* (Kruglyak and Tang, 2000). Interestingly, such co-regulation does have an effect on the rate at which gene order is disrupted in genome evolution in these fungi. Between *S. cerevisiae* and *Candida albicans*, divergently transcribed gene-pairs that are conserved in evolution have a higher probability of being co-regulated than divergently transcribed gene pairs that are disrupted in evolution (Figure 2). Constraints on the promotor region apparently hamper the shuffling process of co-regulated divergently transcribed genes in these fungi. Among the divergently transcribed gene pairs the two fungi share are well known cases like the histone pairs H2A-H2B and H3-H4, but also ones

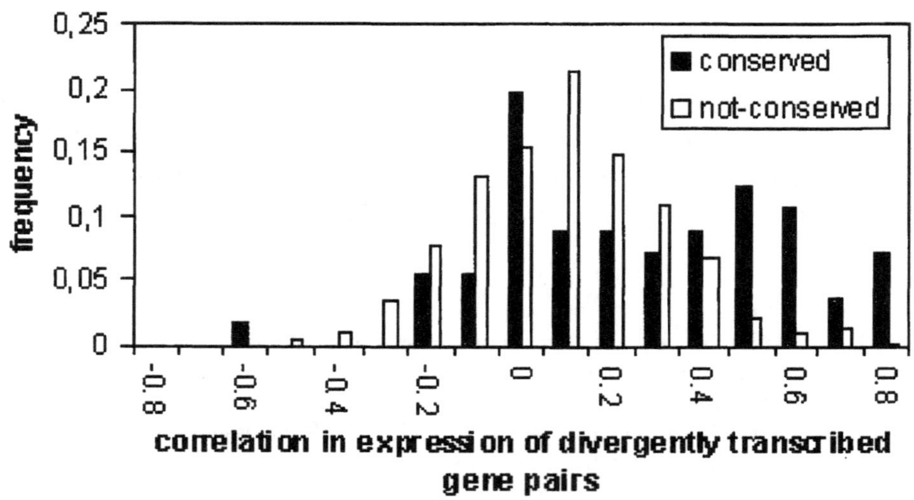

Figure 2. Distribution of the expression correlations of divergently transcribed gene pairs in *Saccharomyces cerevisiae.* The histograms indicate the correlations in gene expression for gene pairs that are (gray) and that are not conserved (white) between *S.cerevisiae* and *Candida albicans.* Gene pairs are considered conserved when they are neighbors in both species, and the relative direction of transcription of the genes to each other is identical in both species. The results show a bimodal distribution for conserved divergently transcribed genes, with a peak at high correlations of co-expression, besides the one at zero. The divergently transcribed gene pairs that are not conserved in *C.albicans* have a unimodal distribution centered round 0.1. The histogram suggests that the rearrangement of divergently transcribed gene pairs between *S.cerevisiae* and *C.albicans* is hampered by overlapping and possibly shared regulatory structures that lie between the genes. For other types of gene-order conservation (genes that are transcribed in the same direction of convergently transcribed gene pairs), no such effect could be observed. Gene expression data are from (Hughes *et al.,* 2000), *C.albicans* genome data are from the Stanford Genome Technology Center (www.sequence-stanford.edu).

like a hexose permease (YJL219W) and an alpha-glucosidase (YJL221C). Such observations suggest the possibility of finding functionally interacting proteins by detecting gene-order conservation in the fungi. Comparisons of the *S.cerevisiae* gene order with that of the more distantly related genomes of *Schizosaccharomyces pombe* and *Encephalotozoon cuniculi* did not reveal significant levels of conservation (data not shown). More, relatively closely related genomes will therefore be needed to do statistical evaluations as the one done in Figure 1b for the fungi.

Reconstructing the Evolution of Iron-Sulfur Cluster Assembly and Identifying (Sub)modules within It

The dramatic increase in the number of available genomes allows us, not only to depict the variation of a pathway, but to actually trace its evolution: *i.e.* when have which proteins been added to the pathway or lost from it, and what is the origin of the added proteins. By doing such analyses we aim to develop a theory of pathway evolution: are there specific processes that recur in the evolution of pathways? Reconstructing pathway evolution involves comparing genomes for the presence of certain orthologous groups of genes, calculating phylogenetic trees to trace the evolution of a specific orthologous group and comparing the presence/absence of genes with experimental data (*e.g.* Huynen *et al.,* 1999). Reconstruction of pathway evolution increases the value of the observed genomic associations for function prediction: *i.e.* rather than observing that two orthologous groups have the same phylogenetic distribution we can actually argue that they have co-evolved. Thereby we handle a classic problem in sequence analysis: discriminating the effects of phylogenetic relatedness from selection. Do two genes occur together because their proteins depend upon each other for their function? Or do they just happen to occur in the same genomes because those genomes are closely related, and therefore have relatively large chance of sharing a set of genes? Given that the intricacies of molecular evolution are hard to delineate with large-scale automated sequence analysis methods, small-scale, manual approaches are required to reconstruct of the evolutionary history of a pathway. This approach to pathway evolution is different from the one in which one analyzes homologies between the proteins in different pathways (Teichmann *et al.,* 2001). Here we actually reconstruct the history of a single pathway that is presently the subject of intensive study: the assembly of 2Fe-2S clusters in Proteobacteria (Zheng *et al.,* 1998) and in mitochondria, (reviewed in Lill and Kispal, 2000). Several proteins have been implicated in assembling the 2Fe-2S clusters and, in the case of mitochondria, exporting them to the cytoplasm (Table 1, Figure 3). The oldest, and phylogenetically most widely distributed part of this pathway is the IscS/IscU pair. IscS, a cysteine desulfurase, is the most essential part of the pathway in *Escherichia coli* (Schwartz *et al.*, 2000; Tokumoto and Takahashi, 2001). IscU, the protein on which the Fe-S clusters are assembled, has a similar wide phylogenetic distribution and the activity of IscS/IscU gene pairs lies both biochemically and phylogenetically at the core of the Fe-S cluster assembly. Other proteins that have a wide distribution are a hypothetical protein in the Bacteria that is orthologous to *S.cerevisiae* ISA2p, and the recently elucidated Fe-S cluster binding, regulatory protein, IscR (Schwartz *et al.*, 2001). From the root of the α-, β- and γ-Proteobacteria on the story that unfolds is highly complex, involving most, if not all, of the primary processes in pathway evolution. Rather than retelling the evolutionary story that is depicted in Figure 4, I will categorize the events into classes, making a distinction, where possible, between modular and non-modular events. Modular events are here defined as events in which multiple proteins from the pathway are involved,

Table 1. The proteins that are involved in Fe-S cluster assembly in the Proteobacteria (left) and in *S.cerevisiae* (right). For a number of proteins, even though their involvement in Fe-S cluster assembly has been established, the exact molecular function is unknown. Proteins on one line are orthologous, a "-" indicates that no orthologous gene for that protein is present in that taxon. A paralogous displacement has occurred between HscA in the Proteobacteria and Ssq1p in the mitochondria. Their orthologs, DnaK and YKL073W respectively do not appear to be involved in Fe-S cluster assembly and have therefore been crossed-out. The evidence for the involvement of Fpr in Proteobacterial Fe-S cluster assembly is tentative.

Proteobacteria	*S.cerevisiae*	Molecular function
IscS	Nfs1p	Cysteine desulfurase (Schwartz *et al.*, 2000)
IscU	Isu1p, Isu2p	Scaffold for Fe-S clusters (Schilke *et al.*, 1999)
IscA	Isa1p	unknown (Kaut *et al.*, 2000; Jensen *et al.*, 2000)
RP063	Isa2p	unknown (Jensen *et al.*, 2000; Pelzer *et al.*, 2000)
HscA	~~YKL073W~~	Bacterial chaperone in Fe-S assembly
~~DnaK~~	Ssq1p	Mitochon. chaperone in Fe-S assembly (Lutz *et al.*, 2001)
HscB	Jac1p	Co-chaperone in Fe-S assembly (Voisine *et al.*, 2001)
Fdx	Yah1p	Reduction of S, Fe-S (Lange *et al.*, 2000)
CyaY	Yfh1p	unknown
Fpr (?)	-	NADP:ferredoxin oxidoreductase (Jung *et al.*, 1999)
-	Arh1	NADP:ferredoxin oxidoreductase (Manzella et. al., 1998)
IscR	-	Regulation of the isc operon (Schwartz *et al.*, 2002)
EC2524	-	unknown
RP667	Nfu1p	unknown (Schilke *et al.*, 1999)
-	Erv1p	Maturation of Fe-S clusters (Lange *et al.*, 2001b)
-	Atm1p	Export of Fe-S clusters

suggesting a functional interaction between them. Note that in principle also non-modular events suggest functional interactions, *e.g.* in the loss of a single protein from a pathway suggests a stronger functional interaction between the remaining proteins than between them and the protein that was lost. Note also that what we regard as modular events maybe actually consist of several non-modular small steps, *e.g.* in the loss of the genes from a pathway, that, in the absence of selection to maintain them, follow each other in rapid succession.

1) Gene Invention (*e.g.* Snel *et al.,* 2002): in multiple instances do genes appear from an (as of yet) unidentifiable source. Using profile based sequence comparison techniques (PSI-Blast, (Altschul *et al.,* 1997)) no homologs could be found (outside their orthologous group) for CyaY/frataxin and for the hypothetical protein that in *E.coli* has number 2524 (EC2524). The origin of

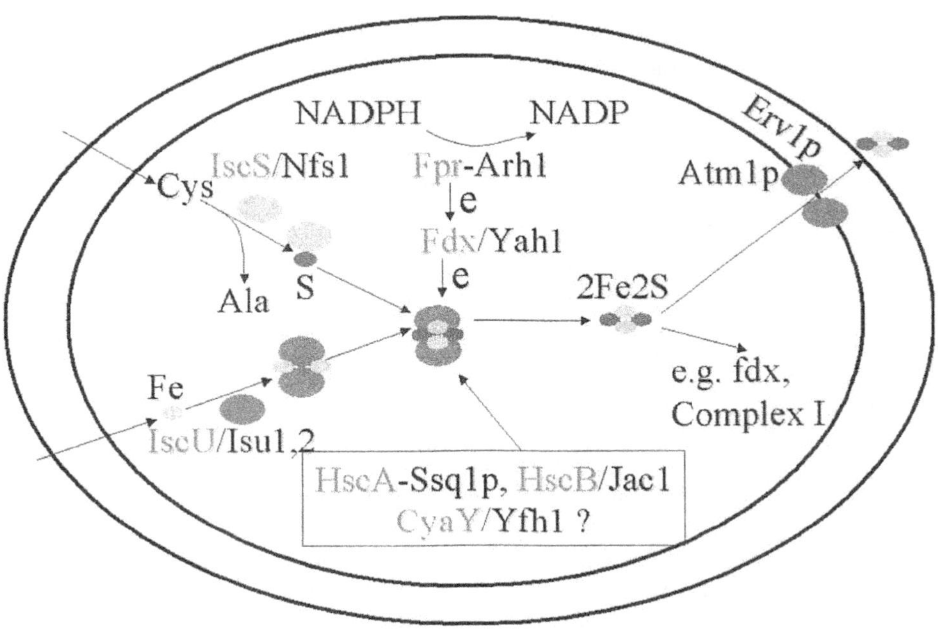

Figure 3. Schematic representation of Fe-S cluster assembly in Proteobacteria and Fe-S cluster assembly and export from the mitochondria of *S.cerevisiae*. Gene displacements are marked with a "-", the Proteobacteria specific proteins are marked in gray. The genes of the proteobacteria and *S.cerevisiae* that are orthologous are separated by a "/". The genes for HscA, HscB and CyaY have a high genomic association are therefore grouped. The picture is largely based on (Lill and Kispal, 2000). The exact order of events in Fe-S cluster assembly on IscU (*e.g.* does the binding and reduction of sulfur take place before or after the binding of iron) has not been resolved (Urbina *et al.,* 2001). ***Colour plate at back of book.***

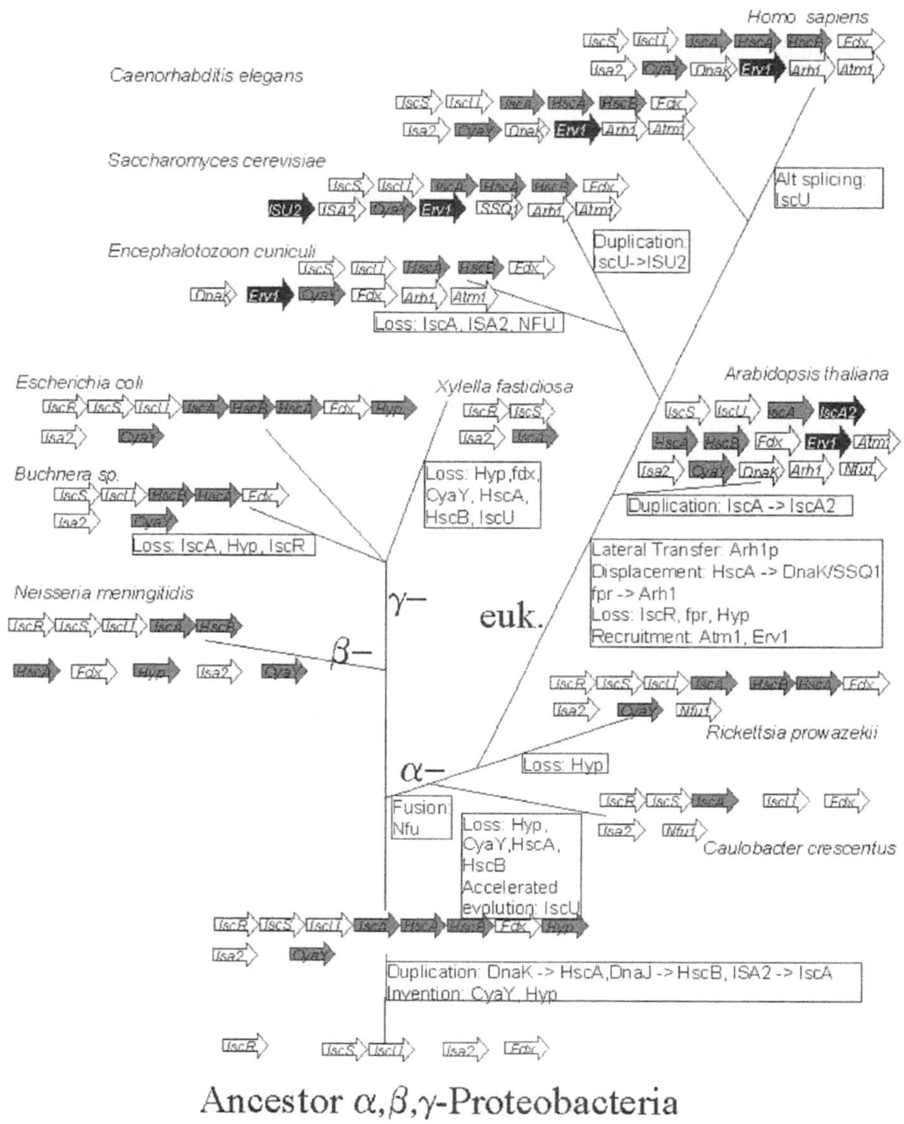

Figure 4. A history of Fe-S cluster assembly. The figure traces the evolution of Fe-S cluster assembly from the ancestor of the α-, β- and γ-Proteobacteria to the eukaryotes (euk.). The results were derived by combining the presence/absence data of genes from the genomes with the phylogenies of individual gene families to determine where gene duplications have occurred. Different types of events that result in the gain and loss of genes are indicated along the branches where they occurred, assuming a most parsimonious evolutionary scenario. We have used bacterial gene names for all the genes that are orthologous to the Bacterial genes from the isc operon or to CyaY/frataxin, and yeast names for the orthologs of genes that were first characterized in *S.cerevisiae*. Genes from the hypothetical gene family of which E.coli gene b2524 is a member are indicated with 'Hyp'. Orthologous groups that within the bacteria also occur outside the Proteobacteria, and that are presumably older than the Proteobacteria are in white, orthologous groups that originated in the Proteobacteria are in gray, orthologous groups that are eukaryote-specific are in black.

such gene families remains an enigma. Structural genomics can potentially solve such cases, as it allows the detection of more distantly related homologs (*e.g.* Ito *et al.,* 2001), but it is very well possible that no homologs for these proteins will ever be found. At least in the case of CyaY/frataxin, solving the structure did not reveal homology with other proteins with known structure (Musco *et al.,* 2000 ; Cho *et al.,* 2000).

2) Gene Duplication (Ohno, 1970): IscA arose from a duplication of a hypothetical gene at the root of the α–, β– and γ–Proteobacteria. An ortholog of this ancestral protein in *Saccharomyces cerevisiae* (Isa2p) is involved in mitochondrial Fe-S cluster assembly. Subsequent gene duplications in this family can be observed in *Arabidopsis thaliana.* IscU has been duplicated in *S.cerevisiae* to ISU1 and ISU2. In *Homo sapiens* an alternative technique to actual gene duplication has been used to obtain multiple, different proteins from one gene: alternative splicing of ISU (Tong and Roualt, 2000).

3) Modular Gene Duplication: The protein pair HscA-HscB arose from duplication of the DnaK-DnaJ pair. Both HscA and HscB are involved in Fe-S cluster assembly (Tokumoto and Takahashi, 2001). The DnaK-DnaJ -> HscA-HscB duplication can thus be regarded as a typical example of two interacting proteins that have been co-duplicated to together assume a function in a new pathway. In this case involving a possible specialization in substrate specificity for IscU (Silberg er al, 2001), and the loss of interaction with a third chaperone, GrpE (Silberg *et al.,* 1998) . In a recent analysis, modular gene duplication (called 'serail recruitment') was considered to play a rather insignificant role in the small molecule metabolism of *E.coli* (Teichmann *et al.,* 2001), including never more than two subsequent enzymatic steps. Note however that in *Pseudomonas aeruginosa* proprionate metabolism, four out of five subsequent steps involve duplication of genes involved in acetate metabolism (Horswill and Escalante-Semerena, 1999; Huynen and Snel, 2000). Modular gene duplication is rather pervasive in signalling pathways, *e.g.* two component signal transduction systems (Koretke *et al.,* 2000) or MAP kinase pathways(Caffrey *et al.,* 1999).

4) Individual Gene Loss: At many instances have individual proteins from the pathway been lost, *e.g.* orthologs of the hypothetical gene EC2524 have independently been lost from *Ricketssia prowazekii*, *Buchnera* sp. and *Pasteurella multocida, Buchnera* has also lost the protein that regulates the isc operon expression, IscR, in what appears to be a more general pattern of the loss of gene regulation in this species (*e.g.* Panina *et al.,* 2001). Genes from the HesB family, *ISA1/IscA* and *ISA2* have been lost multiple times in the evolution, especially from genomes that have undergone extreme size reduction like *Buchnera* sp. and *Encephalotozoon cuniculi*, indicating that their function might not be very essential. Experimental data confirm a non-essential role in Fe-S cluster assembly of the proteins from the HesB family in *S. cerevisiae* (Kaut *et al.*, 2000 ; Jensen and Culotta, 2000 ; Pelzer *et al.*, 2000) and *E. coli* (Tokumoto and Takahashi, 2001) and virtually no effect of deleting EC2425 from *E.coli* (Tokumoto and Takahashi, 2001).

5) Modular Gene Loss (Aravind *et al.*, 2000; Ettema *et al.*, 2001): A large fraction of the pathway has been lost from the proteobacterium *Xylella fastidiosa*, including fdx, HscA, HscB, CyaY, IscU and EC2524. In the lineage leading to the α-bacteria *Caulobacter crescentus* and *Melorhizobium loti* a similar fraction has been lost : HscA, HscB, CyaY and EC2524. The IscU proteins are not lost from these two genomes, but appear to have undergone an accelerated evolution, indicative of loss of function. The two IscU proteins from *M.loti* and the one from *C.crescentus* are less than 20 % identical to their closest sequenced relative, IscU from *R.prowazekii*. This in contrast to the generally much higher level of sequence identity in this orthologous group, which even when comparing IscU from *R.prowazekii* to IscU from the more distantly related *Helicobacter pylori* is 38% (the phylogeny did not reveal any evidence or horizontal gene transfer, data not shown). The coordinated loss of HscA and HscB with the loss or accelerated evolution of IscU is consistent with experimental results that indicate the HscA-HscB have chaperone activity for IscU (Silberg *et al.*, 2001).

6) Gene Recruitment (Jensen, 1976; Copley and Bork, 2000): Atm1p that exports Fe-S clusters from the mitochondrion is derived from an ABC transporter with unknown specificity that is (among others) present in all of the presently sequenced α-Proteobacteria (Huynen *et al.*, 2001b). Although it is not impossible, it is quite unlikely that it functions in these species as Fe-S cluster exporter. The most similar proteins to Atm1p which have experimentally been characterized function as hemolysin exporters (30% identity) (Lally *et al.*, 1991). Atm1p appears to have been recruited for a new function. Another example is the protein Erv1p that is part of a eukaryotic specific gene family and that has a role in the maturation of Fe-S clusters. As we cannot determine whether this family actually existed before the evolution of the mitochondria, Erv1p could also have been an invention: *i.e.* without a function prior to its role in mitochondria.

7) Gene Fusion (Snel *et al.*, 2000): A protein that is involved in Fe-S cluster assembly in *S. cerevisiae*, Nfu1p (Schilke *et al.*, 1999), has only full-length orthologs in the α-Proteobacteria and in the other Eukaryotes. The C-terminal domain of NFU1 is homologous to IscU/NifU and has a much wider phylogenetic distribution than the N-terminal domain. The latter can be found, besides in NFU1 and its orthologs, also as the N-terminal domain in hypothetical proteins from the gram-positive bacteria Bacillus subtilis (YPGR), *Bacillus halodurans* (BH1718) and *Staphylococcus aureus* (SA1262) (data not shown). The specific role of this N-terminal domain remains to be elucidated.

8) Linkage of Genes into One Operon: The combination of evolutionary events that have resulted in the isc operon also includes the linkage of the new and the preexisting genes into a single regulatory and transcriptional unit. This linkage suggests that, not only have the preexisting proteins been linked to new ones in a pathway, also the relations between the preexisting proteins (IscR, IscS, IscU and fdx) themselves have become more direct at this point. However, the link into one operon of IscR and IscS-IscU exists, besides in the Proteobacteria, also in *B.subtilis* and *B.halodurans*. The regulation of IscS-IscU by IscR is thus likely to be more widespread than the Proteobacteria. In contrast, the operon link between fdx, whose ortholog in the eukaryotes, adrenodoxin, donates the electrons for the reduction of Sulfur, and IscS-IscU, is uniquely present in the α-, β- and γ-Proteobacteria. Some species, like the δ–Proteobacteria *Helicobacter pylori* and *Campylobacter jejuni*, that do have a IscS-IscU gene pair do not even have the 2Fe-2S ferredoxin, suggesting that the functional link between the 2Fe-2S ferredoxin and Fe-S cluster assembly is relatively volatile. It should be noted, however, that the absence of an operon link between two genes in a genome does in general not indicate the absence of a functional interaction between the proteins in that species.

9) Gene Displacement (Koonin *et al.,* 1996): HscA has been replaced by SSQ1 in Fe-S cluster assembly in the mitochondria in *S.cerevisiae* (*Lutz et al.,* 2001) HscA is paralogous to the orthologous group to which SSQ1 and the Bacterial DnaK belong (Huynen *et al.,* 2001b). Interestingly, HscA has directly descended from this group, it has thus been replaced by its 'ancestral' proteins. Another, more drastic displacement is that between the Proteobacterial and the mitochondrial ferredoxin reductases. The eukaryotic mitochondrial NADP:ferredoxin oxidoreductase only has orthologs outside the Proteobacteria, in species like *Mycobacterium tuberculosis* (fprB)and *Deinococcus radiodurans* (DR0496). A likely candidate for the Proteobacterial ferredoxin oxidoreductase is *fpr*, which encodes a NADP:ferredoxin reductase that in *Azotobacter vinelandii* is coexpressed with ferredoxin (Jung *et al.,* 1999). 3D structures are available for both types of NADP:ferredoxin oxidoreductases. According to the SCOP classification (Murzin *et al.,* 1995) they fall into different fold classes, implying a drastic, non-homologous gene displacement.

10) Modular Gene Displacement: A step beyond single gene displacement is its modular form: the replacement of multiple subsequent steps in a pathway. In Fe-S cluster assembly we did not encounter such an event. We will therefore refer to another, recently elucidated pathway: the glycolysis in the Pyrococci (Figure 5). Of the ten steps in the glycolysis as we know it in eukaryotes and in the Bacteria, five have been replaced by an enzyme that is not homologous to the Bacterial/Eukaryotic one, and one by an enzyme that is paralogous to the Bacterial/Eukaryotic one. The modularity in this replacement lies in the fact that the displacements are mainly glycolysis specific. They tend to be catabolic rather than anabolic, and to be concentrated in the top (hexamer) part of the glycolysis rather than in the bottom (trimer) part that is less specific for the catabolism of sugar, see (Verhees *et al.,* 2001) and references therein.

Combining all the above evolutionary events allows us to divide the Fe-S cluster assembly into modules. On the one hand, some proteins are gained and lost relatively independently of others: the regulatory protein IscR and the proteins IscA, EC2524 and NFU1p. They therefore are not expected to

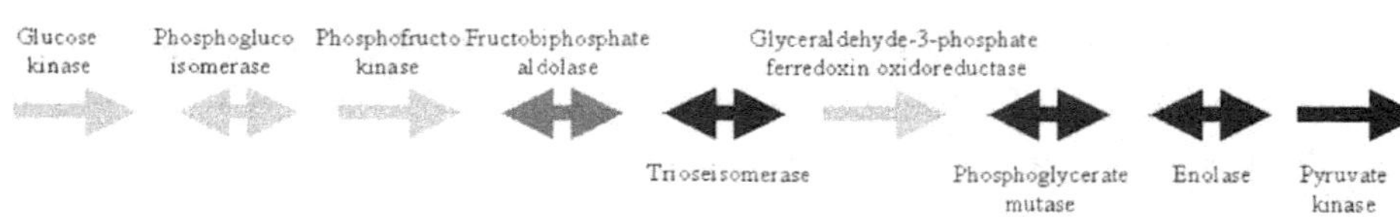

Figure 5. Modular gene displacement in the glycolysis of *Pyrococcus furiosus.* Enzymes in the *P. furiosus* glycolysis that are orthologous to the corresponding *E. coli* enzyme are marked in black, enzymes that are paralogous to the *E.coli* ones in dark-gray and enzymes that are not homologous to the *E.coli* ones are marked in light-gray. The *P. furiosis* glycolysis contains nine enzymes, in contrast to the ten of *E.coli*, because *P. furiosis* Glyceraldehyde phosphate dehydrogenase (GAPOR, Oost *et al.,* 1998) combines glyceraldehyde-3-phosphate dehydrogenase and 3-phosphoglycerate kinase of the *E.coli* glycolysis. Notice that the displacements are glycolysis specific (in contrast to gluconeogenesis) as they tend to catalyze unidirectional reactions and are clustered in the top part of the pathway. For references to the non-homologous displacements see (Verhees *et al.,* 2001). Most of the enzymes of the glycolysis in *P. furiosus* that are orthologous to the ones in *E.coli* have high levels of sequence identity to these enzymes, and are correctly annotated in the databases. The one exception, in a distant ortholog (Dandekar *et al.,* 1999) of the Bacterial Phosphoglycerate mutase in *P. furiosus* does indeed have Phosphoglycerate mutase activity (van der Oost *et al.,* 2002). ***Colour plate at back of book.***

specifically interact with other proteins of the Fe-S cluster assembly. For IscR this makes sense, as it regulates the whole isc operon, rather than specific genes within it. On the other hand, HscA, HscB, IscU and CyaY appear to form a module, while IscU is also part of a module with IscS. The human ortholog of *CyaY*, frataxin, is a well known disease gene. Severely reduced levels of frataxin cause the disease Friedreich's ataxia (Campuzano *et al.,* 1997), which is characterized by degeneration of large sensory neurons and spinocerebellar tracts, cardiomyopathy and increased likelihood of diabetes (Durr *et al.,* 1996). Although the involvement of frataxin in Fe-S cluster assembly has been proposed before, for recent evidence see (Puccio *et al.,* 2001), these results indicate a more specific prediction: CyaY should be involved in the same (sub)process as HscB (Huynen *et al.,* 2001b).

4. Outlook: Challenges and Opportunities of Many Genomes

The rapidly increasing number of sequenced genomes provides challenges and opportunities for the prediction of pathways and the reconstruction of their evolution.

A more quantitative approach to genomic association is becoming feasible and of value (see above). Aside from having statistical thresholds for genomic association, we also need association scores that take into account not only how often are the genes associated, but also how often they are not associated. Furthermore, functional relevance of the combinations of different types of genomic associations are required, not only when those associations are both positive. To give one example: is there any functional information in the fact that, although CyaY/frataxin does co-evolve with proteins involved in Fe-S cluster assembly, their genes never occur within the same operon? Coupling quantitative analyses of genomic association to quantitative data on the types of functional interactions, and to homology-based predictions of possible molecular functions is necessary in order to make predictions more specific than "this set of proteins are somehow functionally interacting", and therewith of practical value (Huynen *et al*, 2000; Wolf *et al.*, 2001a). Part of the challenge is not only to rank the many different predicted interactions between proteins, but also to divide the large networks of predicted or experimental interactions into biologically meaningful modules.

More genomes also increase the resolution at which we can resolve pathway evolution. Biotic systems often are of a baffling complexity, challenging us to explain how they could have evolved from simpler systems that do not appear very useful or functional in themselves. By, based on genomic data, delineating the variation of a single pathway, we observe which intermediates are (apparently) feasible. By subsequently placing this variation on a phylogenetic tree, we can infer the progressive and degenerative evolution of a pathway. With the increase of the number of genomes we can be more confident about claims that a pathway has evolved within one taxon, based on the observation that it is widespread within it and absent outside it. Likewise the accuracy of placing gene duplications or inventions also depends on the number of genomes, *e.g.* until the sequencing of *Rickettsia conorii* (Ogata *et al.,* 2001) the genes of the EC2524 orthologous group had not been observed in the α-Proteobacteria, and the invention of this group would be assumed to have happened in the branch leading to the β and γ-Proteobacteria. As the phylogeny of these genes did not indicate horizontal gene transfer (data not shown), the origin of this orthologous group can now be moved back in time (Figure 4). In general, combining the various types of the data from the many sequenced genomes gives us a more accurate picture not only of the phylogeny of cellular life (Wolf *et al.*, 2001b), but also of evolution at a supra-molecular level.

Acknowledgements

MH thanks Toni Gabaldon for comments on the original manuscript. We apologize to the writers of many relevant papers, including those containing the genome sequences, that we did not refer to.

References

Altschul, S.F., Madden, T.L., Schaffer, A.A., Zhang, G., Zhang, Z., Miller, W., and Lipman, D.J. 1997. Gapped BLAST and PSI-BLAST: a new generation of protein database search programs. Nucl. Acids Res. 25: 3389-3402.

Aravind, L., Watanabe, H., Lipman, D.J., and Koonin, E.V. 2000. Lineage-specific loss and divergence of functionally linked genes in eukaryotes. Proc. Natl. Acad. Sci. USA. 97: 11319-11324.

Boucher, Y., and Doolittle, W.F. 2000. The role of lateral gene transfer in the evolution of isoprenoid biosynthesis pathways. Mol. Microbiol. 37: 703-716.

Caffrey, D.R., O'Neill, L.A., and Shields, D.C. 1999. The evolution of the MAP kinase pathways: coduplication of interacting proteins leads to new signaling cascades. J. Mol. Evol. 49: 567-582.

Campuzano, V., Montermini, L., Lutz, Y., Cova, L., Hindelang, C., Jiralersprong, S., Trottier, Y., Kish, S.J., Faucheux, B., Trouillas, P. *et al.* 1997 Frataxin is reduced in Friedreich ataxia patients and is associated with mitochondrial membranes. Hum. Mol. Genet. 6: 1771-1780.

Cho, S-J., Lee, M.G., Yang, J.K., Lee, J.-Y., Song, H.K., and Suh, S.W. 2000. Crystal structure of *Escherichia coli* CyaY protein reveals a previously unidentified fold for the evolutionarily conserved frataxin family. Proc. Natl. Acad. Sc. U S A. 97: 8932-8937.

Copley R.R., and Bork P. 2000. Homology among (betaalpha)(8) barrels: implications for the evolution of metabolic pathways. J. Mol. Biol. 303: 627-641.

Dandekar, T., Schuster, S., Snel, B., Huynen, M., and Bork, P. 1999. Pathway alignment: application to the comparative analysis of glycolytic enzymes. Biochem. J. 343: 115-124.

Dandekar, T., Snel, B., Huynen, M.A., and Bork, P. 1998. Conservation of gene order: a fingerprint of proteins that physically interact. Trends Biochem. Sci. 23:324-328.

Durr, A., Cossee, M., Agid, Y., Campuzano, V., Mignard, C., Penet, C., Mandel, J.L., Brice, A., and Koenig, M. 1996. Clinical and genetic abnormalities in patients with Friedreich's ataxia. N. Engl. J. Med. 335: 1169-1175.

Enright, A.J., Ilioupoulis, I., Kyrpides, N.C., and Ouzounis C.A.1999. Protein interaction maps for complete genomes based on gene fusion events. Nature 402: 86-90.

Ettema, T., v.d. Oost, J., and Huynen, M.A. 2001. Modularity in the gain and loss of genes: Applications for function prediction. Trends Genet. 17: 485-487.

Eisenberg, D., Marcotte, E.M., Xenarios, I., and Yeates, T.O. 1999. Protein function in the post-genomic era. Nature. 405: 823-826.

Gelfand, M.S., Koonin E.V., and Mironov A.A. 2000. Prediction of transcription regulatory sites in Archaea by a comparative genomic approach. Nucleic Acids Res. 28:695-705.

Goh, C.S., Bogan, A.A., Joachimiak, M., Walther, D., and Cohen F.E. 2000. Co-evolution of proteins with their interaction partners. J. Mol. Biol. 299: 283-293.

Horswill, A.R., and Escalante-Semerena, J.C. 1999. *Salmonella typhimurium* LT2 catabolizes propionate via the 2-methylcitric acid cycle. J. Bacteriol 181: 5615-5623.

Hughes, T.R., Marton, M.J., Jones, A.R., Roberts, C.J., Stoughton, R., Armour, C.D,, Bennett, H.A., Coffey, E., Dai, H., He, Y.D., Kidd, M.J., King, A.M., Meyer, M.R., Slade, D., Lum, P.Y., Stepaniants, S.B.,

Shoemaker, D.D., Gachotte, D., Chakraburtty, K., Simon, J., Bard, M., and Friend, S.H. 2000. Functional discovery via a compendium of expression profiles. Cell. 102: 109-126.

Huynen, M.A., and Bork, P. 1998. Measuring genome evolution. Proc. Natl. Acad. Sci. USA. 95: 5849-5856.

Huynen M.A., Dandekar, T., and Bork, P. 1999. Variation and evolution of the citric acid cycle: a genomic approach. Trends Microbiol. 7: 281-291.

Huynen, M.A., and Snel, B. 2000. Gene and context: integrative approaches to genome analysis. Adv. Protein Chem. 54: 345-379.

Huynen, M.A., Snel, B., Lathe, W. 3rd., and Bork, P. 2000. Predicting protein function by genomic context: quantitative evaluation and qualitative inferences. Genome Res. 10: 1204-1210.

Huynen M.A., Snel B., and Bork P. 2001a. Inversions and the dynamics of eukaryotic gene order. Trends Genet. 17:304-306.

Huynen, M.A., Snel, B., Bork, P., and Gibson, P. 2001b. The phylogenetic distribution of frataxin indicates a role in Fe-S cluster assembly. Hum. Mol. Gen. 10: 2463-2468.

Hynds, P.J., Plucken, H., Westhoff, P., and Robinson, C. 2000. Different lumen-targeting pathways for nuclear-encoded versus cyanobacterial/ plastid-encoded Hcf136 proteins. FEBS Lett. 46: 97-100.

Ito, S., Fushinobu, S., Yoshioka, I., Koga, S., Matsuzawa. H., and Wakagi T. 2001. Structural basis for the ADP-specificity of a novel glucokinase from a hyperthermophilic archaeon. Structure (Camb). 9: 205-214.

Jensen R.A.1976. Enzyme recruitment in evolution of new function. Annu. Rev. Microbiol. 30: 409-425.

Jensen, L.T., and Culotta, V.C. 2000. Role of *Saccharomyces cerevisiae* ISA1 and ISA2 in iron homeostatis. Mol. Cell. Biol. 20: 3918-3927.

Jung, Y.S., Roberts, V.A., Stout, C.D., and Burgess, B.K. 1999. Complex formation between *Azotobacter vinelandii* ferredoxin I and its physiological electron donor NADPH-ferredoxin reductase. J. Biol. Chem. 274: 2978-2987.

Kanehisa M., and Goto S. 2000. KEGG: Kyoto encyclopedia of genes and genomes. Nucleic Acids Res. 28: 27-30.

Kaut, A., Lange, H., Diekert, K., Kispal, G., and Lill, R. 2000. Isa1p is a component of the mitochondrial machinery for maturation of cellular iron-sulfur proteins and requires conserved cysteine residues for function. J. Biol. Chem. 275: 15955-15961.

Koonin, E.V., Mushegian, A.R., and Bork, P. 1996. Non-orthologous gene displacement. Trends Genet.12: 334-336.

Koretke, K.K., Lupas, A.N., Warren, P.V., Rosenberg, M., and Brown J.R. 2000. Evolution of two-component signal transduction. Mol. Biol. Evol.

17: 1956-1970.

Kruglyak, S., and Tang, H. 2000. Regulation of adjacent yeast genes. Trends Genet. 16: 109-111.

Kullback, S. 1959. Information theory and statistics. Wiley, New York.

Lally, E.T., Golub, E.E., Kieba, I.R., Taichman, N.S., Decker, S., Berthold, P., Gibson, C.W., Demuth, D.R., and Rosenbloom, J. 1991. Structure and function of the B and D genes of the *Actinobacillus actinomycetemcomitans* leukotoxin complex. Microb Pathog. 11: 111-121.

Lange, B.M., Rujan, T., Martin, W., and Croteau, R. 2001a. Isoprenoid biosynthesis: the evolution of two ancient and distinct pathways across genomes. Proc Natl Acad Sci USA. 97: 13172-13177.

Lange, H., Kaut, A., Kispal, G., and Lill, R. 2000. A mitochondrial ferredoxin is essential for the biogenesis of cellular iron-sulfur proteins. Proc. Natl. Acad. Sci. USA. 97: 1050-1055.

Lange, H., Lisowsky, T., Gerber, J., Muhlenhoff, U., Kispal. G., and Lill R. 2001b. An essential function of the mitochondrial sulfhydryl oxidase Erv1p/ALR in the maturation of cytosolic Fe/S proteins. EMBO Rep. 2: 715-720.

Lill, R., and Kispal, G. 2000. Maturation of cellular Fe-S proteins: an essential function of mitochondria. Trends Bioch. Sci. 25: 352-356.

Luttgen, H., Rohdich, F., Herz, S., Wungsintaweekul, J., Hecht, S., Schuhr, C.A., Fellermeier, M., Sagner, S., Zenk, M.H., Bacher, A., and Eisenreich, W. 2000. Biosynthesis of terpenoids: YchB protein of *Escherichia coli* phosphorylates the 2-hydroxy group of 4-diphosphocytidyl-2C-methyl-D-erythritol. Proc. Natl. Acad. Sci. USA. 97:1062-1067.

Lutz, T., Westermann, B., Neupert, W., and Herrmann, J.M. 2001. The mitochondrial proteins Ssq1 and Jac1 are required for the assembly of iron sulfur clusters in mitochondria. J. Mol. Biol. 307: 815-825.

Marcotte, E.M., Pellegrini, M., Ng, H.L., Rice, D.W., Yeates, T.O., and Eisenberg, D. 1999. Detecting protein function and protein-protein interactions from genome sequences. Science 285: 751-753.

Manzella, L., Barros, M.H., and Nobrega, F.G. 1998. ARH1 of *Saccharomyces cerevisiae:* a new essential gene that codes for a protein homologous to the human adrenodoxin reductase. Yeast. 14: 839-846.

McGuire, A.M., Hughes J.D., and Church, G.M. 2000. Conservation of DNA regulatory motifs and discovery of new motifs in microbial genomes. Genome Res. 10: 744-757.

Murzin, A.G., Brenner, S.E., Hubbard, T., and Chothia, C. 1995. SCOP: a structural classification of proteins database for the investigation of sequences and structures. J. Mol. Biol. 247: 536-540.

Mushegian, A.R., and Koonin, E.V. 1996. Gene order is not conserved in bacterial evolution. Trends Genet. 12: 289-290.

Musco, G., Stier, G., Kolmerer, B., Adinolfi, S., Martin, S., Frenkiel, T., Gibson T., and Pastore, A. 2000. Towards a structural understanding of Friedreich's ataxia: the solution structure of frataxin. Structure Fold. Des. 257: 507-511.

Ogata, H., Audic, S., Renesto-Audiffren, P., Fournier, P.E., Barbe, V., Samson, D., Roux, V., Cossart, P., Weissenbach, J., Claverie, J.M., and Raoult, D. 2001. Mechanisms of evolution in *Rickettsia conorii* and *R. prowazekii*. Science. 293: 2093-2098.

Ohno, S. 1970. Evolution by gene duplication. Springer-Verlag, New York.

Oost, van der, J., Schut, G., Kengen, S.W., Hagen, W.R., Thomm, M., and de Vos, W.M. 1998. The ferredoxin-dependent conversion of glyceraldehyde-3-phosphate in the hyperthermophilic archaeon *Pyrococcus furiosus* represents a novel site of glycolytic regulation. J. Biol. Chem. 273: 28149-28154.

Overbeek, R., Fonstein, M., D'Souza, M., Pusch, G.D., and Maltsev, N. 1998. Use of contiguity on the chromosome to predict functional coupling. *In Silico* Biology. 1: 93-106.

Overbeek, R., Fonstein, M., D'Souza, M., Pusch, G.D., and Maltsev, N. 1999. The use of gene clusters to infer functional coupling. Proc. Natl. Acad. Sci. USA. 96: 2896-2901.

Panina, E.M., Vitreschak, A.G., Mironov, A.A., Gelfand, M.S. 2001. Regulation of aromatic amino acid biosynthesis in gamma-Proteobacteria. J. Mol. Microbiol. Biotechnol. 3: 529-543.

Pazos, F., and Valencia, A. 2001. Similarity of phylogenetic trees as indicator of protein protein interaction. Protein Eng. 14: 609-614.

Pellegrini, M., Marcotte, E.M., Thompson, M.J., Eisenberg, D., and Yeates, T.O. 1999. Assigning protein functions by comparative genome analysis: protein phylogenetic profiles. Proc. Natl. Acad. Sci. USA. 96: 42885-42888.

Pelzer, W., Muhlenhoff, U., Diekert, K., Siegmund, K., Kispal, G., and Lill, R. 2000. Mitochondrial Isa2p plays a crucial role in the maturation of cellular iron-sulfur proteins. FEBS Lett. 476: 134-139.

Puccio, H., Simon, D., Cossee, M., Criqui-Filipe, P., Tiziano, F., Meiki, J., Hindelang, C., Matyas, R., Rustin, P., and Koenig, M. 2001. Mouse models for Friedreich ataxia exhibit cardiomyopathy, sensory nerve defect and Fe-S enzyme deficiency followed by intramitochondrial iron deposits. Nat. Genet. 27: 181-186.

Schilke, B., Voisine, C., Beinert, H., and Craig, E. 1999. Evidence for a conserved system for iron metabolism in the mitochondria of *Saccharomyces cerevisiae*. Proc. Natl. Acad. Sci. USA. 96: 10206-10211.

Schwartz, C.J., Giel, J.L., Patschkowski, T., Luther, C., Ruzicka, F.J., Beinert, H., and Kiley P.J. 2002. IscR, an Fe-S cluster-containing transcription factor, represses expression of *Escherichia coli* genes encoding Fe-S cluster assembly proteins. Proc Natl Acad Sci USA. *in press*

Schwartz, C.J., Djaman, O., Imlay, J.A., and Kiley, P.J. 2000. The cysteine desulfurase, IscS, has a major role in in vivo Fe-S cluster formation in *Escherichia coli*. Proc. Natl. Acad. Sci. USA. 16: 9009-9014.

Silberg, J. J., Hoff K. G., Tapley T. L., and Vickery L. E. 2001. The Fe/S assembly protein IscU behaves as a substrate for the molecular chaperone Hsc66 from *Escherichia coli*. J. Biol. Chem. 276: 1696-700.

Silberg, J.J., Hoff, K.G., and Vickery, L.E. 1998. The Hsc66-Hsc20 chaperone system in *Escherichia coli*: chaperone activity and interactions with the DnaK-DnaJ-grpE system. J. Bacteriol. 180: 6617-6624.

Snel, B., Lehmann, G., Bork, P., and Huynen M.A. 2000. STRING: a web-server to retrieve and display the repeatedly occurring neighborhood of a gene. Nucleic. Acids. Res. 28: 3442-3444.

Snel, B., Bork, P., and Huynen, M.A. 2002. Genomes in flux: The evolution of archaeal and Proteobacterial gene content. Genome Res. 12: 17-25.

Stec, B., Yang, H., Johnson, K.A., Chen, L., and Roberts, M.F. 2000. MJ0109 is an enzyme that is both an inositol monophosphatase and the 'missing' archaeal fructose-1,6-bisphosphatase. Nat. Struct. Biol. 7: 1046-1050.

Tamames, J., Casari, G., Ouzounis, C., Valencia, A. 1997. Conserved clusters of functionally related genes in two bacterial genomes. J. Mol. Evol. 44: 66-73.

Tatusov, R.L., Natale. D.A., Garkavtsev, I.V., Tatusova, T.A., Shankavaram, U.T., Rao, B.S., Kiryutin, B., Galperin, M.Y., Fedorova, N.D., and Koonin, E.V. 2001. The COG database: new developments in phylogenetic classification of proteins from complete genomes. Nucleic Acids Res. 29: 22-28.

Teichmann, S.A., Rison, S.C., Thornton, J.M., Riley, M., Cough, J., and Chothia C. 2001. The evolution and structural anatomy of small molecule metabolic pathways in Escherichia coli. J. Mol. Biol. 311: 693-708.

Tokumoto, U., and Takahashi, Y. 2001. Genetic Analysis of the isc operon in *Escherichia coli* involved in the biogenesis of cellular Iron-Sulfur proteins. J. Biochem. 130: 63-71.

Tong, W. H., and Rouault, T. 2000. Distinct iron-sulfur cluster assembly proteins exist in the cytosol and mitochondria of human cells. EMBO J. 19: 5692-5700.

Urbina, H.D., Silberg, J.J., Hoff, K.G., Vickery, L.E. 2001. Transfer of sulfur from IscS to IscU during Fe/S cluster assembly. J. Biol. Chem. 276: 44521-44526.

van der Oost, J., Huynen, M.A. and Verhees, C.H. 2002. Molecular characterization of phosphoglycerate mutase in archaea. FEMS Microbiol Lett. 212: 111-120.

Verhees, C.H., Huynen, M.A., Ward, D.E., Schiltze, E., de Vos, W.M., and van der Oost, J. 2001. The phosphoglucoisomerase from the hyperthermophilic archaeon *Pyrococcus furiosus* is a unique enzyme that belongs to the cupin superfamily. J. Biol. Chem. 44: 40926-40932.

Voisine, C., Cheng, Y.C., Ohlson, M., Schilke, B., Hoff, K., Beinert, H., Marszalek, J., and Craig, E. A. 2001. Jac1, a mitochondrial J-type chaperone, is involved in the biogenesis of Fe/S clusters in *Saccharomyces cerevisiae.* Proc. Natl. Acad. Sci. USA. 98: 1483-1488.

Wolf Y.I., Aravind L., Grishin N.V., and Koonin E.V. 1999. Evolution of aminoacyl-tRNA synthetases—analysis of unique domain architectures and phylogenetic trees reveals a complex history of horizontal gene transfer events. Genome Res. 9: 689-710.

Wolf, Y.I., Rogozin, I.B., Kondrashov, A.S., and Koonin E.V. 2001a. Genome alignment, evolution of prokaryotic genome organization, and prediction of gene function using genomic context. Genome Res.11: 356-372.

Wolf, Y.I., Rogozin, I.B., Grishin, N.V., Tatusov, R.L., and Koonin, E.V. 2001b. Genome trees constructed using five different approaches suggest new major bacterial clades. BMC Evol. Biol. 1: 8.

Zheng, L., Cash, V.L., Flint, D.H., and Dean, D.R.1998. Assembly of iron-sulfur clusters. Identification of an *iscSUA-hscBA-fdx* gene cluster from *Azotobacter vinelandii*. J. Biol. Chem. 273: 13264-13672.

From: *Frontiers in Computational Genomics*

7

Using Genomic Context in the Analysis of Multi-Component Systems

Natalie D. Fedorova, Anastasia N. Nikolskaya, and Michael Y. Galperin

Abstract

Multi-domain organization of newly sequenced proteins has long been recognized as a factor that significantly complicates their functional annotation, particularly the automated annotation in the course of large-scale genome sequencing projects. We discuss here some approaches to the analysis of multi-domain (multi-component) systems. We argue that comprehensive annotation of multi-subunit enzymes and complex enzyme systems, such as ABC transporters or members of the two-component signal transduction system, has to combine detailed analysis of their domain composition, operon organization of the corresponding genes, gene content in the particular genome, and phylogenetic analysis. This multi-faceted "genomic context" approach needs to include a human component, *i.e.* a highly-qualified genome biologist who should be able to identify new truly interesting cases and put them into proper context.

Context (noun)
1: the parts of a discourse that surround a word or passage and can throw light on its meaning
2: the interrelated conditions in which something exists or occurs
Etymology: Middle English, weaving together of words, from Latin *contextus* connection of words, coherence, from *contexere* to weave together, from *com-* + *texere* to weave. Date: circa 1568
From *Merriam-Webster Collegiate Dictionary*

1. Introduction

The term "genomic context" was coined by Peer Bork and colleagues in 1998 to describe the use of the non-similarity-based approaches to the analysis of DNA and protein sequences on a whole-genome scale (Bork *et al.*, 1998). Indeed, although primary annotation of new sequences almost always relies on sequence database searches using BLAST, HMMer, or similar tools, it has been repeatedly shown that sequence similarity alone is often insufficient for the identification of the closest relatives, particularly among a group of paralogs (Koski and Golding, 2001; Tatusov *et al.*, 1997). Deducing the exact function of predicted protein can often benefit from taking into account other factors, such as its domain organization, evolutionary history, gene neighborhood, and phyletic pattern (Galperin and Koonin, 2000; Huynen *et al.*, 2000a; b; Huynen and Snel, 2000; Overbeek *et al.*, 1999). These approaches, unified under the term "genomic context", prove useful not only for identification of the general functions of uncharacterized proteins, but also for making functional assignments of complex (*e.g.*, multi-domain) proteins and multi-component systems.

In the course of the analysis of the proteins encoded in completely sequenced microbial genomes we have encountered numerous problems with functional annotation of the two most common multi-component systems in bacteria - the ABC-type transporters and two-component signal transduction systems. Here we consider the domain organization and functional diversity of these two systems and discuss the ways of deducing the exact functions of their components by using genomic context-based methods of sequence analysis.

2. ABC-type Transport Systems

ATP-binding cassette (ABC)-type transporters are membrane-bound protein complexes found universally in all living organisms that use the energy of ATP hydrolysis to transport their substrates across the cytoplasmic membrane (Higgins, 2001; Nikaido and Hall, 1998; Saier, 2000a; b; Tomii and Kanehisa, 1998). A functional ABC-type exporter complex, such as, for example, Na^+

exporter NatAB from *B. subtilis* or heme exporter CcmAB from *E. coli* (Cheng *et al.*, 1997; Schulz *et al.*, 1999), is typically comprised of two membrane channel-forming permease subunits, and two cytoplasmically-located ATPase subunits, encoded by a two-gene operon. ABC-type importers usually include an additional periplasmic (extra-cytoplasmic) ligand-binding protein or domain (PBP). The best-characterized PBPs consist of two β/α globular domains linked by a flexible hinge. They bind their ligands at the bottom of the cleft between these domains, which allows them to discriminate between very similar substrates (Quiocho and Ledvina, 1996; Wolf *et al.*, 1996). Ligand binding induces a conformational change in the protein, which allows the binding protein to interact with other proteins, such as membrane-bound components of the ABC-type transport complexes. This interaction triggers the release of the ligand and its translocation across the cytoplasmic membrane through a pore formed by two permease subunits. The whole process is energized by ATP hydrolysis.

Different subunits of the ABC-type transporters vary in the degree of sequence conservation with ATPase subunits being the most conserved and PBPs and permease subunits showing much less conservation between different families (Holland and Blight, 1999). Nonetheless, there is a general agreement between classifications obtained for permease, ATPase and periplasmic domains/subunits of ABC-type transport systems (Quentin *et al.*, 1999). Typically all subunits of an ABC-type transporter are encoded by a single gene in eukaryotes and by adjacent but separate genes in bacteria and archaea (Quentin *et al.*, 1999; Tomii and Kanehisa, 1998). Since ATPases are usually the most conserved and easily recognizable components of these systems, in most cases identification of certain genes as components of a putative ABC-type transporter relies on the finding of an adjacent ATPase gene.

Experimental studies of various components of the ABC-type systems revealed that in addition to or instead of their role in transport, these proteins could carry out other functions. Many PBPs and some ATPases have been shown to function in signal transduction, virulence, competence, and even transcriptional regulation (Tam and Saier, 1993). Such functional changes can occur without significant changes in amino acid sequences but are often accompanied by changes in the genomic neighborhood and/or protein fusions. In such cases, genes encoding these subunits can be found clustered together with genes encoding proteins involved in these processes and can be recognized through comparative genome analysis. Indeed, components of the predicted ABC-type systems encoded in completely sequenced bacterial, archaeal, and eukaryotic genomes reveal a significant diversity of domain organization, which shows that functional predictions should be based not just on sequence similarity, but also on the genome context and domain/ subunit composition.

ABC Transporter Operons

A functional ABC transporter consists of the ATPase, permease and periplasmic components, which are usually encoded by adjacent genes, forming ABC-type operons (Figure 1). ABC-type operons sometimes include additional genes, for example, those coding for transcriptional regulators, but these proteins are usually too diverse to be helpful in functional annotation. As mentioned before, identification of permease and PBP components of a putative ABC-type transporter relies either on sequence similarity to a previously characterized ABC-type component or on the finding of an adjacent ATPase gene. For example, the *psaA* (pneumococcal surface adhesin) gene that encodes a surface protein in *Streptococcus pneumoniae* is clustered together with the genes encoding a typical ABC-

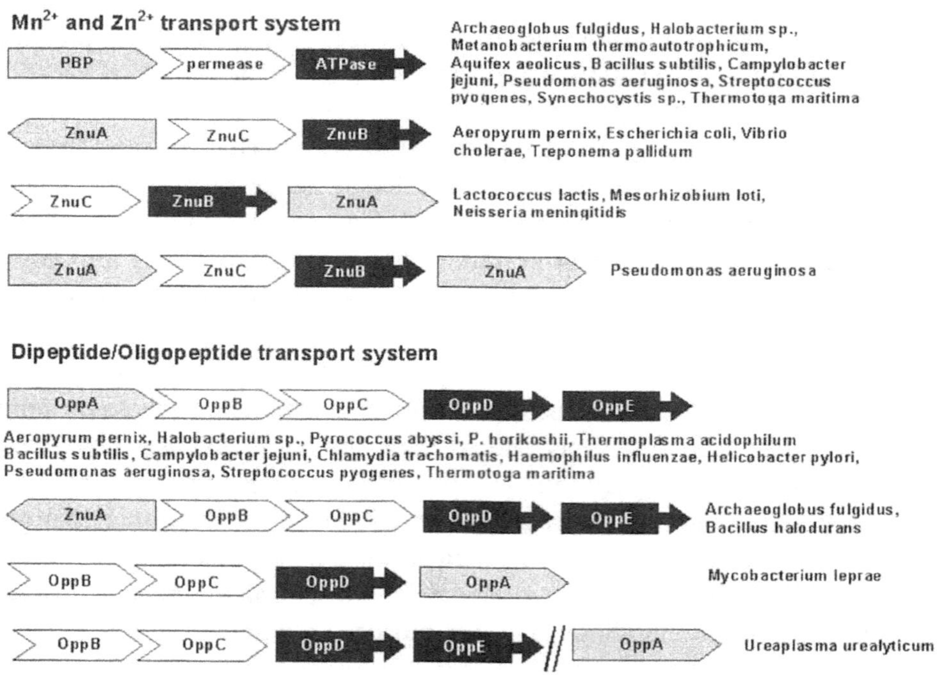

Figure 1. Diversity of operon organization of common ABC-type transporters. The order of the shapes and the direction of the arrows reflect the gene order and orientation in given organisms. The shapes are drawn approximately to scale of the encoded proteins. ***Colour plate at back of book.***

type ATPase and a putative integral membrane protein (Sampson *et al.*, 1994). Based on this observation alone, PsaA could have been predicted to act as a periplasmic solute-binding protein involved in solute import. Indeed, the *psa* operon was recently implicated in Mn^{2+} transport and is homologous to the *adcCBA* operon in *S. pneumoniae* and the *znuABC* operon in *E. coli* that function in Zn^{2+} uptake (Dintilhac *et al.*, 1997; Patzer and Hantke, 1998). PsaA is also homologous to the cell surface antigens of the LraI (lipoprotein receptor antigen) family, found in other streptococci. These surface proteins include fimbrial adhesin FimA in *Streptococcus parasanguis* (Fenno *et al.*, 1989), saliva-binding protein SsaB in *S. sanguis* (Ganeshkumar *et al.*, 1991), lipoprotein ScbA in *S. cristatus* (Correia *et al.*, 1996), coaggregation-mediating adhesin ScaA in *S. gordonii* (Andersen *et al.*, 1993), and endocarditis antigen EfaA in *Enterococcus faecalis* (Lowe *et al.*, 1995). Remarkably, like the *psaA* gene, genes encoding these surface adhesins are located in the vicinity of the genes coding for a predicted ATPase and a permease. Therefore, based on their genomic context, all these surface antigens of the LraI family most likely act as metal ion-binding proteins in ABC-type uptake systems.

Orphan ABC-Type ATPases

There are cases, however, when a gene encoding an ABC-type ATPase is not linked to genes encoding the other two components of an ABC-type transporter. Functional annotation of such ATPases becomes quite complex: some of them might still participate in transport, whereas others might be involved in completely unrelated ATP-dependent processes. For example, several archaeal genomes encode a conserved fusion of a C-terminal ABC-type ATPase domain with an N-terminal acetyltransferase domain. With the exception of the *Aeropyrus pernix* protein APE2272, annotated as "660aa long hypothetical protein", all such proteins (AF1064, MK0520, PAB1929, PF1238, PH1507, see COG2401) have been annotated as ATPase components of an ABC transporter. While this could still turn out to be true, the presence of the N-terminal acetyltransferase domain and the absence of neighboring permease and PBP genes make such functional assignment very shaky. Another orphan ATPase, SufC, has been shown to be involved in Fe^{3+} metabolism and possibly Fe-S cluster assembly (Nachin *et al.*, 2001). Its participation in Fe^{3+} transport, however, does not seem likely. An even more interesting case is the *Bacillus halodurans* protein BH3894 that is currently annotated as "ABC transporter (ATP-binding protein)". This unique protein contains a fusion of an N-terminal ABC-type ATPase with a C-terminal

LytTR-type DNA-binding domain (Nikolskaya and Galperin, 2002) and can be expected to serve as an ATP-regulated (or ATP-dependent) transcriptional regulator.

No ATPase in the Operon

Sometimes the genes for ABC-type ATPases are located outside of the ABC-type operons, which makes the process of functional annotation more complicated. Recently, lipoprotein LplA (lipoprotein-like) was identified as a factor required for starch hydrolysis in *B. subtilis* (Gomez *et al.*, 1994) and is annotated as such in the GenBank. Sequencing of the *B. subtilis* genome revealed that the *lplA* gene is linked to the *lplBC* genes encoding two putative ABC-type sugar permeases and the *lplD* gene encoding a putative glycosyl hydrolase. Proteins LplA, LplB, and LplC are homologous, respectively to the components of AmyEDC, GuaEFG, MsmEAmyDC and UgpBACE transporters, involved in sugar uptake in *Thermoaerobacterium thermosulfurigenes*, *Bacillus stearothermophilus*, *S. mutants* and *E. coli* (Overduin *et al.*, 1988; Sahm *et al.*, 1996; Shulami *et al.*, 1999; Sutcliffe *et al.*, 1993). Of all these transporters, only the *E. coli* UgpBACE transport system has an ATPase gene located in the same operon. Other sugar transporters have "orphan" ATPase genes, that have not been identified yet and are most likely encoded outside ABC-type operons, *i.e.* encoded by "orphan" ATPase genes. We think that the *lplABC* locus most likely encodes another ABC transporter that, along with some starch degrading enzymes, is required for starch utilization.

Missing PBPs

The absence of the PBP-encoding gene in an ABC-type transporter encoding operon would typically imply that the corresponding transporter participates in solute export, but not import. Nevertheless, several well-characterized ABC-type importers also appear to lack the PBPs, which could be due to the non-orthologous gene displacement of the corresponding genes. For example, the ABC-type Co^{2+} transport system CbiMNQO from *Salmonella typhimurium* and several other organisms has a typical organization that includes a short periplasmic protein CbiN, a permease (CbiM, CbiQ) and an ATPase (CbiO) (Roth *et al.*, 1996). However, CbiN has no apparent homologs in the other species (*Synechocystis*, *Haemophilus influenzae*) that still encode the permease and ATPase components of this transport system. This puzzle has been recently resolved when the Co^{2+} transport system of *Actinobacillus pleuropneumoniae* was found to include a predicted periplasmic component CbiK, which was not homologous to CbiN (Bosse *et al.*, 2001). CbiK homologs were found in other genomes, including *H. influenzae* and *Brucella melitensis*, whose *cbiK* genes (HI1624 and BMEI0641, respectively) are linked to other *cbi* genes.

Moonlighting PBP Subunits

An additional complication for the recognition and functional annotation of ABC-type transporters is the fact that PBPs are often recruited for performing other functions that may be unrelated to transport. Phylogenetic analyses and comparisons of domain architectures of various PBP-containing proteins and protein complexes show that their evolution often involves changes in genomic context and/or domain fusions. In most bacterial genomes, there are "extended" PBP families containing many paralogous members that could have acquired additional functions and lost their ability to interact with their permease partners. Some of these proteins are "orphans" that are not encoded in any ABC-type operon, but are instead linked to proteins involved in chemotaxis, sensory transduction, or other regulatory pathways. Functional annotation of these proteins based exclusively on sequence similarity can often lead to poorly justified or even erroneous assignments. On the other hand, analysis of gene neighborhoods (operon structure and domain fusions) can provide valuable data for deducing the function(s) of each particular PBP.

The case that diverges the least from the ABC transporter paradigm is when PBPs with bound ligands also serve as signaling molecules. It is well documented that many PBPs, in addition to participating in transport, can moonlight in chemotaxis or signal transduction pathways. The galactose-binding protein MglB from *E. coli*, for example, acts both as the periplasmic component of the β-methylgalactoside transport system and as galactose chemoreceptor interacting with Trg, the chemotactic signal transducer for galactose (Scholle *et al.*, 1987). Another PBP, ChvE from the plant pathogen *Agrobacterium tumefaciens* is required for chemotaxis and induction of the virulence operon in the plant wound environment (Kemner *et al.*, 1997). The *chvE* gene is part of the *chvE-gguA-gguB* operon coding for a putative ABC-type xylose transport system. It is interesting that its close homologue, ChvE from *Brucella suis*, is essential for uptake of sugars but not for virulence (Alvarez-Martinez *et al.*, 2001). In all these cases, the PBPs are encoded by typical ABC-type operons and are not linked to any signal transduction proteins. This suggests that their primary function is transport and they are just moonlighting as signaling molecules.

PBPs as Full-Time Sensor Proteins (Domains)

While some PBPs participate in signal transduction in addition to their role in transport, others could have lost the transport function altogether. Such cases can be recognized by the absence of the permease- and ATPase-encoding genes next to the PBP-encoding genes. Instead, such PBPs are encoded next to the components of various signal transduction systems, forming likely operons or even fused proteins. For example, the AcfC (VC0841) protein from *V. cholerae* is closely related to sulfate- and molybdate-binding proteins. Nevertheless, its gene does not belong to an ABC transporter operon. Instead, it is adjacent to the gene VC0840 that encodes a typical methyl-accepting chemotaxis protein. AcfC is predicted to function in the biosynthesis of a *V. cholerae* colonization factor (Parsot and Mekalanos, 1992), most likely as a sensor for a still unidentified environmental signal. Likewise, *V. cholerae* gene VCA1033 encodes a close homolog of sugar-binding periplasmic proteins LplA and UgpB, but is clustered together with a gene (VCA1034) that encodes a putative methyl-accepting chemotaxis protein. These two genes are apparently co-expressed, which means that VCA1033 should be annotated as a predicted ligand-binding protein involved in chemotaxis (not transport). A similar situation, when a PBP-like protein is encoded in the same putative operon with a methyl-accepting chemotaxis protein or a histidine kinase, can be found in other organisms, including *Bacillus halodurans* (BH1801-BH1802) and *Halobacterium* sp. (Vng0794g-Vng0793g; Vng1857C-Vng1856G).

In yet another interesting but so far unique system, a PBP-like protein AmiC from *Pseudomonas aeruginosa* (PA3364) is not even exported into the periplasm. It is located in the cytoplasm and serves as the ligand receptor and negative regulator for an amide-inducible aliphatic amidase operon, amiEBCRS. Together with the RNA-binding response regulator AmiR, AmiC forms a control circuit that regulates the expression of amidase AmiE (and their own) in response to inducing amides (Wilson *et al.*, 1996).

On some occasions, PBP-like proteins are not just encoded next to histidine kinases but actually form their periplasmic sensory domains. This is the case, for example, for the *E. coli* histidine kinase EvgS and related proteins in *A. fulgidus* (AF1766), *B. halodurans* (BH2949), and *P. aeruginosa* (PA0600, PA394, and PA2583). Fortunately, detection of the highly conserved histidine kinase domain usually does not pose any problem and allows an easy recognition of such proteins. The problem may arise, however, when PBPs are fused to poorly conserved or poorly characterized signaling domains and comprise the most easily recognizable part of the molecule. For example, *Mycobacterium leprae* gene ML0398 encodes a close homolog of the periplasmic ribose-binding proteins of *E. coli* and *B. subtilis*, which is accordingly annotated as "putative D-ribose-binding protein". However, its nearest neighbor ML0397, starting 113 bp downstream (*i.e.* not necessarily forming an operon) encodes a very unusual fusion protein that consists of a C-terminal ribose permease domain fused to two N-terminal signaling domains, the heme and flavin-binding PAS domain and the GGDEF domain that is believed to have diguanylate cyclase activity (see below). As this region does not encode an ATPase domain, one has to assume that the function of the ML0398-ML0397 pair has something to do with signaling, in addition to transport or even instead of it. The original annotation of ML0397 as "putative transporter protein", while technically correct, misses the biological meaning of this unique transporter-signaller fusion.

PBP-Like Domains in Non-Signaling Proteins

Phylogenetic analysis of individual domains could also help in improving the quality of functional assignments by allowing the identification of relatively recent fusions that might have unusual or duplicate functions. The analysis of YfhD protein from *E. coli*, annotated as putative periplasmic binding transport protein, shows that it is in fact comprised of three domains including, respectively, the N-terminal PBP domain, the central transglycosylase (SLT) domain and a C-terminal low complexity domain. This SLT domain is homologous to that of lytic murein transglycosylases that cleave the glycosidic bond between N-acetylmuramic acid and N-acetylglucosamine in the peptidoglycan and are encoded by bacteriophages, conjugative plasmids and bacterial type III and type IV secretion systems (Dijkstra and Thunnissen, 1994; Mushegian *et al.*, 1996). Since peptidoglycan presents a barrier to transport of large substrates, many macromolecular transport systems have recruited their own specialized transglycosylases that create small localized openings in cell wall peptidoglycan to support the transport complex assembly and/or transport of substrates.

An analysis of the phylogenetic distribution of YfhD protein finds its homologs encoded in genomes of many bacterial pathogens including *S. typhimurium*, *V. cholerae*, *P. multocida*, *Y. pestis*, *H. pylori* and *P. aeruginosa*. All these proteins have the N-terminal PBP domain similar to glutamine-binding protein GlnH and other amino acid-binding proteins involved in ABC-type transport (Hsiao *et al.*, 1996). The presence of this unusual fusion of PBP and SLT domains in many bacterial genomes is quite puzzling, since ABC-type PBPs are typically not involved in DNA/RNA or virulence factor transfer or peptidoglycan metabolism. A quick look at the genomic neighborhood reveals that almost all proteins of the YfhD family are encoded next to predicted cytosine/adenosine deaminase YfhC, phosphoribosylformylglycinamidine synthase PurL and other enzymes of the purine biosynthesis pathway. YfhC-like proteins share the N-terminal Zn^{2+}-binding catalytic region with deoxycytidylate deaminase ComEB required for DNA uptake in *Bacullis subtilis* and a pyrimidine deaminase domain of riboflavin deaminase/reductase RibD/RibG protein of *E. coli* and *B. subtilis*. Taken together, these data allow us to conclude that YfhD and related proteins do not function as substrate-binding components of an ABC-type transporter. They probably act as specialized transglycosylases in assembly of an unknown virulence-related outer membrane structure involved in DNA uptake and/or nucleotide metabolism. Their PBP domains might act as an auxiliary or regulatory subunit.

Another striking example of the necessity of the analysis of domain composition for functional annotation of PBP-containing proteins is the All2911 protein of *Nostoc* sp. PCC7120 (also referred to as *Anabaena* sp. PCC7120). The N-terminal and C-terminal domains of this protein are homologous to the glutamate-binding protein GluH from *E. coli* and related proteins of the same family (Hsiao *et al.*, 1996). As a result, All2911 was originally annotated as a glutamate-binding protein of ABC transport system. However, All2911 also contains a small central MPM (membrane-potassium selective pore-membrane) domain, which is the signature domain for K^+ transport proteins such as Kch and TOK1 (Ketchum *et al.*, 1995). A similar domain composition is found in only one other prokaryotic protein, slr1257 (GluR0) of *Synechocystis* sp., which has been characterized as the first prokaryotic K^+-selective glutamate receptor (Chen *et al.*, 1999), and in numerous eukaryotic ligand-gated ion channel receptors (Mayer *et al.*, 2001; Xue, 1998). GluR0 is thought to be a missing link between the prokaryotic K^+ channels and eukaryotic glutamate-gated transporters (Dingledine *et al.*, 1999). Similarly to All2911, the ligand binding core of GluR0 is homologous to GlnH-like glutamine binding proteins. While this is but one example of the involvement of PBPs in eukaryotic signal transduction, there are many such cases in prokaryotes.

The examples listed above show that confident assignment of PBP-like proteins and ABC-type ATPases as components of ABC transporters requires knowledge of the genome context. Otherwise, annotation of such proteins based solely on their similarity to previously characterized proteins can result in questionable assignments. Even more importantly, such assignments would miss some truly remarkable cases (including those described above), where detailed analysis of the system may lead to important biological insights.

3. Microbial Signal Transduction Systems

Microbial signal transduction systems, traditionally referred to as the "two-component" systems, mediate cellular responses to a variety of environmental stimuli. Although simplest variants of such systems indeed include just two components, a sensor histidine kinase and a DNA-binding response regulator, most signal transduction pathways actually consist of a variety of components that can be found either as stand-alone proteins or in various combinations with each other, forming multi-domain signaling proteins. The modular structure of these proteins and the diversity of their components make signal transduction systems very difficult to annotate, particularly by automated means.

Sensor Kinases and Sensor ATPases

The bacterial cell senses changes in the environment through a network of receptors, of which the most prominent are sensor histidine kinases. A typical sensor kinase is a transmembrane protein that contains an extracellular or membrane-associated sensory domain, one or more transmembrane segments, a cytoplasmic helical linker (HAMP) domain, a His-containing HisKA domain and an ATP-binding HATPase domain (Dutta *et al.*, 1999; Grebe and Stock, 1999; Hoch, 2000; Stock *et al.*, 2000). So-called hybrid kinases additionally contain CheY-like phosphoacceptor (CheY) domains and, sometimes, histidine phosphotransfer (HPt) domains (Dutta *et al.*, 1999; Matsushika and Mizuno, 1998; Mizuno, 1997). Many sensor kinases additionally include ligand-binding PAS and GAF domains (Aravind and Ponting, 1997; Taylor and Zhulin, 1999). In response to the given environmental stimulus (*i.e.* binding of particular ligands or some other physicochemical perturbation), the sensor domain of a given histidine kinase changes its conformation. This change is somehow transferred across the membrane, which causes the sensor kinase to catalyze ATP-dependent autophosphorylation of the conserved His residues, located in its HisK domain, which may transfer the phosphoryl residue further to the HPt and/or CheY domains.

Just like the ATP-binding subunit of the ABC-type transporter typically is the most conserved one and is most commonly used for annotation, recognition of the sensor histidine kinases is typically based on the presence of the C-terminal ATPase domains. Ten years ago, Parkinson and Kofoid have described several conserved motifs, common for most histidine kinases (Parkinson and Kofoid, 1992). In practice, many proteins that lack one or more of such motifs are still being annotated as histidine kinases. A good example is sensor proteins that belong to the HPK8 and HPK10 families (http://www.uni-kl.de/FB-Biologie/AG-Hakenbeck/TGrebe/HPK/Table6.htm) in the classification by Grebe and Stock (Grebe and Stock, 1999). These proteins were always assumed to function as histidine kinases, although neither their autophosphorylation nor phosphorylation of any response regulator by any of these proteins has been ever shown experimentally. Sequence comparisons of these proteins against SMART, Pfam, and COG databases show that they contain N-terminal periplasmic (or membrane-bound) sensor domains and C-terminal ATPase (HATPase in SMART) domains, typical for sensor histidine kinases, but lack the typical histidine kinase phosphoacceptor/dimerization domain (HisKA in SMART). As has been correctly noted for the *P. aeruginosa* FimS sensor protein (Yu *et al.*, 1997), these proteins comprise two separate families of sensors (HPK8 and HPK10 at the above mentioned web site and COG2972 and COG3275 in the COG database). Nevertheless, each of these proteins contains a conserved His residue that could potentially serve as the autophosphorylation site (Grebe and Stock, 1999; Mizuno, 1997). In *Clostridium perfringens* VirS sensor, this conserved His residue has been shown to be essential for signal transduction (Cheung and Rood, 2000). Thus, in spite of the lack of the typical phosphoacceptor/dimerization domain and paucity of experimental data, FimS/LytS-type sensor proteins can still be considered histidine kinases that phosphorylate their cognate response regulators, albeit this might occur without the autophosphorylation step.

An additional argument for the sensor kinase function of these proteins comes from their genome organization: genes that encode these FimS/LytS-type sensors are almost always encoded next to and form predicted operons with genes that encode response regulators. Remarkably, genes that encode sensor proteins and their cognate response regulators are found in operons in both orientations, HK-RR and RR-HK. In *P. aeruginosa fimS-algR* operon, as well as in *B. subtilis lytST*, *Staph.aureus lytSR*, *Staph. lugdunensis agrBA*, *E.coli yehUT* and *ypdAB*, *V. cholerae VC0694-VC0693* and *VCA0851-VCA0850*, *Clostr. acetobutylicum CAC0080-CAC0081*, and many other operons, the gene encoding a FimS-type sensor precedes the gene encoding the response regulator. In contrast, in *Cl. perfringens virRS* and *Cl. acetobutylicum CAC1581-CAC1582* operons, the gene encoding the response regulator precedes the gene encoding the sensor protein, suggesting independent instances of operon assembly, at least in *Clostridium* spp.

These examples show that identification of a novel protein as sensor kinase has to rely on more than just the similarity in the C-terminal HATPase regions. It should include additional evidence, such as presence of an N-terminal extracellular sensor domain, followed by a transmembrane segment(s), presence of additional ligand-binding domains, such as PAS and GAF, presence of the conserved phosphorylation domain, and/or presence of a neighboring gene, encoding a response regulator.

Extracellular Sensor Domains of Kinases and Other Receptors

Identification of the periplasmic (in Gram-positive bacteria - extracellular) ligand-binding sensor domains in histidine kinases is an interesting problem in its own right, due to the extreme diversity of such domains (Table 1). As we have mentioned above, one common type of such domains, referred to as FliY-type in Table 1, is homologous to the periplasmic solute-binding protein components of the ABC-type transporters systems (Tam and Saier, 1993). Many of such domains show highest sequence similarity to ArgT, HisJ, and GlnH, the PBPs of amino acid transport systems and are most likely involved in amino acid binding. Several sensor kinases, including *E. coli* EvgS, mentioned above, and *Bordetella pertussis* BvgS (Beier *et al.*, 1995; Beier *et al.*, 1996), contain duplicated PBP-type domains followed by a transmembrane segment that anchors them to the membrane.

The growth in the number of completely sequenced genomes could have been expected to identify in each novel genome the same set of sensor histidine kinases as in previously characterized model organisms, such as *E. coli* and *B. subtilis*. Indeed, certain sensor kinases, such as CitA, NtrB (GlnL), or UhpB, found in a number of different organisms, recognize the same

Table 1. Conserved domains of the microbial signal transduction systems

Domain name	Length aa	Function	Structure	Domain database entry[a]			Ref.
				SMART	Pfam	COG	
Sensor domains							
Extracellular sensor domains							
FliY (PBP)	~220	Amino acid binding	2lao	PBPb	PF00497	3438	Tam and Saier, 1993
Cache	~80	Small ligand binding	-	-	PF02743	3290[b]	Anantharaman and Aravind, 2000
ERD	~100	Ethylene binding	-	-	-	-	Mount and Chang, 2002
CHASE	~220	Unknown signal	-	-	-	3614	Anantharaman and Aravind, 2001; Mougel and Zhulin, 2001
CHASE2	~240	Unknown signal	-	-	-	4252	Zhulin *et al.*, 2002
CHASE3	~130	Unknown signal	-	-	-	-	Zhulin *et al.*, 2002
CHASE4	~250	Unknown signal	-	-	-	3322	Zhulin *et al.*, 2002
CHASE5	~120	Unknown signal	-	-	-	4250	Zhulin *et al.*, 2002
CHASE6	~240	Unknown signal	-	-	-	3452	Zhulin *et al.*, 2002
CHASE7	~210	Unknown signal	-	-	-	-	Zhulin *et al.*, 2002
Membrane-bound sensor domains							
MHYT	~200	Metal binding?	6TM	-	-	3300	Galperin *et al.*, 2001a
MHYE	~150	Metal binding?	4TM	-		-	Nikolskaya and Galperin, 2002
MASE1	~270	Unknown signal	11TM	-	-	3447	Zhulin *et al.*, 2002
MASE2	~160	Unknown signal	5TM	-	-	-	Zhulin *et al.*, 2002
Cytoplasmic sensor domains							
PAS	~100	Heme, flavin, ATP, and cinnamic acid binding	2phy	PAS PAC	PF00989 PF00785	2202	Taylor and Zhulin, 1999
GAF	~150	cGMP and photo-pigment binding	1f5m	GAF	PF01590	2203	Aravind and Ponting, 1997; Ho *et al.*, 2000
Signal transduction domains							
HAMP	~50	Dimerization?	1joy	HAMP	PF00672	2770	Aravind and Ponting, 1999; Williams and Stewart, 1999
HisKA	~80	Phosphoacceptor,	1joy	HisKA	PF00512		Dutta *et al.*, 1999; Grebe and Stock, 1999; Stock *et al.*, 2000
HATPase	~120	Phosphorylation of HisKA domain	1bxd	HATPase	PF02518	0642[b]	
MCP	~350	Methyl accepting	1qu7	MA	PF00015	0840[b]	Bollinger *et al.*, 1984
HPt	~100	Phosphoacceptor	2a0b	HPT	PF01627	2198	Matsushika and Mizuno, 1998
CheY	~120	Phosphoacceptor	2che	REC	PF00072	0784	Grebe and Stock, 1999; Volz, 1993
Signal output domains							
DNA-binding							
AraC/XylS	~120	DNA binding	1bl0	HTH_AraC	PF00165	2207[b]	Gallegos *et al.*, 1997; Rhee *et al.*, 1998
NarL/FixJ	~120	DNA binding	1rnl	HTH_LuxR	PF00196	2195[b]	Baikalov *et al.*, 1996
Winged helix	~120	DNA binding	1opc	-	PF00486	0745[b]	Martinez-Hackert and Stock, 1997a; b
SARP	~110	DNA binding	-	-	PF03704	3629[b]	Wietzorrek and Bibb, 1997
LytTR	~110	DNA binding	-	-	-	3279[b]	Nikolskaya and Galperin, 2002
RNA-binding							
ANTAR	~80	RNA-binding	1qo0	-	PF03861	3707[b]	Shu and Zhulin, 2002
Enzymatic							
AAA	~300	ATPase	1d2n	AAA	PF00158	2204	Neuwald *et al.*, 1999
GGDEF	~180	Diguanylate cyclase?	-	DUF1	PF01590	2199	Pei and Grishin, 2001; Tal *et al.*, 1998
EAL	~250	Diguanylate phosphodiesterase?	-	DUF2	PF00990	2200	Tal *et al.*, 1998
HD-GYP	~170	Phosphodiesterase	-	-	-	2206	Galperin *et al.*, 1999
CheB	~180	Methylesterase	1chd	-	PF01339	2201[b]	West *et al.*, 1995

[a] Protein domain databases that contain sequence alignments of these signaling domains include SMART (http://smart.embl-heidelberg.de, Schultz *et al.*, 2000), Pfam (http://www.sanger.ac.uk/Software/Pfam, Bateman *et al.*, 2000) , and COGs (http://www.ncbi.nlm.nih.gov/COG, Tatusov *et al.*, 2000).
[b] Members of this COG contain more than one domain

substrates (citrate, nitrite/nitrate, and hexose phosphates, respectively). However, analysis of newly sequenced genomes shows that in different bacteria the same extracellular sensory domains can be found in different transmembrane sensors. A common periplasmic ligand-binding domain, Cache (Ca^{2+} channels, chemotaxis receptors), has been found in histidine kinases, such as the dicarboxylate sensor DctB from *Rhizobium leguminosarum,* and in methyl-accepting chemotaxis proteins (MCPs), such as *Bacillus subtilis* McpB and McpC (Anantharaman and Aravind, 2000; Zhulin, 2001). In addition, the Cache domain has been found in association with cytoplasmic adenylate cyclase domains and with the uncharacterized GGDEF and HD-GYP domains, believed to have diguanylate cyclase (Ausmees *et al.*, 2001; Pei and Grishin, 2001) and phosphodiesterase (Galperin *et al.*, 1999) activity, respectively. While histidine kinases typically control cell metabolism by regulating the level of transcription of certain operons, methyl-accepting chemotaxis proteins act at the level of cellular behavior (chemotaxis) without necessarily changing the cell metabolism. The adenylate cyclases produce cAMP, a secondary messenger that can relieve catabolite repression of many operons, including respiratory and motility systems in many bacteria. Finally, the scarce data on the role(s) of diguanylate cyclases and phosphodiesterases appear to indicate that they control secretion of extracellular proteins and polysaccharides (see below).

The presence of long and relatively well-conserved extracellular sensor domains in diverse membrane receptors poses a serious problem for functional annotation of these proteins, particularly for the automated annotation. A careful analysis of signaling proteins identified in the course of microbial genome sequencing projects shows that many of them are mis-annotated in GenBank. For example, the histidine kinase All3347 and uncharacterized protein Alr0357 from *Nostoc* sp. PCC 7120 (also referred to as *Anabaena* sp. PCC 7120) were both mis-annotated as adenylate cyclases, although their similarity to adenylate cyclases was limited to the common extracellular sensor domain. Curiously, three virtually identical CHASE2-containing proteins were annotated, respectively, as "adenylate cyclase" (All7310), "similar to adenylate cyclase" (All3180) and a "hypothetical protein" (Alr1378), although none of them contained the adenylate cyclase catalytic domain. On the other hand, many transmembrane sensors are being annotated as "unknown", "hypothetical", or "conserved hypothetical" proteins. As we have argued earlier, unless there is a systematic mistake in sequencing, once a protein is conserved across diverse phylogenetic lineages, it is not hypothetical anymore (Galperin, 2001).

So what would be the right way to annotate membrane sensor proteins? In most cases, transmembrane receptors can be annotated based on their C-terminal signal transduction domains. As discussed above, a protein that contains an extracellular sensor domain, a transmembrane segment, a HisKA domain, and an HATPase domain can be safely annotated as a histidine kinase. In the apparent absence of the HisKA domain, one could rely on the presence of the conserved HATPase domain and additional evidence such as the gene neighborhoods. However, when sequence similarity searches do not find either HisKA or HATPase domains, only a careful domain analysis using specialized databases (SMART, Pfam, and COGs) can identify the true function of the novel protein. In most cases, the function can be unequivocally ascribed based on the type of the signal transduction domain attached to the sensor domain. Thus, any extracellular sensor domain attached to the MCP domain forms a methyl-accepting protein, a chemotaxis sensor. Likewise, an extracellular sensor domain attached to the adenylate cyclase domain forms a sensory adenylate cyclase and a sensor domain attached to a diguanylate cyclase or a phosphodiesterase forms, respectively a sensory diguanylate cyclase or a phosphodiesterase. The annotation for proteins that contain both a diguanylate cyclase (GGDEF) and a phosphodiesterase (EAL or HD-GYP) domain becomes slightly more complex, but again, there is nothing there that cannot be resolved by a domain-by-domain analysis. The main complication here occurs when a novel protein lacks a conserved signal transduction domain and the extracellular sensor domain provides the only significant hits in the database. Such cases should be carefully analyzed, as they are bound to result in novel findings. For example, the recently sequenced genome of *Methanosarcina acetivorans* encodes two sensor proteins (MA2076 and MA2922) with a novel archaea-specific signal output domain that belongs to the cupin superfamily and closely resembles bacterial mannose-6-phosphate isomerase. Not only annotation of these proteins as "hypothetical protein (multi-domain)" is exceedingly conservative, it also appears to be vague, as it does not confer the exciting fact that these proteins are definitely sensor proteins that comprise a new class of transmembrane receptors.

Membrane-Bound Sensor Domains

In addition to the extracytoplasmic sensor domains, histidine kinases and other transmembrane receptors may contain membrane-spanning signaling domains. The recently identified MHYT domain consists of six transmembrane segments with conserved Met and His residues located near the outer face of the cytoplasmic membrane. This predicted metal-containing domain was implicated in binding of CO, NO, or a similar ligand (Galperin *et al.*, 2001a). Like CHASE-type domains, the MHYT domain has been found in association with both histidine kinase and diguanylate cyclase/ phosphodiesterase domains (Galperin *et al.*, 2001a). In addition, it has been found in combination with the LytTR-type DNA-binding domain, indicating a direct connection between effector binding and transcriptional regulation

(Nikolskaya and Galperin, 2002). Another recently identified membrane domain, named MHYE, consists of four transmembrane segments with conserved His, Asp, and Glu residues located near the outer face of the cytoplasmic membrane and can also be expected to bind metals (Nikolskaya and Galperin, 2002). Two more conserved membrane-bound sensor domains, MASE1 and MASE2, were identified based on their association with histidine kinase or diguanylate cyclase domains and adenylate cyclase or diguanylate cyclase domains, respectively (Zhulin *et al.*, 2002). It is clear that these domains are just the first examples of the conserved membrane-spanning sensor domains in signal transduction proteins. Because these sensor domains, like extracellular sensor domains, can combine with diverse C-terminal output domains, the corresponding receptors have to be annotated based on their overall domain composition, not just based on a single database hit at their N-terminal region.

Cytoplasmic Ligand-Binding Domains

The PAS domain was originally described as a flavin-binding domain in the *E. coli* chemotaxis receptor Aer and a number of other proteins. The list of ligands that PAS can bind currently includes flavins, heme, cinnamic acid, and ATP (Stephenson and Hoch, 2001; Taylor and Zhulin, 1999). The GAF domain has been shown to bind cGMP and phytochromes (Aravind and Ponting, 1997). In spite of limited sequence similarity, PAS and GAF domains have similar structures (Ho *et al.*, 2000), indicating their common ancestry. Due to their relatively low sequence conservation, PAS and GAF domains alone seldom produce highly statistically significant hits between unrelated proteins. However, some multi-domain proteins contain more than one PAS and/or GAF domain, which complicates their analysis. Although these domains are easily recognized by comparing the query sequence to Pfam, SMART, or COG databases, signaling proteins containing PAS and/or GAF domains are typically annotated based on the functions of other domains found in the same protein, *e.g.* histidine kinase, PP2C-type protein phosphatase, NtrC-type ATPase, or diguanylate cyclase/phosphodiesterase. This certainly makes sense, as those other domains determine the enzymatic activity of the whole protein.

Response Regulators

The response regulators of the two-component system typically consist of an N-terminal CheY-like phosphoacceptor domain (CheY) and a C-terminal effector, or output, domain (Grebe and Stock, 1999; Stock *et al.*, 2000). The most common output domains are the DNA-binding helix-turn-helix (HTH) or winged-helix domains that activate and/or repress transcription of specific target genes (Grebe and Stock, 1999; Martinez-Hackert and Stock, 1997a; Stock *et al.*, 2000).

Recent analyses of the response regulators encoded in complete bacterial genomes, illuminated previously unseen diversity of their output domains. The DNA-binding HTH domains were found to fall into two main categories: the NarL/FixJ family regulators contain a single HTH motif in the middle of a four-helix bundle (Baikalov *et al.*, 1996), whereas regulators of the AraC/XylS family contain two HTH motifs separated by a linker helix and are usually flanked by two additional α-helices (Rhee *et al.*, 1998). A majority of known response regulators, including the well-studied OmpR and PhoB proteins, were found to contain an alternative DNA-binding domain, referred to as "winged helix" or "winged helix-turn-helix" (Martinez-Hackert and Stock, 1997a; b). This domain could have evolved from a typical HTH domain (Aravind and Koonin, 1999). We have also detected a handful of response regulators that combine the CheY domain with the DNA-binding domains of SARP family, which are distantly related to winged-helix domains (Wietzorrek and Bibb, 1997). Finally, response regulators of the ArgA/AlgR/LytR family contain a different type of DNA-binding domain, LytTR (Nikolskaya and Galperin, 2002), while transcriptional regulators of the NtrC family contain a C-terminal Fis-like DNA-binding domain (Klose *et al.*, 1994). In addition to the DNA-binding response regulators, an RNA-binding output domain has been described in *Pseudomonas aeruginosa* response regulator AmiR and related proteins (Shu and Zhulin, 2002).

In certain response regulators, the output domains are enzymatic, such as the methylesterase domain in the chemotaxis response regulator CheB-type, and do not necessarily regulate transcription. In transcriptional regulators of the NtrC family, N-terminal CheY domains and C-terminal DNA-binding Fis-like domains are separated by the AAA-type ATP-binding domains, whose ATPase activity is required for DNA-binding (Hwang *et al.*, 1999). Recent studies of the mechanisms of regulation of cellulose biosynthesis in *Acetobacter xylinum* (Galperin *et al.*, 1999; Tal *et al.*, 1998) identified the signaling molecule as cyclic diguanylate (Ross *et al.*, 1991; Tal *et al.*, 1998) and described proteins involved in the synthesis and hydrolysis of this novel secondary messenger. These proteins had a similar domain organization, consisting of a PAS domain and two novel domains, referred to as GGDEF and EAL domains, which were presumed to function, respectively, as diguanylate cyclase and phosphodiesterase (see Galperin *et al.*, 2001b for a review). Further studies identified an additional enzymatic output domain, HD-GYP, which consists of a moderately conserved phosphodiesterase subdomain and a highly conserved but uncharacterized subdomain, which was assumed to account for the substrate specificity of the former subdomain (Galperin *et al.*, 1999)

While these findings are very interesting and useful for better understanding of microbial regulatory circuits, they additionally complicate the functional annotation of the corresponding proteins. By tradition, all fusion proteins with N-terminal CheY domains and diverse C-terminal output domains are being annotated as response regulators. On the other hand, for consistency, it would be most appropriate to annotate response regulators with enzymatic output domains based on their enzymatic activities, just as it is being done for transmembrane receptors. Thus, a CheY-GGDEF fusion protein could be annotated as "response regulator, putative diguanylate cyclase". Although such an annotation might seem a bit awkward, it would definitely be more appropriate than the current annotations of such proteins in GenBank as "PleD gene product", "putative transcription regulator", "putative response regulator", "regulatory components of sensory transduction system", or even "unknown". Likewise, CheY-EAL and CheY-HD-GYP fusions could be annotated as "response regulator, putative phosphodiesterase", at least until these activities are verified by experimental data, which would allow dropping the qualifier "putative".

Unorthodox Fusions of Signaling Domains

While researchers generally have had little trouble recognizing CheY-GGDEF, CheY-EAL, and CheY-HD-GYP fusions as response regulators even before the abundance of such fusions has become apparent (Hecht and Newton, 1995; Mizuno *et al.*, 1996; Mizuno, 1997), more complex domain fusions still cause a significant confusion. We have already discussed the annotation of transmembrane sensors, consisting of extracytoplasmic or membrane-spanning sensor domains fused with diverse output domains (adenylate cyclase, GGDEF, EAL, and HD-GYP).

Table 2. Diversity of the domain fusions in bacterial signaling proteins

Domain organization	Experimentally studied examples	Examples identified solely from genomic sequences
Transcriptional regulators		
CheY-HTH (AraC-type)	*E. coli* AdiY, AppY, AraC, CelD, EnvY, EutR, FeaR, MarA, MelR, RhaR, RhaS, Rob, SoxS, XylR; *B. subtilis* CitT, ComA, GerE, PhoP, ResD	*E. coli* YbcM, YdeO, YdiP, YeaM, YhiW, YhiX, YidL,YijO, YkgA, YkgD, YneL, YpdC, YqhC; *B. subtilis* YbbB, Ybfl, YbfP, YdeC, YdeE, YesS, YfiF, YisR, YobQ, YtdP
CheY-HTH (NarL-type)	*E. coli* BglJ, EvgA, FimZ, NarL, NarP, RcsB, UhpA, UvrY; *B. subtilis* ComA, DegU	*E. coli* YgeK, YgeL, YhjB, YjjQ; *B. subtilis* Ydfl, YfiK, YhcZ, YocG, YvfU, YvqC, YxjL
CheY-HTH (CitB-type)	*E. coli* CitB, DcuR; *B. subtilis* CitT	*B. subtilis* YufM, YdbG; VC0790, VC1604
CheY-winged helix	*E. coli* ArcA, BaeR BasR, CpxR, CreB, KdpE, OmpR, PhoB, PhoP, RstA, TorR; *B. subtilis* PhoP, ResD,	*E. coli* YgiX, YlcA, YedW, Ybfl, YgeH; *B. subtilis* YbdJ, YcbL, YclJ, YkoG, YrkP, YtsA, YvcP, YvqA, YvrH, YycF, YxdJ
CheY-LytTR	*P. aeruginosa* AlgR, *S. aureus* AgrA	*E. coli* YehT, YpdB; *B. subtilis* LytT, YccH
CheY-SAPR	-	*B. halodurans* BH2016, *D. radiodurans* DR2556
CheY-AAA-Fis	*E. coli* AtoC, GlnG, HydG	*E. coli* YfhA, Cj1024c, HP0703, RP562, CT468, CPn0586, TP0519, BB0763
Simple response regulators		
CheY-GGDEF	*C. crescentus* PleD (CC2462), *Rhizobium leguminosarum* CelR2	BB0419, PA3702, PA4843, VC1086,
CheY-EAL	-	slr1588, PA3947, VC1652
CheY-HD-GYP	*Xanthomonas campestris* RpfG	PA2572, PA4781, slr2100, sll1624, TM0186, TM1147, VC1087, VC1348, VCA0210
CheY-GGDEF-EAL	-	XF0401
Signaling proteins with extracytoplasmic sensor domains		
PBP-GGDEF	-	VC1067, VCA0557
PBP-HD-GYP	-	TM1170
PBP-GGDEF-EAL		Slr2077
CHASE-HD-GYP	-	TM1682, VCA0895
CHASE4-GGDEF	-	VC2285, VC2454, PA0847
CHASE4-GGDEF-EAL	-	PA2072, CC0857
CHASE7-GGDEF-EAL		*E. coli* YhjK
Signaling proteins with cytoplasmic sensor domains		
PAS-GGDEF-EAL	*Acetobacter xylinum* DGC1, PHE	aq_1442
CheY-PAS-GGDEF-EAL	-	slr1305, PA4959, XF2624
GAF-GGDEF	-	*E. coli* YeaP, DRB0044, slr1143, sll0048, PA2771
CheY-GAF-GGDEF	-	slr0687
GAF-GGDEF-EAL	-	VCA0080, VCA0785, PA2567, ML1750
GAF-PAS- GGDEF-EAL	-	PA5017

There is, however, an additional class of signaling proteins where these novel output domains are combined with cytoplasmic ligand-binding domains PAS and/or GAF. There is no doubt that such proteins, too, participate in signal transduction pathways, although the signals sensed by them are not yet clear, and neither are their exact cellular function(s). At this stage we would suggest calling them simply signaling proteins and listing their domain content as in the following examples: *Pseudomonas aeruginosa* PA5017, signaling protein, a PAS-GAF-GGDEF-EAL domain fusion; *Deinococcus radiodurans* DRA0342, signaling protein, a HD-GYP-GGDEF domain fusion (Table 2). Of course, calling these proteins "putative diguanylate cyclase/ phosphodiesterase" would also be entirely appropriate.

In spite of all the variety of domain fusions in signal transduction proteins, there appears to be an underlying uniformity in the order of these domains, which corresponds to the general direction of signal transduction by these proteins, from an N-terminal extracellular or membrane-spanning sensor domain to a transmitter domain (or a chain of domains) to a C-terminal response output domain. Several exceptions notwithstanding, this principle appears to be true for a variety of signaling proteins in bacteria, archaea, and certain eukaryotes and allows one to devise a rational strategy for analysis of such proteins.

In conclusion, it is worth noting that functional annotations described in this chapter can hardly be automated. Although based on simple and fairly straightforward analyses of domain organization and gene neighborhoods, these annotations require a human with certain biological education. While this may be considered a nuisance by some, this reflects a more general principle that to recognize novelty, one has to be aware of the state of the art. If the latest developments are any indication, complex cases of domain fusions would remain outside the domain of automated analysis for years to come.

References

Alvarez-Martinez, M.T., Machold, J., Weise, C., Schmidt-Eisenlohr, H., Baron, C., and Rouot, B. 2001. The *Brucella suis* homologue of the *Agrobacterium tumefaciens* chromosomal virulence operon chvE is essential for sugar utilization but not for survival in macrophages. J. Bacteriol. 183: 5343-5351

Anantharaman, V., and Aravind, L. 2000. Cache - a signaling domain common to animal $Ca2^+$-channel subunits and a class of prokaryotic chemotaxis receptors. Trends Biochem. Sci. 25: 535-537.

Anantharaman, V., and Aravind, L. 2001. The CHASE domain: a predicted ligand-binding module in plant cytokinin receptors and other eukaryotic and bacterial receptors. Trends Biochem Sci. 26: 579-582

Andersen, R.N., Ganeshkumar, N., and Kolenbrander, P.E. 1993. Cloning of the *Streptococcus gordonii* PK488 gene, encoding an adhesin which mediates coaggregation with *Actinomyces naeslundii* PK606. Infect. Immun. 61: 981-987

Aravind, L., and Ponting, C.P. 1997. The GAF domain: an evolutionary link between diverse phototransducing proteins. Trends Biochem. Sci. 22: 458-459

Aravind, L., and Koonin, E.V. 1999. DNA-binding proteins and evolution of transcription regulation in the archaea. Nucleic Acids Res. 27: 4658-4670

Aravind, L., and Ponting, C.P. 1999. The cytoplasmic helical linker domain of receptor histidine kinase and methyl-accepting proteins is common to many prokaryotic signalling proteins. FEMS Microbiol. Lett. 176: 111-116

Ausmees, N., Mayer, R., Weinhouse, H., Volman, G., Amikam, D., Benziman, M., and Lindberg, M. 2001. Genetic data indicate that proteins containing the GGDEF domain possess diguanylate cyclase activity. FEMS Microbiol. Lett. 204: 163-167

Baikalov, I., Schroder, I., Kaczor-Grzeskowiak, M., Grzeskowiak, K., Gunsalus, R.P., and Dickerson, R.E. 1996. Structure of the *Escherichia coli* response regulator NarL. Biochemistry. 35: 11053-110561.

Bateman, A., Birney, E., Durbin, R., Eddy, S.R., Howe, K.L., and Sonnhammer, E.L. 2000. The Pfam protein families database. Nucleic Acids Res. 28: 263-6.

Beier, D., Schwarz, B., Fuchs, T.M., and Gross, R. 1995. In vivo characterization of the unorthodox BvgS two-component sensor protein of Bordetella pertussis. J. Mol. Biol. 248: 596-610

Beier, D., Deppisch, H., and Gross, R. 1996. Conserved sequence motifs in the unorthodox BvgS two-component sensor protein of *Bordetella pertussis*. Mol. Gen. Genet. 252: 169-176

Bollinger, J., Park, C., Harayama, S., and Hazelbauer, G.L. 1984. Structure of the Trg protein: Homologies with and differences from other sensory transducers of *Escherichia coli*. Proc. Natl. Acad. Sci. USA. 81: 3287-3291

Bork, P., Dandekar, T., Diaz-Lazcoz, Y., Eisenhaber, F., Huynen, M., and Yuan, Y. 1998. Predicting function: from genes to genomes and back. J. Mol. Biol. 283: 707-725.

Bosse, J.T., Gilmour, H.D., and MacInnes, J.I. 2001. Novel genes affecting urease acivity in *Actinobacillus pleuropneumoniae*. J. Bacteriol. 183: 1242-1247.

Chen, G.Q., Cui, C., Mayer, M.L., and Gouaux, E. 1999. Functional characterization of a potassium-selective prokaryotic glutamate receptor. Nature. 402: 817-821.

Cheng, J., Guffanti, A.A., and Krulwich, T.A. 1997. A two-gene ABC-type transport system that extrudes Na^+ in *Bacillus subtilis* is induced by ethanol or protonophore. Mol. Microbiol. 23: 1107-1120

Cheung, J.K., and Rood, J.I. 2000. Glutamate residues in the putative transmembrane region are required for the function of the VirS sensor histidine kinase from *Clostridium perfringens*. Microbiology. 146: 517-525.

Correia, F.F., DiRienzo, J.M., McKay, T.L., and Rosan, B. 1996. *scbA* from *Streptococcus crista* CC5A: an atypical member of the *lraI* gene family. Infect. Immun. 64: 2114-2121

Dijkstra, B.W., and Thunnissen, A.M. 1994. 'Holy' proteins. II: The soluble lytic transglycosylase. Curr. Opin. Struct. Biol. 4: 810-813

Dingledine, R., Borges, K., Bowie, D., and Traynelis, S.F. 1999. The glutamate receptor ion channels. Pharmacol Rev. 51: 7-61.

Dintilhac, A., Alloing, G., Granadel, C., and Claverys, J.P. 1997. Competence and virulence of *Streptococcus pneumoniae*: Adc and PsaA mutants exhibit a requirement for Zn and Mn resulting from inactivation of putative ABC metal permeases. Mol. Microbiol. 25: 727-739

Dutta, R., Qin, L., and Inouye, M. 1999. Histidine kinases: diversity of domain organization. Mol. Microbiol. 34: 633-640

Falke, J.J., Bass, R.B., Butler, S.L., Chervitz, S.A., and Danielson, M.A. 1997. The two-component signaling pathway of bacterial chemotaxis: a molecular view of signal transduction by receptors, kinases, and adaptation enzymes. Annu. Rev. Cell. Dev. Biol. 13: 457-512

Fenno, J.C., LeBlanc, D.J., and Fives-Taylor, P. 1989. Nucleotide sequence analysis of a type 1 fimbrial gene of *Streptococcus sanguis* FW213. Infect. Immun. 57: 3527-3533

Gallegos, M.T., Schleif, R., Bairoch, A., Hofmann, K., and Ramos, J.L. 1997. Arac/XylS family of transcriptional regulators. Microbiol. Mol. Biol. Rev. 61: 393-410.

Galperin, M.Y., Natale, D.A., Aravind, L., and Koonin, E.V. 1999. A specialized version of the HD hydrolase domain implicated in signal transduction. J. Mol. Microbiol. Biotechnol. 1: 303-305

Galperin, M.Y., and Koonin, E.V. 2000. Who's your neighbor? New computational approaches for functional genomics. Nat. Biotechnol. 18: 609-613

Galperin, M.Y. 2001. Conserved 'hypothetical' proteins: new hints and new puzzles. Comp. Funct. Genomics. 2: 14-18

Galperin, M.Y., Gaidenko, T.A., Mulkidjanian, A.Y., Nakano, M., and Price, C.W. 2001a. MHYT, a new integral membrane sensor domain. FEMS Microbiol. Lett. 205: 17-23

Galperin, M.Y., Nikolskaya, A.N., and Koonin, E.V. 2001b. Novel domains of the prokaryotic two-component signal transduction systems. FEMS Microbiol. Lett. 203: 11-21

Ganeshkumar, N., Hannam, P.M., Kolenbrander, P.E., and McBride, B.C. 1991. Nucleotide sequence of a gene coding for a saliva-binding protein (SsaB) from *Streptococcus sanguis* 12 and possible role of the protein in coaggregation with actinomyces. Infect. Immun. 59: 1093-1099

Gomez, A., Ramon, D., and Sanz, P. 1994. The *Bacillus subtilis* lipoprotein LplA causes cell lysis when expressed in *Escherichia coli*. Microbiology. 140: 1839-1845

Grebe, T.W., and Stock, J.B. 1999. The histidine protein kinase superfamily. Adv. Microb. Physiol. 41: 139-227

Hecht, G.B., and Newton, A. 1995. Identification of a novel response regulator required for the swarmer- to-stalked-cell transition in *Caulobacter crescentus*. J. Bacteriol. 177: 6223-6229

Higgins, C.F. 2001. ABC transporters: physiology, structure and mechanism—an overview. Res. Microbiol. 152: 205-210

Ho, Y.S., Burden, L.M., and Hurley, J.H. 2000. Structure of the GAF domain, a ubiquitous signaling motif and a new class of cyclic GMP receptor. EMBO J. 19: 5288-5299.

Hoch, J.A. 2000. Two-component and phosphorelay signal transduction. Curr. Opin. Microbiol. 3: 165-170

Holland, I.B., and Blight, M.A. 1999. ABC-ATPases, adaptable energy generators fuelling transmembrane movement of a variety of molecules in organisms from bacteria to humans. J. Mol. Biol. 293: 381-399

Hsiao, C.D., Sun, Y.J., Rose, J., and Wang, B.C. 1996. The crystal structure of glutamine-binding protein from *Escherichia coli*. J. Mol. Biol. 262: 225-242

Huynen, M., Snel, B., Lathe, W., 3rd, and Bork, P. 2000a. Predicting protein function by genomic context: quantitative evaluation and qualitative inferences. Genome Res. 10: 1204-1210

Huynen, M., Snel, B., Lathe, W., and Bork, P. 2000b. Exploitation of gene context. Curr. Opin. Struct. Biol. 10: 366-370.

Huynen, M.A., and Snel, B. 2000. Gene and context: integrative approaches to genome analysis. Adv. Protein. Chem. 54: 345-379.

Hwang, I., Thorgeirsson, T., Lee, J., Kustu, S., and Shin, Y.K. 1999. Physical evidence for a phosphorylation-dependent conformational change in the enhancer-binding protein NtrC. Proc. Natl. Acad. Sci. USA. 96: 4880-4885

Kemner, J.M., Liang, X., and Nester, E.W. 1997. The *Agrobacterium tumefaciens* virulence gene *chvE* is part of a putative ABC-type sugar transport operon. J. Bacteriol. 179: 2452-2458

Ketchum, K.A., Joiner, W.J., Sellers, A.J., Kaczmarek, L.K., and Goldstein, S.A. 1995. A new family of outwardly rectifying potassium channel proteins with two pore domains in tandem. Nature. 376: 690-695.

Klose, K.E., North, A.K., Stedman, K.M., and Kustu, S. 1994. The major dimerization determinants of the nitrogen regulatory protein NtrC from enteric bacteria lie in its carboxy-terminal domain. J. Mol. Biol. 241: 233-245

Koski, L.B., and Golding, G.B. 2001. The closest BLAST hit is often not the nearest neighbor. J. Mol. Evol. 52: 540-542

Lowe, A.M., Lambert, P.A., and Smith, A.W. 1995. Cloning of an *Enterococcus faecalis* endocarditis antigen: homology with adhesins from some oral streptococci. Infect. Immun. 63: 703-706

Martinez-Hackert, E., and Stock, A.M. 1997a. Structural relationships in the OmpR family of winged-helix transcription factors. J. Mol. Biol. 269: 301-312.

Martinez-Hackert, E., and Stock, A.M. 1997b. The DNA-binding domain of OmpR: crystal structures of a winged helix transcription factor. Structure. 5: 109-124.

Matsushika, A., and Mizuno, T. 1998. The structure and function of the histidine-containing phosphotransfer (HPt) signaling domain of the *Escherichia coli* ArcB sensor. J. Biochem. (Tokyo). 124: 440-445

Mayer, M.L., Olson, R., and Gouaux, E. 2001. Mechanisms for ligand binding to GluR0 ion channels: crystal structures of the glutamate and serine complexes and a closed apo state. J. Mol. Biol. 311: 815-836.

Mizuno, T., Kaneko, T., and Tabata, S. 1996. Compilation of all genes encoding bacterial two-component signal transducers in the genome of the cyanobacterium, *Synechocystis* sp. strain PCC 6803. DNA Res. 3: 407-414.

Mizuno, T. 1997. Compilation of all genes encoding two-component phosphotransfer signal transducers in the genome of *Escherichia coli*. DNA Res. 4: 161-168.

Mougel, C., and Zhulin, I.B. 2001. CHASE: an extracellular sensing domain common to transmembrane receptors from prokaryotes, lower eukaryotes and plants. Trends Biochem. Sci. 26: 582-584

Mount, S.M., and Chang, C. 2002. Evidence for the cyanobacterial origin of plant ethylene receptor genes. Plant Physiol. 130: 10-14.

Mushegian, A.R., Fullner, K.J., Koonin, E.V., and Nester, E.W. 1996. A family of lysozyme-like virulence factors in bacterial pathogens of plants and animals. Proc. Natl. Acad. Sci. USA. 93: 7321-7326

Nachin, L., El Hassouni, M., Loiseau, L., Expert, D., and Barras, F. 2001. SoxR-dependent response to oxidative stress and virulence of *Erwinia chrysanthemi*: the key role of SufC, an orphan ABC ATPase. Mol. Microbiol. 39: 960-972

Neuwald, A.F., Aravind, L., Spouge, J.L., and Koonin, E.V. 1999. AAA$^+$: A class of chaperone-like ATPases associated with the assembly, operation, and disassembly of protein complexes. Genome Res. 9: 27-43

Nikaido, H., and Hall, J.A. 1998. Overview of bacterial ABC transporters. Methods Enzymol. 292: 3-20

Nikolskaya, A.N., and Galperin, M.Y. 2002. A novel type of conserved DNA-binding domain in the transcriptional regulators of the AlgR/AgrA/LytR family. Nucleic Acids Res. 30: 2453-2459

Overbeek, R., Fonstein, M., D'Souza, M., Pusch, G.D., and Maltsev, N. 1999. The use of gene clusters to infer functional coupling. Proc. Natl. Acad. Sci. USA. 96: 2896-2901

Overduin, P., Boos, W., and Tommassen, J. 1988. Nucleotide sequence of the ugp genes of *Escherichia coli* K-12: homology to the maltose system. Mol. Microbiol. 2: 767-775

Parkinson, J.S., and Kofoid, E.C. 1992. Communication modules in bacterial signaling proteins. Annu. Rev. Genet. 26: 71-112

Parsot, C., and Mekalanos, J.J. 1992. Structural analysis of the *acfA* and *acfD* genes of *Vibrio cholerae*: effects of DNA topology and transcriptional activators on expression. J. Bacteriol. 174: 5211-5218

Patzer, S.I., and Hantke, K. 1998. The ZnuABC high-affinity zinc uptake system and its regulator Zur in *Escherichia coli*. Mol. Microbiol. 28: 1199-1210

Pei, J., and Grishin, N.V. 2001. GGDEF domain is homologous to adenylyl cyclase. Proteins. 42: 210-216.

Quentin, Y., Fichant, G., and Denizot, F. 1999. Inventory, assembly and analysis of *Bacillus subtilis* ABC transport systems. J. Mol. Biol. 287: 467-484

Quiocho, F.A., and Ledvina, P.S. 1996. Atomic structure and specificity of bacterial periplasmic receptors for active transport and chemotaxis: variation of common themes. Mol. Microbiol. 20: 17-25

Rhee, S., Martin, R.G., Rosner, J.L., and Davies, D.R. 1998. A novel DNA-binding motif in MarA: the first structure for an AraC family transcriptional activator. Proc. Natl. Acad. Sci. USA. 95: 10413-10418.

Ross, P., Mayer, R., and Benziman, M. 1991. Cellulose biosynthesis and function in bacteria. Microbiol. Rev. 55: 35-58

Roth, J.R., Lawrence, J.G., and Bobik, T.A. 1996. Cobalamin (coenzyme B12): synthesis and biological significance. Annu. Rev. Microbiol. 50: 137-181

Sahm, K., Matuschek, M., Muller, H., Mitchell, W.J., and Bahl, H. 1996.

Molecular analysis of the *amy* gene locus of *Thermoanaerobacterium thermosulfurigenes* EM1 encoding starch-degrading enzymes and a binding protein-dependent maltose transport system. J. Bacteriol. 178: 1039-1046

Saier, M.H., Jr. 2000a. Families of transmembrane transporters selective for amino acids and their derivatives. Microbiology. 146: 1775-1795

Saier, M.H., Jr. 2000b. Families of transmembrane sugar transport proteins. Mol. Microbiol. 35: 699-710

Sampson, J.S., O'Connor, S.P., Stinson, A.R., Tharpe, J.A., and Russell, H. 1994. Cloning and nucleotide sequence analysis of *psaA*, the *Streptococcus pneumoniae* gene encoding a 37-kilodalton protein homologous to previously reported *Streptococcus* sp. adhesins. Infect. Immun. 62: 319-324

Scholle, A., Vreemann, J., Blank, V., Nold, A., Boos, W., and Manson, M.D. 1987. Sequence of the *mglB* gene from *Escherichia coli* K12: comparison of wild-type and mutant galactose chemoreceptors. Mol. Gen. Genet. 208: 247-253

Schultz, J., Copley, R.R., Doerks, T., Ponting, C.P., and Bork, P. 2000. SMART: a web-based tool for the study of genetically mobile domains. Nucleic Acids Res. 28: 231-234

Schulz, H., Fabianek, R.A., Pellicioli, E.C., Hennecke, H., and Thony-Meyer, L. 1999. Heme transfer to the heme chaperone CcmE during cytochrome c maturation requires the CcmC protein, which may function independently of the ABC-transporter CcmAB. Proc. Natl. Acad. Sci. USA. 96: 6462-6467

Shu, C.J., and Zhulin, I.B. 2002. ANTAR: an RNA-binding domain in transcription antitermination regulatory proteins. Trends Biochem. Sci. 27: 3-5

Shulami, S., Gat, O., Sonenshein, A.L., and Shoham, Y. 1999. The glucuronic acid utilization gene cluster from *Bacillus stearothermophilus* T-6. J. Bacteriol. 181: 3695-3704

Stephenson, K., and Hoch, J.A. 2001. PAS-A domain of phosphorelay sensor kinase A: a catalytic ATP-binding domain involved in the initiation of development in *Bacillus subtilis*. Proc. Natl. Acad. Sci. USA. 98: 15251-15256

Stock, A.M., Robinson, V.L., and Goudreau, P.N. 2000. Two-component signal transduction. Annu. Rev. Biochem. 69: 183-215

Sutcliffe, I.C., Tao, L., Ferretti, J.J., and Russell, R.R. 1993. MsmE, a lipoprotein involved in sugar transport in *Streptococcus mutans*. J. Bacteriol. 175: 1853-1855

Tal, R., Wong, H.C., Calhoon, R., Gelfand, D., Fear, A.L., Volman, G., Mayer, R., Ross, P., Amikam, D., Weinhouse, H., Cohen, A., Sapir, S., Ohana, P., and Benziman, M. 1998. Three *cdg* operons control cellular turnover of cyclic di-GMP in *Acetobacter xylinum*: genetic organization and occurrence of conserved domains in isoenzymes. J. Bacteriol. 180: 4416-4425

Tam, R., and Saier, M.H., Jr. 1993. Structural, functional, and evolutionary relationships among extracellular solute-binding receptors of bacteria. Microbiol. Rev. 57: 320-346

Tatusov, R.L., Koonin, E.V., and Lipman, D.J. 1997. A genomic perspective on protein families. Science. 278: 631-637

Tatusov, R.L., Galperin, M.Y., Natale, D.A., and Koonin, E.V. 2000. The COG database: a tool for genome-scale analysis of protein functions and evolution. Nucleic Acids Res. 28: 33-36

Taylor, B.L., and Zhulin, I.B. 1999. PAS domains: internal sensors of oxygen, redox potential, and light. Microbiol. Mol. Biol. Rev. 63: 479-506

Tomii, K., and Kanehisa, M. 1998. A comparative analysis of ABC transporters in complete microbial genomes. Genome Res. 8: 1048-59

Volz, K. 1993. Structural conservation in the CheY superfamily. Biochemistry. 32: 11741-11753

West, A.H., Martinez-Hackert, E., and Stock, A.M. 1995. Crystal structure of the catalytic domain of the chemotaxis receptor methylesterase, CheB. J. Mol. Biol. 250: 276-290.

Wietzorrek, A., and Bibb, M. 1997. A novel family of proteins that regulates antibiotic production in streptomycetes appears to contain an OmpR-like DNA-binding fold. Mol. Microbiol. 25: 1181-1184.

Williams, S.B., and Stewart, V. 1999. Functional similarities among two-component sensors and methyl-accepting chemotaxis proteins suggest a role for linker region amphipathic helices in transmembrane signal transduction. Mol. Microbiol. 33: 1093-1102

Wilson, S.A., Wachira, S.J., Norman, R.A., Pearl, L.H., and Drew, R.E. 1996. Transcription antitermination regulation of the *Pseudomonas aeruginosa* amidase operon. EMBO J. 15: 5907-5916.

Wolf, A., Lee, K.C., Kirsch, J.F., and Ames, G.F. 1996. Ligand-dependent conformational plasticity of the periplasmic histidine-binding protein HisJ. Involvement in transport specificity. J. Biol. Chem. 271: 21243-21250.

Xue, H. 1998. Identification of major phylogenetic branches of inhibitory ligand-gated channel receptors. J. Mol. Evol. 47: 323-333.

Yu, H., Mudd, M., Boucher, J.C., Schurr, M.J., and Deretic, V. 1997. Identification of the *algZ* gene upstream of the response regulator *algR* and its participation in control of alginate production in *Pseudomonas aeruginosa*. J. Bacteriol. 179: 187-193.

Zhulin, I.B. 2001. The superfamily of chemotaxis transducers: from physiology to genomics and back. Adv. Microb. Physiol. 45: 157-198

Zhulin, I.B., Nikolskaya, A.N., and Galperin, M.Y. 2002. Novel sensory domains in microbial transmembrane receptors. J. Bacteriol. (in press).

From: *Frontiers in Computational Genomics*

8

Prolegomena to the Evolution of Transcriptional Regulation in Bacterial Genomes

Mikhail S. Gelfand, Olga N. Laikova

Abstract

Availability of many complete bacterial genomes makes it possible to apply the genomic comparative approaches to the analysis of transcriptional regulation and evolution of regulatory interactions. Here we list some observations about co-evolution of regulators and the signals they recognize in DNA; birth, evolution, and degeneration of regulons; influence of horizontal gene transfer on regulation and evolution of ubiquitous regulators; interaction of regulons. Although scattered and incomplete, these observations may provide the first base for development of a general theory of evolution of regulation.

1. Preliminaries

Recent advances in sequencing and analysis of bacterial genomes allow one to ask questions that only a few years ago would be considered pure speculation. Among these questions are metabolic reconstruction and evolution of metabolic pathways, identification of non-orthologous replacements, evolution of whole genomes, in particular, positional analysis of the operon structure, statistical analysis of genomes and identification of replication origins and horizontally transferred genes, and, last but not least, prediction of binding sites for transcription regulatory factors.

Comparative approaches were used both in large scale studies (McGuire *et al.*, 2000; McGuire and Chruch, 2000; McCue *et al.*, 2001; Terai *et al.*, 2001) and in analysis of individual systems (Mironov *et al.*, 1999; Stojanovich *et al.*, 1999; Gelfand *et al.*, 2000a; Rodionov *et al.*, 2000, 2001a, 2001b; Makarova *et al.*, 2001; Panina *et al.*, 2001; Tan *et al.*, 2001), for reviews see (Gelfand, 1999; Gelfand *et al.*, 2000b). Although it is probably too early to speak about generalities of the evolution of transcriptional regulation, some patterns start to emerge. Here we try to list some of the observations.

The plan of the paper is as follows. First, we briefly describe the basics of the comparative analysis of regulation. Then we consider co-evolution of regulators and their binding signals using the LacI family of transcriptional repressors as the main example. Several examples of emerging regulons are presented in the next section. Two closely related topics, regulators present in (almost) all bacterial genomes and regulation of horizontally transferred genes, are considered in the last two sections.

We concentrate here on bacterial transcriptional regulation. This is caused by several reasons. Firstly, the regulatory systems in prokaryotes are much simpler than in eukaryotes, especially multicellular ones. Indeed, in most cases it is not too much of oversimplification if one speaks about regulons, that is, sets of genes regulated by one transciptional factor. Usually one can ignore, in the first approximation, the interaction between factors binding to sites within one regulatory region: even if this interaction is essential (as in the CytR-CRP system of *Escherichia coli* or regulation of σ^{54} promoters of Gram-negative bacteria), it is relatively simple to take it into account in the course of computer analysis. The situation in eukaryotes with their huge regulatory cassettes bound by dozens of transcription factors is much more complicated.

Secondly, in many cases the identification of individual sites is simpler due to the larger size, higher information content (degree of sequence conservation), and the symmetry (complementary-palindromic or repeat structure) of the recognition signals. Although in most cases this is not sufficient to derive reliable recognition rules, rough genome-scale predictions provide a useful starting point for subsequent comparative analysis. Thirdly, there simply are more data: complete genomes of varying degree of relatedness.

The majority of examples in this paper have not been experimentally verified, and usually when we write about binding sites, this means ***candidate*** binding sites. However, the consistency of predictions makes it likely that most of them are indeed correct.

2. Comparative Analysis of Regulation

The main assumption of the comparative approach to prediction of regulatory sites is based on the idea that regulons are not random sets of genes, but form physiologically or metabolically logical structures. Thus if bacterial genomes contain orthologous transcription factors, the corresponding regulons can be assumed to be conserved. Even if one has a bad, insufficiently specific recognition rule for regulatory sites, simultaneous analysis of several genomes and the use of the above assumption yields the following consistency check: true sites occur upstream of orthologous genes, whereas false positives are scattered at random. Note that the recognition rules can be different in the considered genomes; indeed, as we will see below, often orthologous factors seem to possess slightly different recognition signals.

If the genomes are so close that there is considerable residual conservation of non-coding regions (*e.g. E. coli* and *Salmonella* spp.), the consistency filter is less selective. Indeed, in this case occurrences of candidate sites upstream of orthologous genes are not independent and could be due to the overall similarity rather than evolutionary conservation. However, the binding sites of transcription factors often show higher degree of stability than surrounding regions and can be seen in multiple alignments as runs of exact matches. The analysis of such conservation is widely used in eukaryotic genomics under the name "phylogenetic footprinting" (Tagle *et al.*, 1988; Gumuchio *et al.*, 1996), for more examples and references see (Duret and Bucher, 1997; Hardison *et al.*, 1997; Gelfand, 1999; Wasserman *et al.*, 2000; Dubchak *et al.*, 2000; Wasserman *et al.*, 2000). An attempt to formalize this approach and apply it to bacterial genomes was presented in (Stojanovich *et al.*, 1999), and the corresponding software was described in (Florea *et al.*, 2000).

Another modification arises during analysis of systems subject to horizontal transfer. In this case it is often difficult to resolve the orthology relationships. Thus it is sufficient to require that candidate sites occur upstream of closely homologous genes in however distant species. Of course, the orthologous regulators should be present in all considered genomes. Several such situations are treated in more detail below.

3. Co-Evolution of Regulators and Signals

Although it seems almost obvious that closely homologous transcription factors should have similar DNA binding activities, and this is supported by numerous experiments where a factor introduced on a plasmid complements the disrupted native factor, this is not always the case. In many cases the signals recognized by orthologous factors show subtle, but consistent differences. This does not necessarily mean that cross-recognition is lost completely, but only that the affinities of individual nucleotides to the DNA-binding surface of the protein, and hence the positional nucleotide frequencies, are different.

It is well established that positions conserved in an alignment of sites from one genome that can be bound by the same transcription factor are the positions most sensitive to mutations: changes in these positions disrupt the protein binding and the regulatory activity of the site. This conforms well to the evolutionary dogma: what is conserved, is functionally important. The structural basis for this also is clear: in a recent study of bacterial transcriptional factors with known structure of protein-DNA complexes, it was shown that the conserved bases are those that form the majority of DNA-protein contacts (Mirny and Gelfand, 2001a).

On the other hand, outside closely related taxonomic groups there is little direct correlation between evolutionary trees of factors and regulated systems (Otsuka *et al.*, 1996) and the evolution of structural and regulatory genes is independent (de Lorenzo and Pérez-Martín, 1996). Without experimental data or positional analysis of the operon structure (Dandekar *et al.*, 1998; Overbeek *et al.*, 1999; Yada *et al.*, 1999; Salgado *et al.*, 2000; Ermolaeva *et al.*, 2001; Rogozin *et al.*, 2001), it is not possible to link factors and regulons: for instance, it is not known what factor regulates the purine regulon of *Pyrococcus* spp. identified by computational analysis (Gelfand *et al.*, 2000a).

Still, there is a loose tendency of factors from one family to regulate related metabolic subsystems. Thus, the repressors from the LacI family are usually concerned with sugar utilization (*e.g.* in *E. coli*: AscG, EbgR, FruR, GalR, GalS, GntR, LacI, MalI, RbsR, RafR, TreR, but also CytR and PurR; see Figure 1), the same is true for the DeoR family (*e.g.* in *E. coli*: AgaR, DeoR, FucR, GatR, GlpR, SrlR), a more functionally diverse LysR family contains many global activators and regulators of amino acid synthesis (*e.g.* in *E. coli*: IciA, OxyR, CysB, DsdC, GcvA, IlvY, LeuO, LysR, MetR, TdcR) as well as specific regulators of catabolism in Pseudomonas (de Lorenzo and

Pérez-Martín, 1996) , the AraC family contains activators of catabolic operons in various organisms, the FUR family contains regulators of ion (ferric iron and zink) uptake in many different species, etc.

Given a phylogenetic tree of related transcription factors and functional assignments of factors to metabolic systems, one can tentatively identify homogeneous groups defined as branches where all factors have the same cellular role in different genomes, and the branching order is congruent to the species tree. One may expect cross-recognition between factors from one group, although detailed analysis sometimes reveals subtle differences between the recognition signals (see below the PurR case).

Analysis of positions conserved within functional groups of transcription factors demonstrates that the conserved alignment positions concentrate within the DNA-recognition surface and the ligand-binding pocket (Mirny and Gelfand, 2001b), by the way, in eukaryotes the situation is similar with the addition of protein-protein interaction surfaces (Lichtarge *et al.*, 1997).

Consider in detail the co-evolution of some transcription factors from the LacI family and their DNA signals (Figure 1). As the number of binding sites for any particular factor is low, it is difficult to distinguish between random fluctuations and changes in specificity of base pairs at specific positions. However, if the signal is palindromic, as is the case for LacI-family factors that bind DNA as dimers, the regular changes occur at symmetric positions, somewhat reminiscent of complementary substitutions preserving RNA secondary structure. This observation provides a base for consistency check, which is not absolute, but useful.

There are three groups of sucrose repressors in γ-proteobacteria. One group contains ScrR repressors from Enterobacteriaceae (*Salmonella typhimurium* plasmid, *Klebsiella pneumoniae*, *Erwinia amylowora*, *Yersinia enterocolitica*) and *Pasteurella multocida*. The recognition signals are slightly different: **tg**CTAAA**C**CG**G**TTTAG**CA** for the former and **AA**CTAAA**A**CG**T**TTTAG**TT** for the latter (differing positions are highlighted). The two other groups are CscR from a different set of Enterobacteriaceae (*E. coli*, *K. pneumoniae*) and *Pseudomonas syringae*, and ScrR from Vibrionaceae (*Vibrio cholerae* and *Vibrio alginolyticus*) that forms a common branch with trehalose repressors TreR from Enterobacteriaceae and Vibrionaceae (*Vibrio cholerae*) and sucrose repressors from several Gram-positive bacteria (*Bacillus halodurans*, *Bacillus cereus*, *Staphylococcus aureus*, *Staphylococcus xylosus*, *Clostridium difficile*, *Clostridium beijerinckii*). The branching order of the

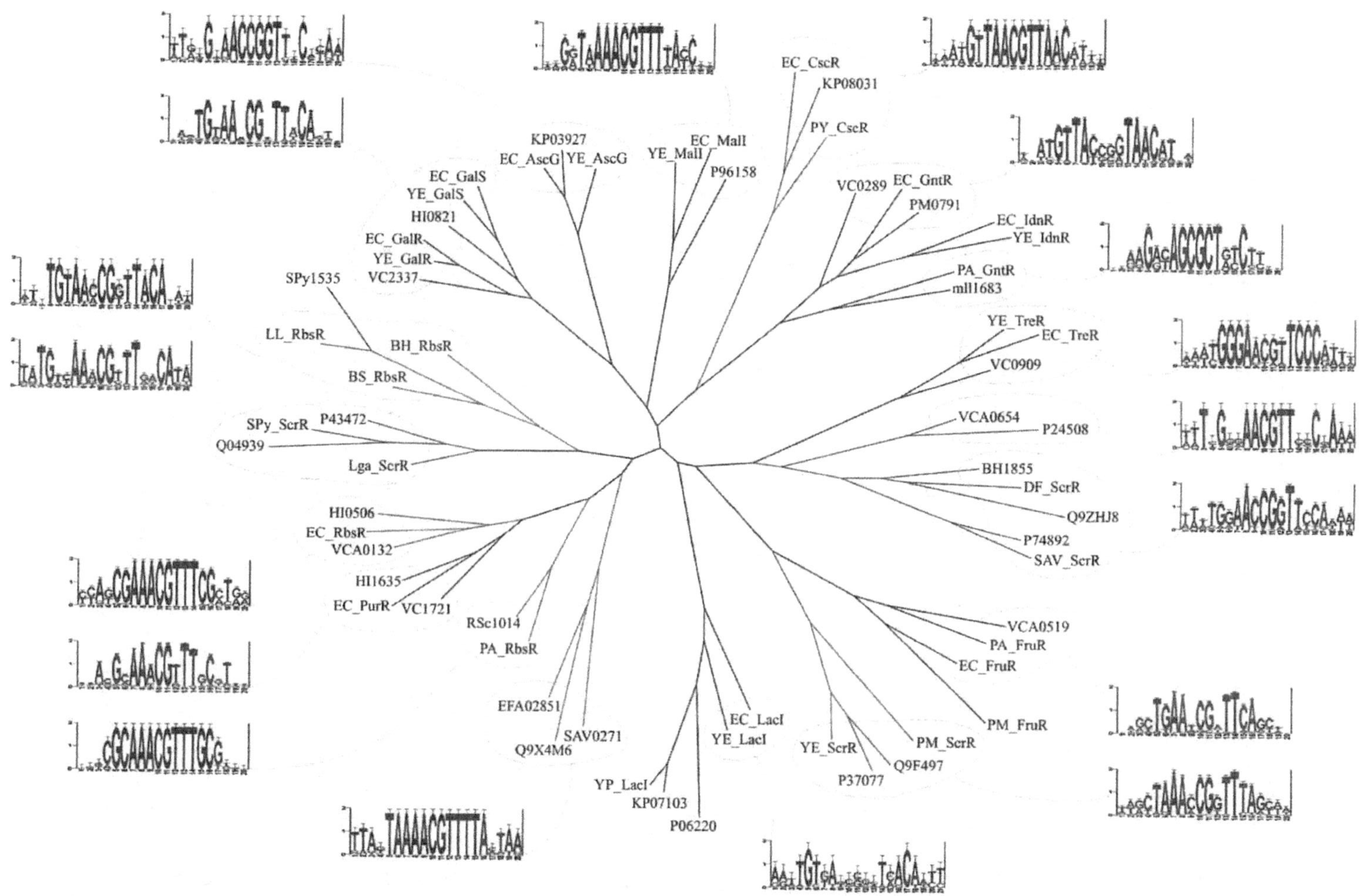

Figure 1. LacI family of transcriptional repressors. Cellular roles of repressors (regulated metabolic subsystems) are shown. Sequence logos are given for presumably homogeneous site sets.

latter three subgroups (ScrR from Vibrionaceae / TreR / sucrose regulators from the Bacillus/Clostridium group) is uncertain. Note that *K. pneumoniae* contains repressors from two groups at two different loci (*scrK-scrYABR* and *cscR-cscBArafY*, where R, K, Y denote repressors, a kinase, and porins respectively, *scrB* and *cscA* encode invertases specific for sucrose-6-P and sucrose respectively, and finally *scrA* and *cscB* encode transporters from the PTS and MFS familes respectively. The recognition signals are aaawG**TT**AACGTT**AA**Cwwww for CscR, wwwtG**GG**AACGTT**CC**C**ATT**T for TreR, and wwTTG**ca**AACGTT**CC**C**GAA**a for ScrR (Vibrionaceae). There are some common features in all three signals (stronger conservation of the right half-site), as for the consensus, the CscR signal seems to be an outgroup, since the TreR and ScrR (Vibrionaceae) signals look closer to each other.

The genes encoding the purine biosynthesis pathway in the core γ-proteobacteria are regulated by the purine repressor PurR whose recognition signal is AcGcAAA**CG**TTTgCgT with minor differences between Enterobacteriaceae, Pasteurellaceae and *V. cholerae* (D. Ravcheev, personal communication). A related repressor, RbsR, regulates the ribose utilization operon through binding to the site AGCGAAACGTTTCGCT; in Pasteurellaceae and *V. cholerae* the signal is ATCGAAACGTTTCGAT. Now, in four *Pseudomonas* spesies there is a repressor orthologous both to PurR and RbsR. The recognition signal of this repressor is closer to the PurR one: AAAACGTTTT. However, this repressor is co-localized with the ribose utilization genes and is likely to be the ribose repressor. A possible scenario is that following the duplication of the ancestral repressor in the branch leading to the core γ-proteobacteria, one repressor (PurR) changed the ligand-binding specificity, whereas the other (RbsR) changed the recognition signal. At that, note that the mode of action of PurR and RbsR is different: the former binds to its operator in the presence of the co-repressor (as there is no need to transcribe the purine synthesis genes in the presence of purines), whereas the latter is bound in the absence of the inducer (the ribose utilization genes are not transcribed when no ribose is available).

Similar observations can be made for other regulator families. For instance, the xylose regulators of Gram-positive bacteria from the Bacillus/Clostridium group belong to the ROK family (Rodionov *et al.*, 2001b). They form two large subfamilies (repressors from the *Bacillus* spp., *Staphylococcus xylosus*, *Enterococcus faecalis* and *Clostridium difficile*, and from *Lactobacillus pentosus* and *Lactobacillus brevis*) and one separate branch formed by the regulator from *Clostridium acetobutylicum*. The recognition signals of the first two groups are GTTwGTTtatnnnataAACwAAC and

GTTgGTTGnnnnnnnCAACcAAC respectively, similar to the consensus of experimentally determined sites of *Bacillus subtilis*. The signal of *C. acetobutylicum* is ACTTTTTAAAnnnnnTTTAAAAAGT, reminiscent of the NagC signal STTATTTnnnnnnnnnAAATAAS of *E. coli*. In addition, the xylose metabolism in *Lactococcus lactis* is regulated by an activator from the AraC family.

4. Creation and Evolution of Regulons

The process of creation of new regulatory interactions is difficult to observe, as most existing bacteria are highly adapted to their ecological niches. However, an elegant review (de Lorenzo and Pérez-Martín, 1996) describes currently evolving systems of xenobiotic catabolism in *Pseudomonas* species. Metabolic systems from different strains catalyzing the same compounds recruit transcription factors from different families. At that, a newly formed operon is brought under control of a pre-existing factor regulating a related metabolic pathway. At a later stage, the binding sites are fine-tuned to preclude cross-regulation by factors from one family, the specificity of regulators to effectors increases, and new regulons are formed each responding to a specific compond. Various stages of this process are observed in the phenol, naphthalene, and octane catabolism pathways.

New factors often arise by duplications. Then either one factor recognizes a subset of sites and thus regulates a sub-pathway, or some sort of a regulatory cascade is formed. A striking example is that of GalS and GalR regulators of *E. coli*. They bind the same set of operators upstream of *mgl*, *gal* and *galP* operons. At that, the affinity of the operators to GalS decreases in that order, as it increases for GalR (Geanacopoulos and Adhya, 1997). In addition, GalS regulates its own gene, whereas transcription of GalR is thought to be constitutive. Finally, the binding of GalS to DNA is an order of magnitude more sensitive to galactose than that of GalR. The duplication of the regulators seems to be specific to Enterobacteriaceae, as only one regulator exists in Pasteurellaceae and Vibrionaceae. However, GalS has been lost by *Yersinia pestis*, although both regulators can be found in closely related *Yersinia enterocolitica*.

Similarly, UxuR and ExuR repressors of *E. coli* are 49% identical and regulate related metabolic pathways of glucuronate and galacturonate utilization respectively; ExuR also regulates the gene *exuT* encoding transporter of both sugar acids; the first step of both pathways, isomerization to tagaturonate and fructuronate respectively, is catalized by UxaC whose gene also is regulated by ExuR. At the first glance it looks likely that the galacturonate pathway and its regulator is the primary one. However, in *Salmonella typhi*, only the glucuronate pathway and its regulator UxuR are present; only UxuR is also present in *Haemophilus influenzae*. The binding sites for the two regulators are indistinguishable computationally, and more data, both experimental and genomic, are needed to reconstruct the evolution of these pathways and their regulation (Rodionov *et al.*, 2000).

The L-rhamnose utilization system of *E. coli* is encoded by two operons, one of which is responsible for transport, and another one for metabolism. Both are regulated by an AraC-family activator RhaS. The latter forms an operon with a homologous activator RhaR, and this operon, *rhaSR* is activated by RhaR and CRP (Holcroft and Egan, 2000). Thus there arises a controlled positive feedback loop. This arrangement is conserved in Enterobacteriaceae. The rhamnose loci of *Listeria monocytogenes* and *Listeria innocua* contain AraC-family regulators distant from RhaS/RhaR, whereas the homologous loci of *B. subtilis* and *Bacillus halodurans* contain regulators from the DeoR family, however in the latter case it is not obvious that these loci are concerned in the rhamnose utilization, as they lack the gene for the essential enzyme RhaD. A more interesting situation is that of *Enterococcus faecium* and *Enterococcus faecalis* where the rhamnose loci contain two AraC-family regulators in a different arangement: >*R2*><*R1*<> *rhaBADyiiL*> and <*R1*<> *rhaBADyiiL*>>*R2*> respectively (angle brackets show the direction of transcription).

One more mode of the regulon evolution is expansion. The most striking example is that of the FruR regulon in enteric bacteria. It is a global regulator of sugar metabolism in *E. coli* where it represses at least sixteen operons (Ramseier *et al.*, 1995). However, in related groups of genomes, Vibrionaceae and Pseudomonadaceae, it behaves as a garden variety sugar transcriptional regulator from the LacI family, regulating only itself and the fructose catabolism operon (transcribed divergently from the *fruR* gene) and, in pseudomonads, one more operon containing a porin gene. The situation in the fourth well-represented group of gamma-proteobacteria, Pasteurellaceae, is obscure: there are no *fru* genes in *Haemophilus ducreyi*, no *fruR* with preserved *fru* operon in *Haemophilus influenzae*, and both *fruR* and structural *fru* genes, but in separate loci with problematic FruR candidate sites, in *Pasteurella multocida* and *Actinobacillus actinomycetemcomitans*. An intermediate situation is observed in *Yersinia pestis*, a closer relative of *E. coli* belonging to the same family Enterobacteriaceae. In *Y. pestis*, FruR is a global regulator, but some operons that are under the FruR control in *E. coli* have no upstream FruR binding sites. Of course, an alternative explanation could be that the state of the global regulon is the ancestral one, and it has been retained in *E. coli* and, partially, *Y. pestis*, and degraded in other genomes. However, the "globalization" of the FruR regulon looks more likely, as it is confirmed by comparison with other LacI-related repressors of sugar metabolism operons (see above).

A very similar case is that of the Lrp regulon (Friedberg *et al.*, 2001). In *E. coli*, it is a global regulon, as Lrp, present in a concentration of 3,000 dimers per cell, affects expression of dozens of operons. The orthologous factor of *H. influenzae*, LrfB, is present in a concentration of 130 dimers per cell, and seems to regulate only few genes.

Another example of more or less recent global regulons seems to be provided by the catabolite repression regulons of Gram-positive and Gram-negative bacteria. Indeed, both systems are functionally very close, but the exact mode of regulation is quite different. At that, the catabolite regulator of gamma-proteobacteria, CRP, has been recruited from the FNR family, whose other members regulate the switches between aerobic and anaerobic metabolism (Sawers *et al.*, 1997). The regulator of the Bacillus/Clostridium group, CcpA, is a member of the LacI family and thus this case resembles the FruR one from the previous paragraph (Miwa *et al.*, 2000). Thus it is likely that these regulons emerged after divergence of the ancestors of the Bacillus/Clostridium group and proteobacteria, and the global regulation formed when local regulators of sugar metabolism were already in place. As the result, the sugar catabolism operons fell under dual regulation: the dependence of their expression on the absence of the preferred source, glucose, is mediated by CRP or CcpA, and the dependence on availability of the respective sugar compound, by the local regulator. Thus the overall functional or, rather, metabolic similarity of the catabolite repression regulatory systems is likely due to convergence (D. Rodionov, personal communication). In this respect it is also noteworthy that CRP regulates many regulator genes in *E. coli* forming regulatory cascades (Thieffry *et al.*, 1998).

We have already mentioned the opposite process, that of loss of regulatory interactions. It may happen on the level of individual sites or entire regulons. One example of a clearly degrading site is the TrpR-binding site upstream of the *aroH* gene in *Yersinia pestis* and *Erwinia herbicola*, Figure 2 (Panina *et al.*, 2001).

```
EC gggccttagtcgccgaATGTACTAGAGAACTAGTGCATtagcttatttttttgttatcatg
ST ggggcttaaccgcaggATGTACTAGTAAACTAGTTTAAtggattggtttt-gttatcatg
   ::: ::::  ::: : :::::::::  :::::::  ::: : ::  ::::: :::::::::
   gGG cT A tCGccG ATGTacTa   aa tA t    t g t   T ttt     t aTg
    ::  ::: ::: :: ::::  : :      :              ::   ::::    :
YP agggatgagtcgtcggATGTTTTAACTAAATATTTTCAtgagtgattatttgcggtttata
EH gggcctgactcgccgcATGTACTGATGGGTAACCGGCGctgaactgtaaa-gcgg-cgctg
```

Figure 2. Evolution of the TrpR-binding site upstream of the *aroH* gene in four genomes of Enterobacteriaceae. EC: *Escherichia coli*, ST: *Salmonella typhimurium*, YP: *Yersinia pestis*; EH: *Erwinia herbicola*. The first two lines: capitals – TrpR-binding sites; the last two lines: capitals – degraded TrpR-binding sites. Boldface – positions, conforming to the consensus. Middle line: capitals – positions conserved in all four sequences, lower case – positions, conserved in three sequences. Colons – positions conserved between EC and ST or between YP and EH.

A converse case is retention of a site for an absent regulator. Probably it indicates a very recent loss of the regulator, like in the case of *rbsDK* operon of *Yersinia pestis*. The *rbsDK* upstream region is conserved between *Y. pestis* and other enteric bacteria (where the operon is *rbsDACBK*) despite the absence of the regulator gene *rbsR*. Notably, the conserved region covers the RbsR-binding site (Figure 3).

A similar example is provided by the Nac regulator of several nitrogen-assimilation operons in enterobacteria (Reitzer and Schneider, 2001). The histidase operon *hut* is activated by Nac in *Klebsiella aerogenes*, but is not expressed in *Salmonella typhimurium*, as the latter lacks Nac. However, there is a potential Nac-binding site upstream of the *hut* operon of *S. typhimurium*,

```
EC  tcaccaTGTAAAACGTTTCGAGgttgatcacatttccgtaacgtcacgatggttttcccaact
ST  tccgctAGCGAAACGTTTCGACggcgatcacaattctgttacgtcatgatggttttatgaaca
TY  tccgctAGCGAAACGTTTCGACggcgatcacaattctgttacgtcatgatggttttatgaaca
KP  ctgcctCGCGAAACGTTTCGATggcgatcacatttctctcttctggtgatggttttct--gct
YP  atcgctAGCAAAGTGTTTCGGTggcgatcacaatttcactaaatgaggtttgtcttct--act
              **  ******  *  ******* **        *   * * ** **     *

EC  cagtcaggattaaactgtgggtcAGCGAAACGTTTCGCTgatggagaaaaa-------aatga
ST  catccggggttacactgcggc-cAGCGAAACGTTTCGCTagtggagcagaa------aaatga
TY  catccagggttacactgcggc-cAGCGAAACGTTTCGCTagtggagcagaa------aaatga
KP  cacacattgataataattattttAGCGAAACGTTTCGCTagtggagcaaagagaaaaaatga
YP  gccgtttttctaaactccttgttAGCGAAACGTTTCGCTcttggagtagat-------catga
              **           ****************  ***** * *         ****
```

Figure 3. Evolution of the RbsR-binding site upstream of the *rbsD* gene in five genomes of Enterobacteriaceae. EC: *Escherichia coli*, ST: *Salmonella typhimurium*, TY: *Salmonella typhi*, KP: *Klebsiella pneumoniae*, YP: *Yersinia pestis*. Capitals – RbsR-binding sites. Italics – CRP-binding sites. Boldface – positions, conforming to the consensus of RbsR or CRP. Underlined: start codon.

```
putP_EC  caaccgcaaaaaatgtgagagagagtgcaacctGATGAAAAATAGTGTcgctgagcact
putP_TY  caaccgcaaaaaatgtgagagagagtgcaacctGGAGAAAAATAGTATccctctgcaat
putP_KP  caacctttaaaaacgtgagagagagtgcaacctCTAACAAATTGACGTttgcgccgtc
         *****   ***** *****************     *** *    *       *

hutU_KP  ggcgtcagcaaaacgtgattcgtgacgcaatATAACAAAATTGTATcatttctgtta
hutU_TY  tgctgccggaaaacgtgattgctgacacaatATAACAAAATTGTATcattttgtta
          **  * * ***********  **** ************************ *****
```

Figure 4. Evolution of the Nac-binding sites upstream of *hut* and *put* genes in three genomes of Enterobacteriaceae. EC: *Escherichia coli*, TY: *Salmonella typhi*, KP: *Klebsiella pneumoniae*. Capitals – Nac-binding sites; italics – CRP-binding sites; boldface – positions, conforming to the CRP consensus.

and the operon can be activated by Nac from *K. aerogenes* (Best and Bender, 1990). Similarly, there are conserved Nac-binding sites upstream of the proline transporter gene *putP* in *E. coli* and *K. pneumoniae* (having Nac) and *S. typhi* (not having Nac), Figure 4. In all these cases the Nac sites are adjacent to conserved CRP-binding sites.

An example of recent loss of a global regulator is the σ^{54} sigma-factor of *Neisseria* spp. (Buck *et al.*, 2000). These genomes still contain *rpoN* pseudogenes, and some genes retain the σ^{54}-specific promoters (Laskos *et al.*, 1998). It might be interesting to do a systematic inventory of the σ^{54} regulatory systems in β-proteobacteria, aligning promoters of all orthologs of σ^{54}-regulated genes.

5. Horizontal Transfer and Ubiquitous Regulons

Analysis of complete genomes established the horizontal transfer as a major aspect of bacterial evolution (Paul, 1999; Koonin *et al*, 2001). The genes acquired by horizontal transfer may either bring the regulator with them, or become members of already existing regulons.

The first case is probably the simplest, and it has been observed many times, *e.g.* in the drug resistance systems traveling on mobile elements, transposons or plasmids (*e.g.* Arthur and Courvalin, 1993; Quintiliani and Courvalin, 1994; Nwosu, 2001; Carattoli, 2001). The structural operon and the regulator often form a divergently transcribed cassette and may integrate into the host chromosome, forming a self-contained locus. The same applies also to many metabolic systems. A nice example is provided by the urease genes of Enterobacteriaceae that are located in plasmids in *Providencia stuartii*, *E.*

coli and *Salmonella* spp., and in the chromosome of *Proteus mirabilis* (Thomas and Collins, 1999) and *Vibrio parahaemoliticus* (Park *et al.*, 2000); these genes are regulated by UreR, an activator from the AraC family. Noteworthy, in a related bacterium, *P. vulgaris*, there is no UreR, and the urease operon is regulated by a σ^{54} promoter (Moersdorf and Kaltwasser, 1990).

Indeed, the divergent arrangement of the regulator and structural operons may be due not only to functional requirements, when binding of the regulator to the same site in the intergenic region activates the structural promoter and represses the regulator gene itself (Gralla and Collado-Vides, 1996), but may be evidence of the history of lateral transfers. The latter explanation is even more likely for the loci where the regulator gene and the structural operon are in convergent or collinear order as, *e.g.*, in the case of restriction-modification (RM) cassettes where all three possible relative arrangements of the RM-operon and the control gene encoding a transcription factor have been observed (Bart *et al.*, 1999).

An example of a system that is likely horizontally transferred between Gram-positive and Gram-negative bacteria, although it is not clear in what direction, is the experimentally uncharacterized locus of *E. coli*, *Listeria monocytogenes*, *Listeria innocua* and *Enterococcus faecium* consisting of *ycjMb1310ycjOPQb1314ycjSTUV* (in the *E. coli* nomenclature) and probably regulated by the *ycjW* gene product. Indeed, not only the regulators are very close (50-60% identity), but the binding sites are similar (Figure 5). The composition of the locus suggests that its function might be utilization of an oligosaccharide: b1310, YcjO, YcjP, and YcjV are the four components of an ABC transporter (the binding protein, two transmembrane proteins and the ATPase respectively), YcjM and YcjT are similar to oligosaccharide

```
EFA02102  atgct--aTTGACAgtaagcgttttccataaTATGAT-caaatcgtTGGGAACGCTCCCA
lin2973   gtg-t---TTGACAgaaagcgctttccatacTATAAT-tcttttgTGGGAGCGCTCCCG
EC_ycjM   gtgctcaaTTGACAgtttattttctgcggagTAGTCTctcgtttcaTGGGACCGCTACCA
           ** *   *******       * * *  * **   *     *   ***** **** **

          aaaaagaa---aggggaatccgatgtatacaaa
          gttaggagg-taaatgatgacgaatttaagaaa
          c-ggaaaggcaacatgaaacagaaaattacgga
                *    *   **    **     *   *
```

Figure 5. Transcription factor binding sites upstream of a candidate oligosaccharide utilization operon. EC: *Escherichia coli*, lin: *Listeria innocua*, EFA: *Enterococcus faecium*. Capital italics – candidate promoter –35 and –10 boxes. Capitals – candidate binding site of YcjW. Boldface – positions conforming to the palindromic consensus of the YcjW-box. Underlined – putative start codons.

phosphorylases, YcjU is a putative phosphatase or phosphomutase, and YcjQ and YcjS are putative oxidoreductases.

The structure of the loci is slightly different in *E. coli* and Gram-positive bacteria:

`[ ] >ycjMb1310ycjOPQb1314ycjSTUV>> ompG><ycjW<`

and

`<ycjW<[ ] >ycjMb1310ycjOPQb1314ycjSTlin2965ycjU> // ycjV`

here the angle brackets denote the direction of transcription, square brackets [ ] denote the binding sites, and // shows that in *Listeria* spp. and *E. faecium* the ATPase gene *ycjV* is located separately. Location of the regulator *ycjW* gene relative to the structural operon is different: they are transcribed divergently in *Listeria* spp., and *E. faecium*, whereas in *E. coli* they are separated by a porin gene and transcribed convergently. There is also an additional gene in the *Listeria* spp. and *E. faecium* genomes (*lin2965* in *L. innocua*), encoding a protein similar to glycerate kinases.

Sometimes the horizontal transfer is evident, but the binding site is no longer recognizable. For instance, the *trpBA* operon of *Chlamidia trachomatis* is clearly regulated by TrpR whose gene is located immediately upstream. No other *Chlamidia* genome contains orthologs of any of these genes, and the only other group of genomes containing homologs of TrpR is gamma-proteobacteria. However, attempts to identify a candidate TrpR-binding site upstream of *trpBA* or *trpR* operons using the gamma-proteobacterial recognition rule failed (Panina *et al.*, 2001).

A different situation arises when a gene falls into an already existing regulon. The simplest case is that when the transferred gene integrates into a regulated operon. A more interesting situation is when a gene or an operon has a binding site for an ubiquitous regulator that has the same binding signal in many bacterial genomes. One example is provided by the *umuC/uvrX* family of genes regulated by homologous factors, LexA in Enterobacteriaceae and DinR in the *Bacillus/Clostridium* group. These factors have diverged DNA-binding domains and their recognition signals are different, but conserved within the respective taxonomic groups. The *umuDC* operons are often encountered on plasmids and always preceded by candidate LexA binding sites (SOS-boxes); similarly, *uvrX* belongs to prophage regions in *Bacillus subtilis* and is flanked by transposon genes in *Enterococcus faecalis*, and in all such cases it is preceded by candidate DinR binding sites (Cheo-boxes) (Permina *et al.*, 2000).

A more universal factor is the iron response regulator FUR, whose binding signal is conserved between Gram-positive and Gram-negative bacteria, although some differences can be observed (E.M. Panina, personal communication). The iron utilization operons form large loci that are different even in closely related genomes and similar in distant relatives, and thus likely have been frequently transferred laterally.

6. Evolution of Interacting Regulons

Although the bacterial gene regulation is nowhere as complicated as the eukaryotic one, still there are numerous examples of interacting regulatory systems. Sometimes several factors regulate one functional system, and the mode of regulation of individual genes differs in related genomes.

Thus, the aromatic amino acid metabolism in the *Bacillus/Clostridium* group is regulated by two RNA-based systems. The TRAP protein senses the presence of excess tryptophan and inhibits the expression of the tryptophan operon by a termination-antitermination mechanism (Babitzke and Gollnick, 2001). T-boxes sense lack of amino acids via the presence of uncharged tRNAs and prevent formation of terminators, thus switching on the expression of the aminoacyl-tRNA synthetase and biosynthetic operons (Chopin *et al.*, 1998; Pelchat and Lapointe, 1999). Some transporter operons are also regulated using these mechanisms (A.G.Vitreschak, personal communication). In addition, there seem to be at least two transcription-based systems (E.M. Panina, personal communication). What is interesting, the aromatic amino acid metabolism operons are shuffled between TRAP, tryptophan-specific and tyrosine-specific T-boxes, and transcription regulators, and the exact mode of regulation of any particular gene can involve multiple different mechanisms even in closely related genomes (Panina *et al.*, in preparation).

Another example is provided by the heat shock regulation. It involves one ubiquitous regulator (HrcA) and several genome-specific ones, in particular, the heat-shock sigma-factor σ^{32} in beta- and gamma-proteobacteria and CtsR in bacilli. The binding signal of HrcA, called CIRCE, is an inverted repeat of 9 nucleotides with a 9 base pair spacer `TTAGCACT-N9-GAGTGCTAA`. It is one of the most conserved signals both within genomes, with at most 2 mismatches with the consensus at each half-site (Segal and Ron, 1996), and between genomes, as it is the same in such diverse taxonomic groups as Firmicutes, Proteobacteria, Cyanobacteria, Chlamidiae and Spirochaetes (Segal and Ron, 1006; Gelfand, 1999). In most genomes, HrcA regulates only the chaperone genes, and other heat-shock genes are regulated by other factors. The simplest situation is that of mycoplasmas, where HrcA is the only transcription factor with identifiable function; there it regulates not only chaperone genes, but also the genes of heat-shock proteases such as *lon* and *clpB* (Gelfand, 1999). In *Clostridium acetobutylicum*, HrcA regulates itself, as there is a CIRCE element upstream of the operon *hrcA-grpE-dnaKJ* (Narberhaus *et al.*, 1992). Similarly, the *rpoH* gene, encoding the heat-shock sigma-factor σ^{32}, is autoregulated in *Caulobacter crescentus* (Reisenauer *et al.*, 1996) and regulated by another sigma factor, σ^{28}, in *E. coli* (Gross, 1996). The most interesting situation is that of beta-proteobacteria such as *Bordetella pertussis*, where there are candidate CIRCE elements upstream of *rpoH*, and candidate heat-shock promoters upstream of *hrcA,* yielding a regulatory loop (E.A. Permina, personal communication).

An example of a transcription factor participating in several different systems is the arginine repressor ArgR/AhrC. This factor regulates the genes for arginine metabolism in several bacterial lineages, binding to palindromic sites that occur in pairs upstream of biosynthetic operons, and as single boxes upstream of catabolic and transport operons (Makarova *et al.*, 2001). In all these cases ArgR interacts with the main sigma-factor σ^{70}. In addition, in *S. typhimurium*, ArgR interacts with another sigma-factor σ^{54} at a promoter upstream of *astCADBE* (Lu and Abdelal, 1999). Finally, ArgR has been described as the XerC protein involved in recombination of the *E. coli* plasmid ColE1 (Stirling *et al.*, 1988), where it binds an ARG box within the recombination origin *cer* (Guhathakurta and Summers, 1995).

7. Conclusions

However scattered and incomplete, the observations listed above form some background for further generalization. We see that regulation is a constantly evolving aspect of bacterial genomes. New regulons arise from duplication of regulators with subsequent divergence of signals and re-distribution of regulated genes into emerging regulons, recruitment of regulators for new systems, and horizontal transfer. Many regulatory systems are redundant, which provides the cell with an opportunity for fine tuning of the gene expression levels and metabolic rates. On the other hand, many regulatory interactions seem to be relics of past states (optimized to previous envirnoments?) or random genomic events. Although core regulons tend to be conserved in related genomes, the details may be quite different. Another emerging pattern of the evolution of regulation seems to be re-distribution of genes and operons between regulatory systems.

Of course, as more data are available, these preliminary points will coalesce into something more fundamental and substantial. These data include both additional genomes and experimental information, including verification of predicted sites, analysis of cross-regulation, understanding of structural aspects of protein-DNA interactions. An important gap in theoretical models is the absence of techniques for analysis of evolution of (sets of) regulatory sites that would allow one to distinguish between random fluctuations and changes in DNA signals recognized by transcription factors.

And, of course, somewhere on the horizon is computational analysis of regulatory interactions in eukaryotes.

Acknowledgements

We are grateful to Vadim Brodyansky, Anna Gerasimova, Ekaterina Kotelnikova, Ekaterina Panina, Elizaveta Permina, Dmitry Ravcheev, Dmitry Rodionov and Alexei Vitreschak who shared with us many insightful observations, to Vsevolod Makeev, Andrei Mironov and Alexandra Rakhmaninova for useful discussions, and to Eugene Koonin, Kira Makarova and Leonid Mirny for collaboration. The studies cited here were partially supported by grants from the Howard Hughes Medical Institute (55000309), the Ludwig Institute of Cancer Research (CRDF RB0-1268), Russian Fund of Basic Research (00-15-99362), and INTAS (99-1476),

References

Arthur, M., and Courvalin, P. 1993. Genetics and mechanisms of glycopeptide resistance in enterococci. Antimicrob. Agents Chemoter. 37: 1563-1571.

Babitzke, P., and Gollnick, P. 2001. Posttranscription initiation control of tryptophan metabolism in Bacillus subtilis by the trp RNA-binding attenuation protein (TRAP), anti-TRAP, and RNA structure. J. Bacteriol. 183: 5795-5802.

Bart, A., Dankert, J., and van der Ende, A. 1999. Operator sequences for the regulatory proteins of restriction modification systems. Mol. Microbiol. 31: 1277-1278.

Best, E.A., and Bender, R.A. 1990. Cloning of the *Klebsiella aerogenes nac* gene, which encodes a factor required for nitrogen regulation of the histidine utilization (*hut*) operons in *Salmonella typhimurium*. J. Bacteriol. 192: 7043-7048.

Buck, M., Gallegos, M.-T., Studholme, D.J., Guo, Y., and Gralla, J.D. 2000. The bacterial enhancer-dependent σ^{54} (σ^{N}) transcription factor. J. Bacteriol. 182: 4129-4136.

Carattoli, A. 2001. Importance of integrons in the diffusion of resistance. Vet. Res. 32: 243-259.

Chopin, A., Biaudet, V., and Ehrlich, S.D. 1998. Analysis of the *Bacillus subtilis* genome sequence reveals nine new T-box leaders. Mol. Microbiol. 29: 662-664.

Dandekar, T., Snel, B., Huynen, M.A., and Bork, P. 1998. Conservation of gene order: a fingerprint of physically interacting proteins. Trends Biochem. Sci. 23: 324-328.

Dubchak, I., Brudno, M., Loots, G.G., Pachter, L., Mayor, C., and Rubin, E.M. 2000. Active conservation of noncoding sequences revealed by three-species comparisons. Genome Res. 10: 1304-1306.

Duret, L., and Bucher, P. 1997. Searching for regulatory elements in human noncoding sequences. Curr. Opin. Struct. Biol. 7: 399-406.

Ermolaeva, M.D., White, O., and Salzberg, S.L. 2001. Prediction of operons in microbial genomes. Nucleic Acids Res. 29: 1216-1221.

Florea, L., Riemer, C., Schwartz, S., Zhang, Z., Stojanovich, N., Miller, W., and McClelland, M. 2000. Web-based visualization tools for bacterial genome alignments. Nucleic Acids Res.28: 3486-3496.

Friedberg, D., Midkiff, M., and Calve, J.M. 2001. Global versus local regulatory roles for Lrp-related proteins: *Haemophilus influenzae* as a case study. J. Bacteriol. 183: 4004-4011.

Geanacopoulos, M., and Adhya, S. 1997. Functional characterization of roles of GalR and GalS as regulators of the *gal* regulon. J. Bacteriol. 179: 228-234.

Gelfand, M.S. 1999. Recognition of regulatory sites by genomic comparison. Res. Microbiol. 150: 755-771.

Gelfand, M.S., Koonin, E.V., and Mironov, A.A. 2000a. Prediction of transcription regulatory sites in Archaea by a comparative-genomic approach. Nucleic Acids Res. 28: 695-705.

Gelfand, M.S., Novichkov, P.S., Novichkova, E.S., and Mironov, A.A. 2000b. Comparative analysis of regulatory patterns in bacterial genomes, Briefings in Bioinformatics. 1: 357-371.

Gralla J.D., and Collado-Vides, J. 1996. Organization and function of transcription regulatory elements. *Escherichia coli* and *Salmonella*. Cellular and Molecular Biology. Neidhardt, F.C., ed. ASM Press, Washington, DC. 1232-1245.

Gross, C.A. 1996. Function and regulation of heat-shock proteins. *Escherichia coli* and *Salmonella*. Cellular and Molecular Biology. Neidhardt, F.C., ed. ASM Press, Washington, DC. 1382-1399.

Guhathakurta, A., and Summers, D. 1995. Involvement of ArgR and PepA in the pairing of ColE1 dimer resolution sites. Microbiology. 141: 1163-1171.

Gumucio, D.L., Shelton, D.A., Zhu, W., Millinoff, D., Gray, T., Bock, J.H., Slightom, J.L., and Goodman, M. 1996. Evolutionary strategies for the elucidation of *cis* and *trans* factors that require the developmental switching programs of the β-like globin genes. Mol. Phylogenet. Evol. 5: 18-32.

Hardison, R.C., Oeltjen, J., and Miller, W. 1997. Long human-mouse sequence alignments reveal novel regulatory elements: A reason to sequence the mouse genome. Genome Res. 7: 759-766.

Holcroft, C.C., and Egan S.M. 2000. Interdependence of activation at *rhaSR* by cyclic AMP receptor protein, the RNA polymerase alpha subunit C-terminal domain, and RhaR. J. Bacteriol. 182: 6774-6782.

Koonin, E.V., Makarova, K.S., and Aravind L. 2001. Horizontal gene transfer in prokaryotes: quantification and classification. Annu. Rev. Microbiol. 55: 709-742.

Laskos, L., Dillard, J.P., Seifert, H.S., Fyfe, J.A.M., and Davies, J.K. 1998. The pathogenic neisseriae contain an inactive *rpoN* gene and do not utilize the *pilE* σ^{54} promoter. Gene. 208: 95-102.

Lichtarge, O., Yamamoto, K.R., and Cohen, F.E. 1997. Identification of functional surfaces of the zinc binding domains of intracellular receptors. J. Mol. Biol. 274: 325-337.

de Lorenzo, V., and Pérez-Martín, J. 1996. Regulatory noise in prokaryotic promoters: how bacteria learn to respond to novel environmental signals. Mol. Microbiol. 19: 1177-1184.

Lu, C.D., and Abdelal, A.T. 1999. Role of ArgR in activation of the *ast* operon, encoding enzymes of the arginine succinyltransferase pathway in *Salmonella typhimurium*. J. Bacteriol. 181: 1934-1938.

Makarova, K.S., Mironov, A.A., and Gelfand, M.S. 2001. Conservation of the arginine repressor DNA binding signal in all bacterial lineages. Genome Biology. 2: research0013.1-0013.8.

Mirny, L.A., and Gelfand, M.S. 2001a. Structural analysis of conserved base-pairs in protein-DNA complexes. Nucleic Acids Res. 30: 1704-1711.

Mirny, L.A., and Gelfand, M.S. 2001b. Using orthologous and paralogous proteins to identify specificity-determining residues in bacterial transcription factors. J. Mol. Biol. 321: 7-20.

Mironov, A.A., Koonin, E.V., Roytberg, M.A., and Gelfand, M.S. 1999. Computer analysis of transcription regulatory patterns in completely sequenced bacterial genomes. Nucleic Acids Res. 27: 2981-2989.

Miwa, Y., Nakata, A., Ogiwara, A., Yamamoto, M., Fujita, Y. 2000. Evaluation and characterization of catabolite-responsive elements (cre) of Bacillus subtilis. Nucleic Acids Res. 28: 1206-1210.

McCue, L.A., Thompson, W., Carmack, C.S., Ryan, M.P., Lie, J.S., Derbyshire, V., and Lawrence, C.E. 2001. Phylogenetic footprinting of transcription factor binding sites in proteobacterial genomes. Nucleic Acids Res. 29: 774-782.

McGuire, A.M., and Church, G.M. 2000. Predicting regulons and their cis-regulatory motifs by comparative genomics. Nucleic Acids Res. 28: 4523-4530.

McGuire, A.M., Hughes, J.D., and Church, G.M. 2000. Conservation of DNA regulatory motifs and discovery of new motifs in microbial genomes. Genome Res. 10: 744-757.

Moersdorf, G., and Kaltwasser, H. 1990. Cloning of genes encoding urease from *Proteus vulgaris* and sequencing of the structural genes. FEMS Microbiol. Lett. 66: 67-74.

Narberhaus F., Giebeler, K., and Bahl, H. 1992. Molecular characterization of the *dnaK* gene region of *Clostridium acetobutylicum*, including *grpE*, *dnaJ*, and a new heat shock gene. J Bacteriol. 174: 3290-3299.

Nwosu, V.C. 2001. Antibiotic resistance with particular reference to soil microorganisms. Res. Microbiol. 152: 421-430.

Otsuka, J., Watanabe, H., and Mori K.T. 1996. Evolution of transcriptional regulation system through promiscuous coupling of regulatory proteins with operons; suggestion from protein sequence similarities in *Escherichia coli*. J. Theor. Biol. 178: 183-204.

Overbeek, R., Fonstein, M., D'Souza, M., Pusch, G.D., and Maltsev, N. 1999. The use of gene clustering to infer functional coupling. Proc. Natl. Acad. Sci. USA 96: 2896-2901.

Panina, E.M., Vitreschak, A.G., Mironov, A.A., and Gelfand, M.S. 2001. Regulation of aromatic amino acid biosynthesis in gamma-proteobacteria. J. Mol. Microbiol. Biotechnol. 3: 529-543.

Park,K.S., Iida,T., Yamaichi,Y., Oyagi,T., Yamamoto,K., and Honda,T. 2000. Genetic characterization of DNA region containing the *trh* and *ure* genes of *Vibrio parahaemolyticus*. Infect. Immun. 68: 5742-5748.

Paul, J.H. 1999. Microbial gene transfer: An ecological perspective. J. Molec. Microbiol. Technol. 1: 45-50.

Pelchat, M., and Lapointe, J. 1999. Aminoacyl-tRNA synthetase genes of *Bacillus subtilis*: organization and regulation. Biochem. Cell. Biol. 77: 343-347.

Permina, E.A., Mironov, A.A., and Gelfand, M.S. 2000. Comparative approach to analysis of regulation in complete genomes: SOS repair in *Bacillus subtilis*. Proc. 2nd Int. Conf. on Bioinformatics of Genome Regulation and Structure BGRS'2000. 2: 22-23.

Quintiliani R. Jr., and Courvalin, P. 1994. Conjugal transfer of the vancomycin resistance determinant *vanB* between enterococci is associated with the movement of large genetic elements from chromosome to chromosome. FEMS Microbiol. Lett. 119: 359-364.

Reisenauer, A., Quon, K., and Shapiro, L. 1996. Regulation of a heat shock sigma32 homolog in *Caulobacter crescentus*. J Bacteriol. 178: 1919-1927.

Ramseier, T.M., Bledig, S., Michotey, V., Feghali, R., and Saier, M.H. Jr. 1995. The global regulatory protein FruR modulates the direction of carbon flow in *Escherichia coli*. Mol. Microbiol. 16: 1157-1169.

Reitzer, L., and Schneider, B.L. 2001. Metabolic context and possible physiological themes of σ^{54}-Dependent genes in *Escherichia coli*. Microbiol. Mol. Biol. Rev. 65: 422-444.

Rodionov, D.A., Mironov, A.A., Rakhmaninova, A.B., and Gelfand, M.S. 2000. Transcriptional regulation of transport and utilization systems for hexuronides, hexuronates and hexonates in gamma purple bacteria. Mol. Microbiol. 38: 673-683.

Rodionov, D.A., Gelfand, M.S., Mironov, A.A., and Rakhmaninova, A.B. 2001a. Comparative approach to analysis of regulation in complete genomes: multidrug resistance systems in gamma-proteobacteria. J. Mol. Microbiol. Biotechnol. 3: 319-324.

Wolf, Y.I., Rogozin, I.B., Kondrashov, A.S., and Koonin E.V. 2001. Genome alignment, evolution of prokaryotic genome organization, and prediction of gene function using genomic context. Genome Res. 11: 356-372.

Salgado, H., Moreno-Hagelsieb., G., Smith, T.F., and Collado-Vides, H. 2000. Operons in *Escheriachia coli*: gneomics analyses and predictions. Proc. Natl. Acad. Sci. 97: 6652-6657.

Sawers, G., Kaiser, M., Sirko, A., and Freundlich, M. 1997. Transcriptional activation by FNR and CRP: reciprocity of binding-site recognition. Mol. Microbiol. 23: 835-845.

Segal, R., and Ron, E.Z. 1996. Regulation and organization of the groE and dnaK operons in Eubacteria. FEMS Microbiol. Lett.138: 1-10.

Stirling, C.J., Szatmari, G., Stewart, G., Smith, M.C., and Sherratt, D.J. 1988. The arginine repressor is essential for plasmid-stabilizing site-specific recombination at the ColE1 *cer* locus. EMBO J. 7: 4389-4395.

Stojanovic, N., Florea, L., Riemer, C., Gumucio, D., Slightom, J., Goodman, M., Miller, W., and Hardison, R. 1999. Comparison of five methods for finding conserved sequences in multiple alignments of gene regulatory regions. Nucleic Acids Res. 27: 3899-3910.

Tagle, D.A., Koop, B.F., Goodman, M., Slightom, J.L., Hess, D.L., and Jones, R.T. 1988. Embryionic ε and γ globin genes of a prosimian primate (*Galago crassicaudatus*) nucleotide and amino acid sequences, developmental regulation and phylogenetic footprints. J. Mol. Biol. 203: 439-455.

Tan, K., Moreno-Hagelsieb, G., Collado-Vides, J., and Stormo, G.D. 2001. A comparative genomics approach to prediction of new members of regulons. Genome Res. 11: 566-584.

Terai, G., Takagi, T., and Nakai, K. 2001. Prediction of co-regulated genes in *Bacillus subtilis* on the basis of upstream elements conserved across three closely related species. Genome Biology. 2: research0048.1-0048.12.

Thieffry, D., Huerta, A.M., Pérez-Rueda, E., and Collado-Vides, J. 1998. From specific gene regulation to genomic networks: a global analysis of transcriptional regulation in *Escherichia coli*. BioEssays. 20: 433-440.

Thomas, V.J., and Collins, C.M. 1999. Identification of UreR binding sites in the Enterobacteriaceae plasmid-encoded and *Proteus mirabilis* urease gene operons. Mol. Microbiol. 31: 1417-1428.

Wasserman, W.W., Palumbo, M., Thompson, W., Fickett, J.W., and Lawrence, C.E. 2000. Human-mouse genome comparisons to locate regulatory sites. Nat. Genet. 26: 225-228.

Yada, T., Nakao, M., Totoki, Y., and Nakai, K. 1999. Modeling and predicting transcriptional units of *Escherichia coli* genes using hidden Markov models. Bioinformatics. 15: 987-993.

From: *Frontiers in Computational Genomics*

9

Experimental RNomics

Alexander Hüttenhofer and Jürgen Brosius

Abstract

A major milestone of genome projects, identification of all genes, can only be achieved when genes encoding non-messenger RNAs including longer mRNA-like RNAs that lack open reading frames and small non-messenger RNAs (snmRNAs) are not ignored. RNAs play a variety of important roles in different compartments and functions of the cell. We review biomathematical identification of snmRNA candidates as well as experimental approaches in various organisms from bacteria to man. As Experimental RNomics is unbiased in identifying novel snmRNAs, it is suitable to identify even novel classes of RNAs, such as micro RNAs (miRNAs). Combined approaches promise a further plethora of additional genome encoded RNAs and their roles in gene regulation.

1. Introduction

Small non-messenger RNAs (snmRNAs) do not encode proteins but have various cellular functions ranging from regulation of replication, transcription, RNA processing or translation to metabolic roles (reviewed by Inouye and Delihas, 1988; Wassarman *et al.*, 1999) on their own or in complex with proteins that are bound to the RNA (see also Eddy, 2001; Erdmann *et al.*, 2001a; Erdmann *et al.*, 2001b; Mattick, 2001; Wassarman *et al.*, 1999). Thus, they are able to form complexes with proteins (ribonucleoprotein complexes,

RNPs) in cellular compartments as diverse as the nucleolus or dendritic processes of nerve cells. Despite early efforts to identify and characterize such snmRNAs (Hindley, 1967; Larsen *et al.*, 1967; Prestayko and Busch, 1968; Vesco and Penman, 1969; Weinberg *et al.*, 1967; Weinberg and Penman, 1968; Zieve and Penman, 1976), these efforts remained restricted to RNAs that were present as distinct bands when total RNA was separated on denaturing polyacrylamide gels and visualized by staining with *e.g.* ethidium bromide. As only few scientists in those days entertained ideas that non-messenger RNAs could exert important functions apart from templating, structurally supporting, or assembling the predominant protein macromolecules, even identified RNA molecules were largely ignored. Such negligence might explain the fact that it took decades between the first identification of, for example, 10Sa RNA in *E. coli* and discovery of its function in translation (Keiler *et al.*, 1996; Tu *et al.*, 1995). The function of one of the most prominent nuclear RNA bands in mammalian cell extracts, 7SK RNA (Zieve and Penman, 1976), only recently has been elucidated (Nguyen *et al.*, 2001; Yang *et al.*, 2001).

2. RNomics: Computational Versus Experimental Approaches

When in 1995 the first sequences of entire (bacterial) genomes became available (Fleischmann *et al.*, 1995; Fraser *et al.*, 1995), it became obviouus, that annotation and finding of novel protein coding regions could in general be achieved by more or less routine biomathematical approaches. In contrast, annotation - even of orthologues of known small non-messenger RNAs (snmRNAs) - was often lacking (Brosius, 1996). In the meantime, a few computational approaches have been undertaken to investigate the content of non-coding RNAs in the genomes of various model organisms (Olivas *et al.*, 1997; Rivas *et al.*, 2001). Most of these computational approaches rely on identification of conserved sequence motifs found in non-messenger RNAs. However, this severely restricts the identification of entirely novel snmRNA species exhibiting yet unknown sequence motifs. In addition, the search for promotor or termination signals to identify snmRNAs is hampered by the low sequence conservation of these motifs and/or their frequent occurrence as, for example, tRNA-derived retronuons in any parts of the genome (a nuon is any definable stretch of nucleic acid, a retronuon is one that had been reverse transcribed from any RNA and its cDNA reinserted into the genome, Brosius and Gould, 1992).

Alternatively, some computational approaches rely on folding predictions, assuming that non-coding RNAs might exhibit extensive secondary structures by intra-molecular base pairing (Carter *et al.*, 2001). This exludes, however, less structured or completely unstructured snmRNAs. In addition, most computational searches, when applied to archaeal and bacterial genomes almost exclusively focus on intergenic regions of genomes. Thereby, antisense snmRNAs to known ORFs, snmRNAs processed from the sense strand of ORFs, or snmRNAs which are located within a wrongly assigned ORF would remain undetected. According to our own unpublished data in the archaeon *Archaeoglobus fulgidus*, the latter scenario might, in fact, be grossly overlooked in these approaches (Tang *et al.*, 2002).

As for eukaryotic model organisms, computational approaches are even more difficult to perform. The search for snmRNAs within restricted intergenic regions, as successfully applied for example to Bacteria and Archaea, is severely impeded since most eukaryotic genomes consist almost entirely out of intergenic or interexonic regions with only a few genes interspersed. For example, the recently released first draft of the human genome (Lander *et al.*, 2001; Venter *et al.*, 2001) shows that about 75% of the genome is not transcribed at all. Other estimates (Mattick, 2001) indicate that as much as two thirds of the human genome is being transcribed and many of the transcripts serve a function as RNA only as a mere 1.5 % of the genome is being translated into proteins. Additionallly, as mentioned before, the existence of so called pseudogenes and retronuons greatly interferes with the identification of *bona fide* snmRNA genes:

Small RNAs frequently serve as templates for reverse transcriptase (Brosius, 1999). The resulting cDNAs integrate on a more or less random basis back into the genome, a process termed retoposition. As putative flanking promoter elements are missing, they represent mostly inactive copies. Tens, hundreds, up to thousands of copies of snmRNAs are being generated. Selected snmRNAs serve as extremely efficient RNA templates for retroposition and reach copy numbers between tens and hundreds of thousands. Such snmRNA pseudogenes are then termed SINEs (short interspersed repetitive elements). A most notable case is exemplified by up to one million copies of Alu elements in primate genomes. The plethora of snmRNA pseudogenes makes it difficult to identify potential snmRNA genes based on, for example, internal RNA polymerase III promoter elements. There are millions of such elements in the genome, but only a few are actively transcribed genes. Additional external promoter elements are ill defined, as of yet, and cannot be used to discriminate between *e.g.* a retronuon and a *bona fide* actively transcribed gene. On the other hand, as many snmRNAs potentially serve as templates for retronuons identification of even smaller retronuon families may ultimately lead to the identification of their founder RNA, often a functional snmRNA. Of course, experimental approaches are again necessary in order to discriminate the active snmRNA gene from the respective pseudogenes.

In order to fill the conceptual as well as the intergenic gaps, we decided to take an experimental approach (Hüttenhofer *et al.*, 2001) akin to the established EST approaches for finding and characterizing novel mRNAs and/or alternativel processed forms thereof (ERNSs = expressed RNA sequences). ESTs generated from poly(A)$^+$ mRNAs have proven to be indispensable tools for basic and applied research including identification of protein coding genes, functional analysis and correlation of gene variation with disease (Adams *et al.*, 1993; Wiemann *et al.*, 2001). In contrast, not a single experimental effort addresses the class of non-messenger RNAs. Due to the lack of open reading frames, it would be very difficult to identify all RNA genes on a given stretch of genomic sequence of any organism. An estimate of up to 1000 genes and their RNA products may thus remain undetected in a mammalian genome. The functions, interactions in cellular circuits and roles in diseases (see below) and pharmacogenomics of snmRNAs would thus remain unknown and our understanding of the functioning of a cell would be incomplete. In addition, these small RNAs can serve as ideal targets for drugs, such as antibiotics.

Developments in the discovery of snmRNAs in *Escherichia coli*, the best studied organism and still a preferred model organism due to its relatively small genome, are a an excellent example, how many genes could be missed in the genomes that are being sequenced and annotated: Meanwhile, via a combination of biomathematical predictions with subsequent experimental confirmation at least two dozen novel snmRNAs were discovered over the last few years (Argaman *et al.*, 2001; Wassarman *et al.*, 2001). Still, there might be no end in sight as predicted by a most recent biomathematical study (Rivas *et al.*, 2001) and our own experimental ERNSs approach in *E. coli,* where we identified several additional candidates for snmRNAs (V. Bartels *et al.*, unpublished). The biomathematical approach of identifying candidates for snmRNAs that so successfully had been applied to *E. coli* will also be useful if applied to other Bacteria, Archaea and Eukarya that contain simple genomes. As mentioned before, this strategy will be complicated in organisms whose genomes are rich in retronuons, especially SINEs (short interspersed repetitive elements) that are derived from snmRNAs (mainly tRNAs or SRP RNA). Their internal structural elements including promoter sequences will yield up to hundreds of thousands of false positives. A categorical elimination of retronuons is also problematic, as two of the most intriguing snmRNAs, BC1 RNA in rodents and BC200 RNP in anthropoid primates are retronuons derived from $tRNA^{Ala}$(DeChiara and Brosius, 1987; Martignetti and Brosius, 1993b) and a monomeric Alu transcript (Kuryshev *et al.*, 2001; Martignetti and Brosius, 1993a; Skryabin *et al.*, 1998), respectively. Yet these RNAs are expressed almost exclusively in neurons and transported into their dedritic processes and may have intriguing functions in the nervous system (Brosius, 1999; Brosius and Tiedge, 1995a; Brosius and Tiedge, 2001; Tiedge and Brosius, 1996; Tiedge *et al.*, 1993; Tiedge *et al.*, 1991).

Even experimental approaches such as our Experimental RNomics search for ERNSs which initially identifies candidates for snmRNAs, need to be further corroborated by additional experiments, including mapping on the genome, RNA blots etc. A major problem are ERNSs that are derived from degraded large RNAs. While ribosomal RNA sequences can easily be excluded, even when they slip through our initial screening, ERNSs generated from degraded mRNAs can lead to uncertainty, especially since annotation of genes in genomic sequences can be incomplete and/or false. Even if an ERNS maps to a portion of a bona fide $poly(A)^+$ mRNA, it could be that a stable processing product corresponds to a functional snmRNA. If a major class of snmRNAs, small nucleolar RNAs (snoRNAs), is being processed from intronic sequences of mRNA precursors (hnRNAs), we cannot rule out a corresponding mechanism that generates snmRNAs from 5' or 3' UTRs or even coding regions of mature mRNAs.

In the following, we will exclusively focus on identification of novel snmRNAs by the Experimental RNomics approach rather than computational RNomics. As mentioned above, computational approaches are biased towards motifs, structures or locations of snmRNA genes. Only few approaches have employed the Experimental RNomics method so far. In the following we will review Experimental RNomics searches in model organisms including mouse, human, fruit fly and nematode. Excellent reviews on identification and function of novel snmRNAs by computational or other approaches can be found in the recent literature (Eddy, 1999; Eddy, 2001; Erdmann *et al.*, 2001a; Erdmann *et al.*, 2001b; Mattick, 2001; Wassarman *et al.*, 1999).

3. Experimental RNomics

The Method

As compared to conventional EST approaches, the main difference to our strategy (Figure 1) is the source and treatment of RNA. First, we isolate small RNAs in the size range of 50-500 nucleotides that usually is not included (*i.e.* it will be not be present in poly(A)$^+$ mRNA). In particular, the small

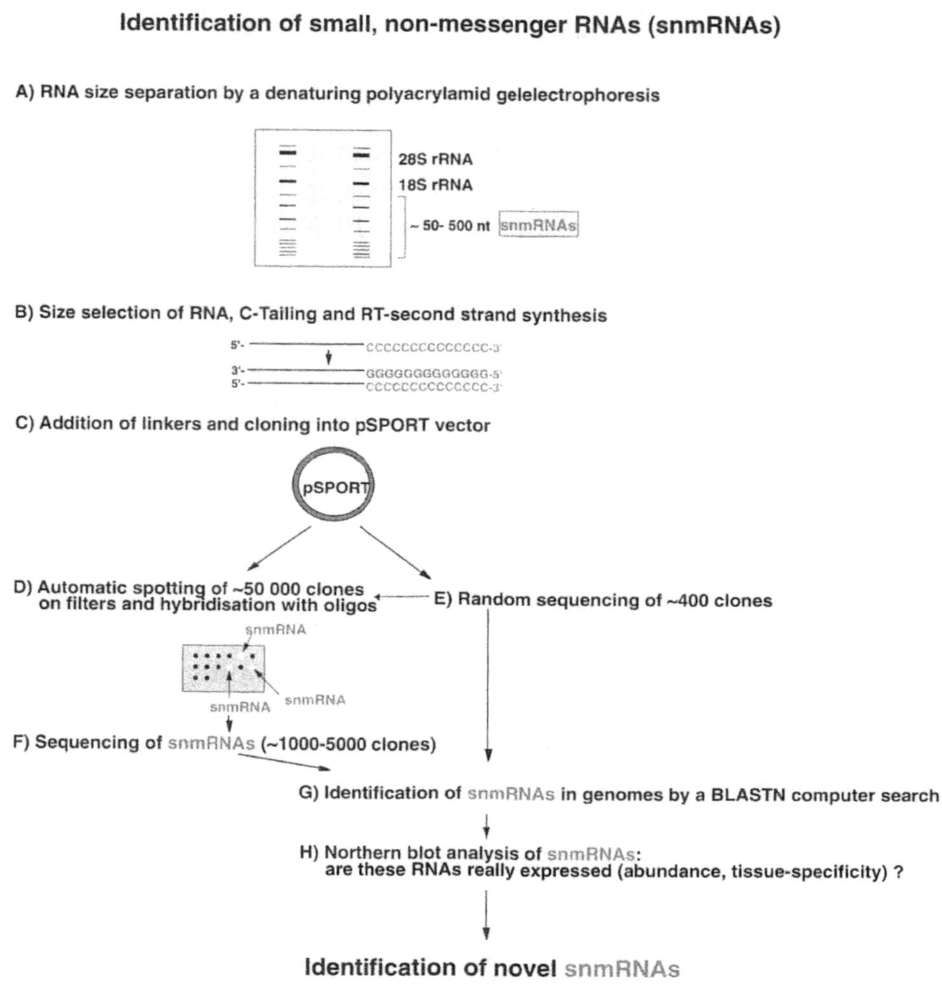

Figure 1. Identification scheme of candidates for small non-messenger RNA (snmRNAs) in model organisms. Gel-purified bulk small RNA is C-tailed and reverse transcribed. Resulting cDNAs are cloned into the pSPORT plasmid vector. After identification and exclusion of clones derived from known RNAs by hybridization of arrays, remaining cDNAs are sequenced. Expression of candidates for snmRNAs is confirmed by Northern blot analysis. ***Colour plate at back of book.***

RNA fraction had been counterselected in most previous EST approaches. RNA isolation is achieved by size separation of total RNA (either from the entire organism at different developmental stages or from an individual organ) on denaturing polyacrylamide gelelectrophoresis. Mostly, this size-selected RNA lacks polyadenylated tails. As a first step to generate cDNA from this RNA fraction, we add tails of oligo(C) to the RNA *in vitro* (DeChiara and Brosius, 1987). Subsequently, C-tailed RNAs are reverse transcribed employing an oligo(dG)-containing primer. After second strand synthesis and ligation of linker sequences, cDNAs are cloned into a standard vector system (pSPORT1/GibcoBRL) generating a plasmid cDNA library.

Usually we pick between 10.000 and 50.000 colonies from such a cDNA library (dependent on the genome size of the model organism) and by an automated robotic system, spot DNAs from these clones on filters in high density arrays (in collaboration with the German Resource Center, Berlin). In parallel, we sequence about 400 randomly selected clones from the cDNA library to identify the most abundant, known snmRNAs. These usually correspond to small ribosomal RNAs, such as 5S or 5.8S rRNAs, degradation fragments of the large rRNAs, tRNAs, snRNAs or 4.5S RNA. Subsequently, we hybridize the filter arrays with labelled oligonucleotides matching the most abundant known RNAs to reduce redundancy while sequencing our cDNA clones in an attempt to exclude clones corresponding to known snmRNAs. From each library, between 10.000 and 50.000 clones are screened. By computer aided analysis, signals obtained in the filter hybridization are ranked and between 1000 and 5000 cDNA clones exhibiting the lowest hybridization scores are sequenced. By this selective hybridisation procedure, we significantly increase the amount of candidates for novel snmRNA species (Figure 1).

Experimental RNomics: Alternative Strategies

As an alternative to C-tailing of RNAs, other groups have employed a different method for cloning small RNA molecules. Instead, two short identical phosphorylated RNA-linkers are ligated to the 5' and 3' ends of snmRNAs employing T4 RNA Ligase (Lagos-Quintana *et al.*, 2001; Lau *et al.*, 2001). Subsequently, ligated RNAs are transcribed into cDNA by RT-PCR using oligonucleotides derived from the RNA-linker sequences. The disadvantage of this method is a rather inefficient ligation step of the RNA-linkers to the potential snmRNA, the possibility of oligomerization of RNA linkers and the failure of linker attachment to modified (*e.g.* 5') termini. The advantage of this method is, however, that often full length cDNA clones can be obtained.

Expression Analysis of Candidates for snmRNAs

In addition to searching for matches of cDNA clones with other sequences including genomic DNA, we examine candidates for novel snmRNAs by Northern blot analysis. Thus, we can determine their size, abundance, potential precursors, degradation products, and - if RNA samples from different tissues and/or at different developmental stages are used - their spatial and temporal expression patterns. It is not always possible to detect a signal in the RNA blot. This might be due to low expression levels of some of the snmRNAs in a given tissue, not excluding high-level expression that is restricted to a rare cell type. Another explanation could be that the cDNA is derived from a degraded larger RNA (mRNA or hnRNA) and is not detectable on our polyacrylamide gel system that is optimized for separation of small RNAs. In some cases, we were able to detect snmRNAs, whose expression could not be confirmed by Northern blot analysis, by application of RT-PCR (Hüttenhofer *et al.*, 2001).

In general, Northern blot analysis indicated a larger size when compared to the length of respective cDNAs isolated by our experimental RNomics approach. This is explained by our method of library construction (see above). As a result, the very 5' ends of snmRNAs are usually not represented in the cDNA clones. Concerning lengths, sequences of snmRNA candidates resemble more closely ESTs (expressed sequence tags) of mRNAs and are therefore designated as ERNSs (expressed RNA sequences, see above). Often, larger bands observed on RNA blots also correspond to precursor RNAs or other larger forms.

4. Experimental RNomics in Mouse Brain: Identification of 201 Candidates for Novel snmRNAs

Our study represented the first unbiased survey of the population of small non-messenger RNA species in a mammalian tissue, providing the basis for a comprehensive understanding of genomic, cellular and organismal function (Hüttenhofer *et al.*, 2001). From mouse brain, we identified so far a total of 201 different expressed RNA sequences potentially encoding novel snmRNA species (Figure 2). This number represents 22 % of all cDNA clones analysed

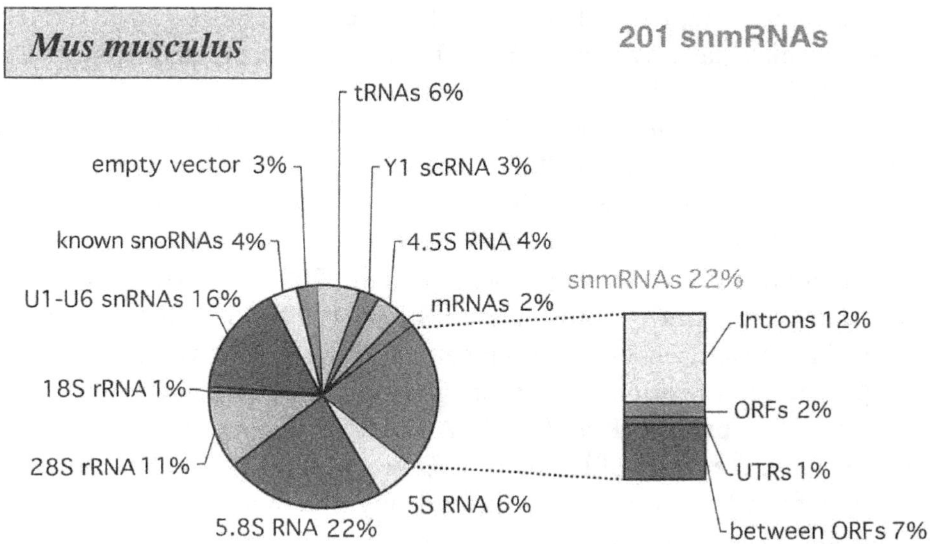

Figure 2. Sequence analysis of a mouse brain cDNA library containing candidates for snmRNAs (Hüttenhofer *et al.*, 2001). Indicated is the total number of novel snmRNA candidates identified (upper right), as well as the number of all RNAs species identified in our screen (in %). The location of genes on the mouse genome encoding novel snmRNA candidates is shown on the right. ***Colour plate at back of book.***

(not counting clones that were sequenced multiple times). Of these 54% (12% of the total) were located in introns, 32% (7% of the total) in intergenic regions, 5% (1% of the total) within annotated 5' or 3' UTRs of mRNAs and 9% (2% of the total) were derived from annotated protein coding regions. The predominant location of snmRNAs genes within introns can be explained by the fact that the majority of these snmRNA belong to the class of small nucleolar RNAs (snoRNAs) which, in vertebrates, are usually cotranscribed in hnRNAs and located in introns. Subsequently, these snmRNAs are processed from their respective introns by splicing and exo- as well as endonucleolytic trimming (Bachellerie *et al.*, 2000; Weinstein and Steitz, 1999).

snoRNAs: An Ever Growing Class of RNA Molecules with Expanding Functions

Based on sequence and structural motifs, 113 of the 201 RNAs can be assigned to either C/D box or H/ACA box subclass of snoRNAs (for reviews see: Bachellerie *et al.*, 2000; Balakin *et al.*, 1996; Smith and Steitz, 1997;

Weinstein and Steitz, 1999). Each snoRNA subclass includes a small number of RNA species that are required for defined pre-rRNA processing steps and are essential for viability. The vast majority of snoRNAs, however, serve a role in posttranscriptional modification of rRNA nucleotides. It is well documented that each 2'-O-ribose methylation or pseudouridylation in eukaryal rRNA (about 200 modifications per ribosome) is selected by a cognate snoRNA belonging to either the C/D box or H/ACA box family, respectively (for reviews see: Bachellerie *et al.*, 2000; Balakin *et al.*, 1996; Smith and Steitz, 1997; Weinstein and Steitz, 1999). In both cases, this is exerted through antisense elements located within snoRNAs complementary to a specific modification site within ribosomal RNAs. This results in the formation of a specific rRNA/snoRNA duplex spanning the modification site. Recently, the range of action of C/D box methylation guide snoRNAs has been found to significally extend beyond their role in ribosome biogenesis: Several snoRNA species were reported that predicted a role in directing 2'-O-ribose methylations in U6 snRNA, through similar RNA/RNA interactions (Ganot *et al.*, 1999; Tycowski *et al.*, 1998). Until our studies, it was assumed that C/D or H/ACA box snoRNAs are, just as their targets, expressed ubiquitously.

Of the 201 novel snmRNA candidates in mouse (Hüttenhofer *et al.*, 2001), 30 merely represented mouse homologs of previously identified human snoRNAs (Figure 3). Importantly, 83 corresponded to entirely novel snoRNAs, 49 of the C/D box and 34 of the H/ACA box type. We identified 24 C/D box- and 19 H/ACA box snoRNAs that are likely to guide rRNA modification, as evidenced by their complementary sequences to rRNA. Among the remainder of novel snoRNAs, we identified for the first time four C/D box snoRNAs and four H/ACA box snoRNAs predicted to direct modifications within U2, U4 or additional modifications in U6 snRNAs (Figure 3). Furthermore, 32 snoRNAs from either class lacked antisense elements for rRNAs or snRNAs. Therefore, additional snoRNA targets have to be considered, such as hnRNAs, mRNAs or other nucleic acid targets. Moreover, seven of 32 snoRNAs, lacking a snRNA or rRNA target, are not ubiquitously expressed in mouse tissues, but are specific to brain (Figure 3).

This might lead to further studies to identify tissue-specifically expressed snoRNAs in other tissues than brain. Interestingly, three of the identified brain-specific snoRNAs map to human chromosome 15q11-q13 (Cavaillé *et al.*, 2000). Deletions of larger fragments including the snoRNA clusters and other paternally expressed genes is associated with the neurodevelopmental Prader-Willi-Syndrome (PWS). Disruption of more

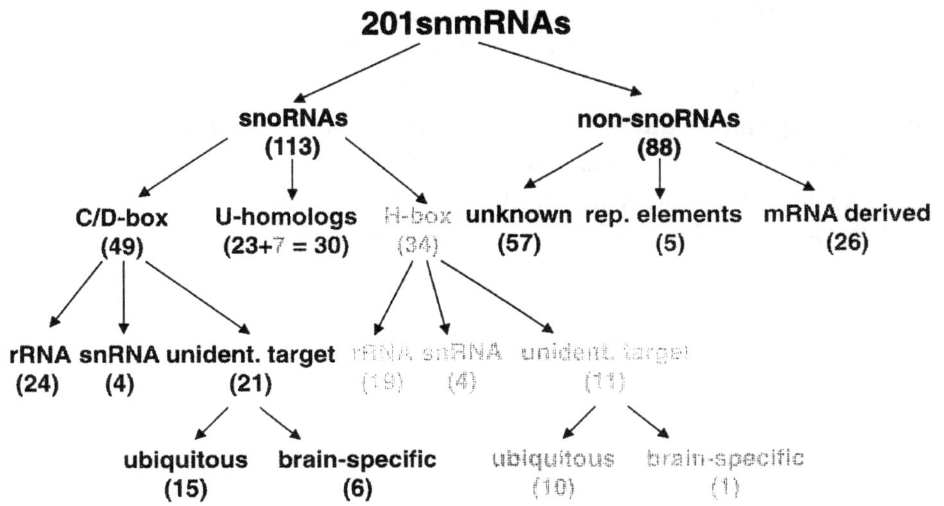

Figure 3. Grouping and functional assignment of 201 novel snmRNAs identified in mouse brain. ***Colour plate at back of book.***

telomerically located genes that are maternally expressed causes Angelman Syndrome. For an excellent review, see Nicholls and Knepper, 2001. One of the brain-specific snoRNAs (MBII-52) has been suggested to target serotonin receptor 5-HT_{2C} hnRNA or mRNA that, in turn, is specifically expressed in brain (Cavaillé *et al.*, 2000). This may be indicative of a novel function of snoRNAs, namely the regulation of gene expression by binding to and/or modifying mRNAs or their hnRNA precursors via their antisense elements. In this case, it would dramatically expand the proposed target sites for snoRNAs.

MBII-52 and MBII-85 snoRNAs are among the most abundant RNAs expressed in the brain, and one should further consider that they are expressed only in neurons. A sole regulatory function is more difficult to reconcile with the high abundance of MBII-52 and MBII-85 snoRNAs. An alternative function of MBII-52, MBII-85 and similarly expressed snoRNAs, could be regulating not only their own imprinted expression but also that of neighbouring genes via interaction with their own genomic locus by RNA/DNA hybridization. Perhaps binding of the snoRNAs precludes binding of other macromolecules, such as proteins, to DNA or vice versa. Alternatively, snoRNA binding guides, in analogy to RNA, methylation of bases in DNA or guides demethylation or perhaps prevents one or the other.

It is also interesting to note that most copies of MBII-52 and MBII-85 snoRNAs harbour in their presumed antisense elements upstream from box D at least one CpG dimer. Already, it is known that RNA molecules can play a role of in regulating the expression of genes. For example, the non-protein coding RNA transcript Xist is involved in X chromosome inactivation (for a review see: (Kelley and Kuroda, 2000). It would not surprise, if MBII-52, MBII-85 and similar snoRNAs were involved in gene regulation by interacting with DNA. Of course, some of the variants of MBII-52 and MBII-85 snoRNAs still could additionally function in processing or editing of hnRNA/mRNA.

Non-snoRNA Species in Mouse Brain

Of the 88 RNAs not belonging to either snoRNA subclass, at least 26 are probably derived from truncated hnRNAs or mRNAs. Short interspersed repetitive elements (SINEs) are located on five RNA sequences and may represent rare examples of transcribed SINEs. The remaining RNA species could neither be assigned to any snmRNA class nor to a part of a larger hnRNA/mRNA, as of yet. At this stage, it is difficult to speculate about the function of potential snmRNAs of the non-snoRNA type. Even small RNA species derived from hnRNAs or mRNAs might regulate the expression of mRNAs by yet unknown mechanisms, especially when located within their 5' or 3' UTRs. Their detection sets the stage for direct experimental testing of these hypotheses.

5. Experimental RNomics Approaches in *Drosophila melanogaster*, *C. elegans* and *Homo sapiens*

Very recently, other experimental RNomics approaches have been performed in different model organisms, namely *Drosophila melanogaster*, *C. elegans* and *Homo sapiens* (Lagos-Quintana *et al.*, 2001; Lau *et al.*, 2001) (Lee and Ambros, 2001). However, although unbiased towards specific sequence or structural motifs of RNAs, all three studies were strongly geared towards very small RNAs since the respective groups were exclusively interested in the identification of siRNAs (small interfering RNAs) or miRNAs (micro RNAs, see below). siRNAs, stRNAs (small temporal RNAs) or miRNAs usually range from 21 to 23 nt and hence RNAs bracketing that size were cloned.

siRNAs

siRNAs are usually generated from long double stranded RNAs by an enzyme termed Dicer (Bernstein *et al.*, 2001; Grishok *et al.*, 2001b; Hutvagner *et al.*, 2001; Ketting *et al.*, 2001). Dicer is thought to cleave double stranded RNA templates into short siRNAs which exhibit a length from 21 to 23 nt. These double stranded short RNA-oligomers, in turn, are able to target a certain mRNA region complementary to siRNAs which results in specific cleavage of the mRNA at this site. This mechanism of mRNA degradation has been termed RNA interference (RNAi)(Fire *et al.*, 1998) and is thought to have evolved as a mechanism for antiviral defense by targeting certain double stranded RNA viruses (Elbashir *et al.*, 2001; Hamilton and Baulcombe, 1999; Ketting *et al.*, 1999; Sharp, 2001; Tabara *et al.*, 1999; Waterhouse *et al.*, 2001; Wu-Scharf *et al.*, 2000).

stRNAs

Another species of small RNA molecules, designated as stRNAs has been identified in *C. elegans*. The first two identified members of this class, lin-4 and let-7 (Reinhart *et al.*, 2000; Wightman *et al.*, 1993) are expressed only at certain larval stages of *C. elegans* development. The single-stranded stRNAs are able to downregulate translation of distinct target mRNAs via the 3' untranslated region (UTR) elements of mRNAs that are complementary to each regulatory stRNA to specify the temporal progression of cell fate (Slack *et al.*, 2000; Wightman *et al.*, 1993).

The common denominator between stRNAs and siRNAs is their size (about 22 nt) as well as a similar mechanism to generate these RNA species from a larger precursor: Here too, the enzyme Dicer, (related to RNase III) is thought to process stRNAs from longer double stranded RNA precursors (Grishok *et al.*, 2001a; Hutvagner *et al.*, 2001). In case of stRNAs these RNA precursors are stable stem-loop structures of 60-70 nt in size. The 22 nt stRNAs are processed from the stem portion of the precursor (Grishok *et al.*, 2001a; Hutvagner *et al.*, 2001). In contrast to siRNAs, only one of the strands of the RNA helix is processed and stable while siRNAs are acting as double stranded RNAs. However, it is unclear whether the siRNA-endonuclease complex is containing double-stranded siRNA. The second difference between stRNAs and siRNAs is their functional mode. While siRNAs are able to degrade specific mRNAs by the RNAi mechanism (see above), stRNA presumably do not degrade their target mRNAs but apparently exert their function at the translational level by binding to the 3'UTRs of mRNAs.

miRNAs

The group of Tuschl aimed to clone endogenous siRNAs to investigate more closely the mechanism of RNAi function *in vivo* (Lagos-Quintana *et al.*, 2001) while the groups of Bartels and Ambros sought to identify stRNAs or miRNAs (Lau *et al.*, 2001; Lee and Ambros, 2001). In their approaches, total RNA from the respective organisms was separated on a denaturing polyacrylamid gel. RNAs sized from about 18 to 26 nucleotides were excised and cloned using either the RNA-linker/ligation method (Lagos-Quintana *et al.*, 2001; Lau *et al.*, 2001), see above) or the RNA-tailing approach (Lee and Ambros, 2001), see above). With respect to the sizes of RNA species selected, these approaches are complementary to our study in the mouse, where RNAs in the range of ~50 to 500 nt are the starting material of cDNA libraries. In the study of Tuschl's lab, 19 novel RNA species were cloned from *Homo sapiens* (*e.g.* Hela cells) and 14 novel RNAs from the fruit fly, *Drosophila melanogaster* (Lagos-Quintana *et al.*, 2001). In the second study by the Bartels group, 55 novel RNAs from the nematode *C. elegans* were isolated (Lau *et al.*, 2001) while in a third study by the lab of Ambros (Lee and Ambros, 2001), 15 novel RNAs were detected in this organisms. Ten of these were also identified by the Bartels group.

In all cases, precursor RNAs could be identified resembling stem-loop structures from which miRNAs could be processed in analogy to stRNAs. Some but not all of these novel identified RNAs were developmentally regulated (as has been shown for siRNAs), as demonstrated by Northern blot analysis. Hence, these novel RNA species, in total 93 from three different model organisms, expand the class of stRNAs dramatically. Since only some of the novel RNAs were developmentally regulated, the novel RNAs were not designated as stRNAs but as miRNAs. It was proposed that miRNAs, in analogy to stRNAs regulate the translation of certain target mRNAs via complementarity to their 3' UTR (Lagos-Quintana *et al.*, 2001; Lau *et al.*, 2001; Lee and Ambros, 2001; Ruvkun, 2001). In addition, it was estimated, that the 93 novel miRNAs might only reflect the "tip of the iceberg" and that maybe hundreds if not thousands, of miRNAs exist that are able to regulate gene expression (Ambros, 2001; Grosshans and Slack, 2002; Ruvkun, 2001).

It is interesting to point out that our approach has not turned up, as of yet, such precursors, which lie within the size range of our selection scheme (Hüttenhofer *et al.*, 2001). Perhaps simply difficulties in the reverse transcription or the presumed instability of such precursors makes them rare cDNA species. Alternatively they might just be rare in adult brain tissue.

Nevertheless, in our approaches covering model organisms such as *D. melanogaster* and *C. elegans*, which covers different developmental stages, we also failed to identify putative precursors of miRNAs.

6. Experimental RNomics: Lessons To Be Learned

What lessons can be learned from these few selected Experimental RNomics approaches performed thus far? First, we can expect many more surprises concerning snmRNAs than ever anticipated. Novel classes of snmRNAs are already rapidly emerging exhibiting novel functions. From the studies performed, it appears that the mammalian cell contains at least four different classes of highly abundant snmRNAs exhibiting different sizes: first, the class of miRNAs/stRNAs/siRNAs with an average size of about 22 nt, second the class of tRNAs, involved in protein biosynthesis, with an average size of about 75 nt, third, C/D box snoRNAs which are usually in the size range of 50-100 nt and fourth, H/ACA box snoRNAs which are sized between 100 and 200 nt. All classes have been shown to contain numerous representatives and the numbers of novel members are still growing.

Second, novel functions are being ascribed to already known classes of snmRNAs. In particular, this applies to the class of snoRNAs which was known, up to now, to interact exclusively with rRNA or snRNAs, via antisense elements located within snoRNAs. In our study, 32 novel members of snoRNAs appeared (seven of them only brain-specifically expressed) that did not exhibit antisense elements to any known non-messenger RNA. In one case, a (brain-specific) snoRNA, designated as MBII-52 snoRNA, exhibited 18-nucleotide complementarity to the serotonin receptor 2C pre-mRNA (Cavaillé *et al.*, 2000) at a site where alternative splicing (Canton *et al.*, 1996) and editing (Burns *et al.*, 1997) was observed within the mRNA. This immediately suggests a function of snoRNAs other than modification of rRNAs or snRNAs, namely the regulation of gene expression on a posttranscriptional level. A similar way of action is proposed for stRNAs and miRNAs which, in some cases, were shown to interact with the 3' UTR of certain mRNAs thereby regulating their translation (Ruvkun, 2001).

It thus appears that apart from proteins, cells contain numerous small RNA molecules (snRNAs, stRNAs, miRNAs and maybe even snoRNAs) able to bind to and interact with mRNAs via an antisense element, thus regulating gene expression (Figure 4). In addition, certain snmRNAs from *E. coli*, such as MicF, DicF or DsrA RNAs, have been shown to interact with the translation

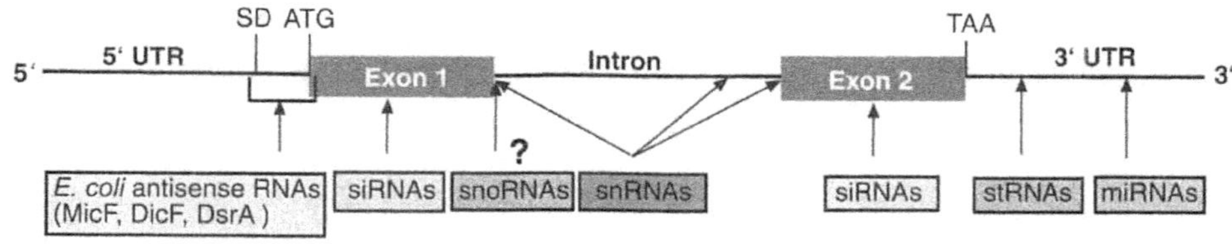

Figure 4. Targeting of bacterial or eukaryal mRNAs or pre-mRNAs by snmRNAs thereby regulating gene expression. It should be noted that binding of a snoRNA to the splice-site of a eukaryal pre-mRNA or involvement in RNA editing has recently been proposed by our group (Cavaillé *et al.*, 2000) as two of several alternatives for a functional role. They remain to be experimentally sustained or disproven. ***Colour plate at back of book.***

initiation region of target mRNAs via an antisense element as well. By this mechanism the snmRNAs regulate the translation of their target mRNAs (for a review see: Wassarman *et al.*, 1999 and Figure 4). It might very well be that all these snmRNAs reflect powerful remnants of the "early RNA or RNP world", for example, able to target mRNAs at different sites in a highly specific manner. Additionally, unlike proteins, they have not to be translated into proteins and hence can serve as a fast switch for the cell to rapidly react to certain environmental changes. On the other hand, the concept of young recruits to the family of functional RNAs and novel regulatory or structural strategies in the cell is not out of the question anymore (Brosius and Gould, 1992; Brosius and Tiedge, 1995a; Brosius and Tiedge, 1995b).

7. Outlook

With the completion of more genomic sequences the power of phylogenetic analysis that, for example, has revolutionized RNA structure predictions (Gutell *et al.*, 1994; Gutell *et al.*, 1986; Noller *et al.*, 1981; Woese *et al.*, 1980), efforts in detecting and/or evaluating candidates for novel snmRNAs must also encompass phylogenetic comparison. Coding regions as well as regulaory regions are usually conserved at significantly higher levels than intergenic or intronic sequences. Studies in that direction are already underway (Chen *et al.*, 2001; Frazer *et al.*, 2001; Hardison *et al.*, 1997; Shabalina *et al.*, 2001) and should also be applicable to finding non-protein coding RNAs. This phylogenetic biomathematical approach may be hampered – at least in some phyla. Comparing stretches from the mouse and human genomes, (Shabalina *et al.*, 2001) found that conserved DNA stretches that may encompass as much as 15% of the genomes, about an order of magnitude larger than the estimated 1.5% coding regions. The significance of these conserved sequences is not clear yet.

Currently, we are analysing cDNA libraries encoding snmRNAs from five additional model organsims, namely the archaeon *Archaeoglobus fulgidus*, the bacterium *Escherichia coli*, the nematode *Caenorhabditis elegans*, the fruitfly *Drosophila melanogaster* and the plant *Arabidopsis thaliana*. By our Experimental RNomics approach, we have already identified well over 500 different snmRNA species in these model organisms. In the eukaryotic organisms, *Arabidopsis thaliana* and *Drosophila melanogaster*, we are also searching for developmentally regulated as well as tissue-specifically expressed snmRNAs. The comparison of novel snmRNAs between species should result in the identification of novel classes of snmRNAs. Despite the value of phylogenetic analysis in finding additional genes, analysis of large genomes with respect to their snmRNA content will always require experimental verification, if not initial approaches. In addition, the challenging part of our investigation has merely begun: identification of the biological function of these snmRNAs in the cell.

References

Adams, M.D., Soares, M.B., Kerlavage, A.R., Fields, C., and Venter, J.C. 1993. Rapid cDNA sequencing (expressed sequence tags) from a directionally cloned human infant brain cDNA library. Nat. Genet. 4: 373-380.

Ambros, V. 2001. microRNAs: tiny regulators with great potential. Cell. 107: 823-826.

Argaman, L., Hershberg, R., Vogel, J., Bejerano, G., Wagner, *E.G.*, Margalit, H., and Altuvia, S. 2001. Novel small RNA-encoding genes in the intergenic regions of Escherichia coli. Curr. Biol. 11: 941-950.

Bachellerie, J.P., Cavaillé, J., and Qu, L.H. 2000. Nucleotide modifications of eukaryotic rRNAs: the world of small nucleolar RNAs revisited. ASM Press, Washington, DC.

Balakin, A.G., Smith, L., and Fournier, M.J. 1996. The RNA world of the nucleolus: two major families of small RNAs defined by different box elements with related functions. Cell. 86: 823-834.

Bernstein, E., Caudy, A.A., Hammond, S.M., and Hannon, G.J. 2001. Role for a bidentate ribonuclease in the initiation step of RNA interference. Nature. 409: 363-366.

Brosius, J. 1996. More Haemophilus and Mycoplasma genes. Science. 271: 1302-1304.

Brosius, J. 1999. RNAs from all categories generate retrosequences that may be exapted as novel genes or regulatory elements. Gene. 238: 115-

134.

Brosius, J., and Gould, S.J. 1992. On "genomenclature": a comprehensive and respectful taxonomy for pseudogenes and other "junk DNA". Proc. Natl. Acad. Sci. USA. 89: 10706-10710.

Brosius, J., and Tiedge, H. 1995a. Neural BC1 RNA: Dendritic localization and transport. In: Localized RNAs. H.D. Lipshitz, ed. R.G. Landes, Austin, TX, p. 289-330.

Brosius, J., and Tiedge, H. 1995b. Reverse transcriptase: mediator of genomic plasticity. Virus Genes. 11: 163-179.

Brosius, J., and Tiedge, H. 2001. Neuronal BC1 RNA: Intracellular transport and activity-dependent modulation. In Richter, D. (ed.. Cell polarity and subcellular RNA localization. Springer Verlag, Berlin, pp. 129-138.

Burns, C.M., Chu, H., Rueter, S.M., Hutchinson, L.K., Canton, H., Sanders-Bush, E., and Emeson, R.B. 1997. Regulation of serotonin-2C receptor G-protein coupling by RNA editing. Nature. 387: 303-308.

Canton, H., Emeson, R.B., Barker, E.L., Backstrom, J.R., Lu, J.T., Chang, M.S., and Sanders-Bush, E. 1996. Identification, molecular cloning, and distribution of a short variant of the 5-hydroxytryptamine2C receptor produced by alternative splicing. Mol. Pharmacol. 50; 799-807.

Carter, R.J., Dubchak, I., and Holbrook, S.R. 2001. A computational approach to identify genes for functional RNAs in genomic sequences. Nucleic Acids Res. 29: 3928-3938.

Cavaillé, J., Buiting, K., Kiefmann, M., Lalande, M., Brannan, C.I., Horsthemke, B., Bachellerie, J.P., Brosius, J., and Huttenhofer, A. 2000. Identification of brain-specific and imprinted small nucleolar RNA genes exhibiting an unusual genomic organization. Proc. Natl. Acad. Sci. USA. 97: 14311-14316.

Chen, R., Bouck, J.B., Weinstock, G.M., and Gibbs, R.A. 2001. Comparing vertebrate whole-genome shotgun reads to the human genome. Genome Res. 11: 1807-1816.

DeChiara, T.M., and Brosius, J. 1987. Neural BC1 RNA: cDNA clones reveal nonrepetitive sequence content. Proc. Natl. Acad. Sci. USA. 84: 2624-2628.

Eddy, S.R. 1999. Noncoding RNA genes. Curr. Opin. Genet. Dev. 9: 695-699.

Eddy, S.R. 2001. Non-coding RNA genes, and the modern RNA world. Nat. Rev. Genet. 2: 919-929.

Elbashir, S.M., Lendeckel, W., and Tuschl, T. 2001. RNA interference is mediated by 21- and 22-nucleotide RNAs. Genes Dev. 15: 188-200.

Erdmann, V.A., Barciszewska, M.Z., Hochberg, A., de Groot, N., and Barciszewski, J. 2001a. Regulatory RNAs. Cell Mol. Life Sci. 58: 960-977.

Erdmann, V.A., Barciszewska, M.Z., Szymanski, M., Hochberg, A., de Groot, N., and Barciszewski, J. 2001b. The non-coding RNAs as riboregulators. Nucleic Acids Res. 29: 189-193.

Fire, A., Xu, S., Montgomery, M.K., Kostas, S.A., Driver, S.E., and Mello, C.C. 1998. Potent and specific genetic interference by double-stranded RNA in Caenorhabditis elegans. Nature. 391: 806-811.

Fleischmann, R.D., Adams, M.D., White, O., Clayton, R.A., Kirkness, E.F., Kerlavage, A.R., Bult, C.J., Tomb, J.F., Dougherty, B.A., Merrick, J.M., *et al*. 1995. Whole-genome random sequencing and assembly of Haemophilus influenzae Rd. Science. 269: 496-512.

Fraser, C.M., Gocayne, J.D., White, O., Adams, M.D., Clayton, R.A., Fleischmann, R.D., Bult, C.J., Kerlavage, A.R., Sutton, G., Kelley, J.M. *et al*. 1995. The minimal gene complement of Mycoplasma genitalium. Science. 270: 397-403.

Frazer, K.A., Sheehan, J.B., Stokowski, R.P., Chen, X., Hosseini, R., Cheng, J.F., Fodor, S.P., Cox, D.R., and Patil, N. 2001. Evolutionarily conserved sequences on human chromosome 21. Genome Res. 11: 1651-1659.

Ganot, P., Jady, B.E., Bortolin, M.L., Darzacq, X., and Kiss, T. 1999. Nucleolar factors direct the 2'-O-ribose methylation and pseudouridylation of U6 spliceosomal RNA. Mol. Cell Biol. 19: 6906-6917.

Grishok, A., Pasquinelli, A.E., Conte, D., Li, N., Parrish, S., Ha, I., Baillie, D.L., Fire, A., Ruvkun, G., and Mello, C.C. 2001a. Genes and mechanisms related to RNA interference regulate expression of the small temporal RNAs that control C. elegans developmental timing. Cell. 106: 23-34.

Grishok, A., Pasquinelli, A.E., Conte, D., Li, N., Parrish, S., Ha, I., Baillie, D.L., Fire, A., Ruvkun, G., and Mello, C.C. 2001b. Genes and mechanisms related to RNA interference regulate expression of the small temporal RNAs that control C. elegans developmental timing. Cell. 106: 23-34.

Grosshans, H., and Slack, F.J. 2002. Micro-RNAs: small is plentiful. J. Cell Biol. 156: 17-22.

Gutell, R.R., Larsen, N., and Woese, C.R. 1994. Lessons from an evolving rRNA: 16S and 23S rRNA structures from a comparative perspective. Microbiol. Rev. 58: 10-26.

Gutell, R.R., Noller, H.F., and Woese, C.R. (1986. Higher order structure in ribosomal RNA. EMBO J. 5: 1111-1113.

Hamilton, A.J., and Baulcombe, D.C. 1999. A species of small antisense RNA in posttranscriptional gene silencing in plants. Science. 286: 950-952.

Hardison, R.C., Oeltjen, J., and Miller, W. 1997. Long human-mouse sequence alignments reveal novel regulatory elements: a reason to sequence the mouse genome. Genome Res. 7: 959-966.

Hindley, J. 1967. Fractionation of ^{32}P-labelled ribonucleic acids on polyacrylamide gels and their characterization by fingerprinting. J. Mol. Biol. 30: 125-136.

Hüttenhofer, A., Kiefmann, M., Meier-Ewert, S., O'Brien, J., Lehrach, H., Bachellerie, J.-P., and Brosius, J. 2001. RNomics: an experimental approach that identifies 201 candidates for novel, small, non-messenger RNAs in mouse. EMBO J. 20: 2943-2953.

Hutvagner, G., McLachlan, J., Pasquinelli, A.E., Balint, E., Tuschl, T., and Zamore, P.D. 2001. A cellular function for the RNA-interference enzyme Dicer in the maturation of the let-7 small temporal RNA. Science. 293: 834-838.

Inouye, M., and Delihas, N. 1988. Small RNAs in the prokaryotes: a growing list of diverse roles. Cell. 53: 5-7.

Keiler, K.C., Waller, P.R., and Sauer, R.T. 1996. Role of a peptide tagging system in degradation of proteins synthesized from damaged messenger RNA. Science. 271: 990-993.

Kelley, R.L., and Kuroda, M.I. 2000. Noncoding RNA genes in dosage compensation and imprinting. Cell. 103: 9-12.

Ketting, R.F., Fischer, S.E., Bernstein, E., Sijen, T., Hannon, G.J., and Plasterk, R.H. 2001. Dicer functions in RNA interference and in synthesis of small RNA involved in developmental timing in C. elegans. Genes Dev. 15: 2654-2659.

Ketting, R.F., Haverkamp, T.H., van Luenen, H.G., and Plasterk, R.H. 1999. Mut-7 of C. elegans, required for transposon silencing and RNA interference, is a homolog of Werner syndrome helicase and RNaseD. Cell. 99: 133-141.

Kuryshev, V.Y., Skryabin, B.V., Kremerskothen, J., Jurka, J., and Brosius, J. 2001. Birth of a gene: locus of neuronal BC200 snmRNA in three prosimians and human BC200 pseudogenes as archives of change in the Anthropoidea lineage. J. Mol. Biol. 309: 1049-1066.

Lagos-Quintana, M., Rauhut, R., Lendeckel, W., and Tuschl, T. 2001. Identification of novel genes coding for small expressed RNAs. Science. 294: 853-858.

Lander, E.S., Linton, L.M., Birren, B., Nusbaum, C., Zody, M.C., Baldwin, J., Devon, K., Dewar, K., Doyle, M., FitzHugh, W., Funke, R., Gage, D., Harris, K., Heaford, A., Howland, J., Kann, L., Lehoczky, J., LeVine, R., McEwan, P., McKernan, K., Meldrim, J., Mesirov, J.P., Miranda, C., Morris, W., Naylor, J., Raymond, C., Rosetti, M., Santos, R., Sheridan, A., Sougnez, C., Stange-Thomann, N., Stojanovic, N., Subramanian, A., Wyman, D., Rogers, J., Sulston, J., Ainscough, R., Beck, S., Bentley, D., Burton, J., Clee, C., Carter, N., Coulson, A., Deadman, R., Deloukas, P., Dunham, A.,

Dunham, I., Durbin, R., French, L., Grafham, D., Gregory, S., Hubbard, T., Humphray, S., Hunt, A., Jones, M., Lloyd, C., McMurray, A., Matthews, L., Mercer, S., Milne, S., Mullikin, J.C., Mungall, A., Plumb, R., Ross, M., Shownkeen, R., Sims, S., Waterston, R.H., Wilson, R.K., Hillier, L.W., McPherson, J.D., Marra, M.A., Mardis, E.R., Fulton, L.A., Chinwalla, A.T., Pepin, K.H., Gish, W.R., Chissoe, S.L., Wendl, M.C., Delehaunty, K.D., Miner, T.L., Delehaunty, A., Kramer, J.B., Cook, L.L., Fulton, R.S., Johnson, D.L., Minx, P.J., Clifton, S.W., Hawkins, T., Branscomb, E., Predki, P., Richardson, P., Wenning, S., Slezak, T., Doggett, N., Cheng, J.F., Olsen, A., Lucas, S., Elkin, C., Uberbacher, E., Frazier, M., *et al.* 2001. Initial sequencing and analysis of the human genome. Nature. 409: 860-921.

Larsen, C.J., Galibert, F., Lelong, J.C., and Boiron, M. 1967. [Characterization of a metabolically stable RNA of low molecular weight in the nuclear fraction of KB cells]. CR Acad. Sci. Hebd. Seances Acad. Sci. D. 264: 1523-1526.

Lau, N.C., Lim, L.P., Weinstein, *E.G.*, and Bartel, D.P. 2001. An abundant class of tiny RNAs with probable regulatory roles in Caenorhabditis elegans. Science. 294: 858-862.

Lee, R.C., and Ambros, V. 2001. An extensive class of small RNAs in Caenorhabditis elegans. Science. 294: 862-864.

Martignetti, J.A., and Brosius, J. 1993a. BC200 RNA: a neural RNA polymerase III product encoded by a monomeric Alu element. Proc. Natl. Acad. Sci. USA. 90: 11563-11567.

Martignetti, J.A., and Brosius, J. 1993b. Neural BC1 RNA as an evolutionary marker: guinea pig remains a rodent. Proc. Natl. Acad. Sci. USA. 90: 9698-9702.

Mattick, J.S. 2001. Non-coding RNAs: the architects of eukaryotic complexity. EMBO Rep. 2: 986-991.

Nguyen, V.T., Kiss, T., Michels, A.A., and Bensaude, O. 2001. 7SK small nuclear RNA binds to and inhibits the activity of CDK9/cyclin T complexes. Nature. 414: 322-325.

Nicholls, R.D., and Knepper, J.L. 2001. Genome organization, function, and imprinting in Prader-Willi and Angelman syndromes. Annu. Rev. Genomics Hum. Genet. 2: 153-175.

Noller, H.F., Kop, J., Wheaton, V., Brosius, J., Gutell, R.R., Kopylov, A.M., Dohme, F., Herr, W., Stahl, D.A., Gupta, R., and Waese, C.R. 1981. Secondary structure model for 23S ribosomal RNA. Nucleic Acids Res. 9: 6167-6189.

Olivas, W.M., Muhlrad, D., and Parker, R. 1997. Analysis of the yeast genome: identification of new non-coding and small ORF-containing RNAs. Nucleic Acids Res. 25: 4619-4625.

Prestayko, A.W., and Busch, H. 1968. Low molecular weight RNA of the chromatin fraction from Novikoff hepatoma and rat liver nuclei. Biochim. Biophys. Acta. 169: 327-337.

Reinhart, B.J., Slack, F.J., Basson, M., Pasquinelli, A.E., Bettinger, J.C., Rougvie, A.E., Horvitz, H.R., and Ruvkun, G. 2000. The 21-nucleotide let-7 RNA regulates developmental timing in Caenorhabditis elegans. Nature. 403: 901-906.

Rivas, E., Klein, R.J., Jones, T.A. and Eddy, S.R. 2001. Computational identification of noncoding RNAs in E. coli by comparative genomics. Curr. Biol. 11: 1369-1373.

Ruvkun, G. 2001. Molecular biology. Glimpses of a tiny RNA world. Science. 294: 797-799.

Shabalina, S.A., Ogurtsov, A.Y., Kondrashov, V.A., and Kondrashov, A.S. 2001. Selective constraint in intergenic regions of human and mouse genomes. Trends Genet. 17: 373-376.

Sharp, P.A. 2001. RNA interference—2001. Genes Dev. 15: 485-490.

Skryabin, B.V., Kremerskothen, J., Vassilacopoulou, D., Disotell, T.R., Kapitonov, V.V., Jurka, J., and Brosius, J. 1998. The BC200 RNA gene and its neural expression are conserved in Anthropoidea (Primates). J. Mol. Evol. 47: 677-685.

Slack, F.J., Basson, M., Liu, Z., Ambros, V., Horvitz, H.R., and Ruvkun, G. 2000. The lin-41 RBCC gene acts in the C. elegans heterochronic pathway between the let-7 regulatory RNA and the LIN-29 transcription factor. Mol. Cell. 5: 659-669.

Smith, C.M., and Steitz, J.A. 1997. Sno storm in the nucleolus: new roles for myriad small RNPs. Cell. 89: 669-672.

Tabara, H., Sarkissian, M., Kelly, W.G., Fleenor, J., Grishok, A., Timmons, L., Fire, A. and Mello, C.C. 1999. The rde-1 gene, RNA interference, and transposon silencing in C. elegans. Cell. 99: 123-132.

Tang, T.-H., Bachellerie, J.-P., Rozhdestvensky, T., Bortolin, M.-L., Huber, H., Drungowski, M., Elge, T., Brosius, J., and Hüttenhofer, A. 2002. Identification of 86 candidates for small non-messenger RNAs from the archaeon *Archaeoglobus fulgidus*. Proc. Natl. Acad. Sci. USA. 99: 7536-7541.

Tiedge, H., and Brosius, J. 1996. Translational machinery in dendrites of hippocampal neurons in culture. J. Neurosci. 16: 7171-7181.

Tiedge, H., Chen, W., and Brosius, J. 1993. Primary structure, neural-specific expression, and dendritic location of human BC200 RNA. J. Neurosci. 13: 2382-2390.

Tiedge, H., Fremeau, R.T., Weinstock, P.H., Arancio, O., and Brosius, J. 1991. Dendritic location of neural BC1 RNA. Proc. Natl. Acad. Sci. USA.

88: 2093-2097.

Tu, G.F., Reid, G.E., Zhang, J.G., Moritz, R.L., and Simpson, R.J. 1995. C-terminal extension of truncated recombinant proteins in Escherichia coli with a 10Sa RNA decapeptide. J. Biol. Chem. 270: 9322-9326.

Tycowski, K.T., You, Z.H., Graham, P.J., and Steitz, J.A. 1998. Modification of U6 spliceosomal RNA is guided by other small RNAs. Mol. Cell. 2: 629-638.

Venter, J.C., Adams, M.D., Myers, E.W., Li, P.W., Mural, R.J., Sutton, G.G., Smith, H.O., Yandell, M., Evans, C.A., Holt, R.A., Gocayne, J.D., Amanatides, P., Ballew, R.M., Huson, D.H., Wortman, J.R., Zhang, Q., Kodira, C.D., Zheng, X.H., Chen, L., Skupski, M., Subramanian, G., Thomas, P.D., Zhang, J., Gabor Miklos, G.L., Nelson, C., Broder, S., Clark, A.G., Nadeau, J., McKusick, V.A., Zinder, N., Levine, A.J., Roberts, R.J., Simon, M., Slayman, C., Hunkapiller, M., Bolanos, R., Delcher, A., Dew, I., Fasulo, D., Flanigan, M., Florea, L., Halpern, A., Hannenhalli, S., Kravitz, S., Levy, S., Mobarry, C., Reinert, K., Remington, K., Abu-Threideh, J., Beasley, E., Biddick, K., Bonazzi, V., Brandon, R., Cargill, M., Chandramouliswaran, I., Charlab, R., Chaturvedi, K., Deng, Z., Di Francesco, V., Dunn, P., Eilbeck, K., Evangelista, C., Gabrielian, A.E., Gan, W., Ge, W., Gong, F., Gu, Z., Guan, P., Heiman, T.J., Higgins, M.E., Ji, R.R., Ke, Z., Ketchum, K.A., Lai, Z., Lei, Y., Li, Z., Li, J., Liang, Y., Lin, X., Lu, F., Merkulov, G.V., Milshina, N., Moore, H.M., Naik, A.K., Narayan, V.A., Neelam, B., Nusskern, D., Rusch, D.B., Salzberg, S., Shao, W., Shue, B., Sun, J., Wang, Z., Wang, A., Wang, X., Wang, J., Wei, M., Wides, R., Xiao, C., Yan, C., *et al.* 2001. The sequence of the human genome. Science. 291: 1304-1351.

Vesco, C., and Penman, S. 1969. The cytoplasmic RNA of HeLa cells: new discrete species associated with mitochondria. Proc. Natl. Acad. Sci. USA. 62: 218-225.

Wassarman, K.M., Repoila, F., Rosenow, C., Storz, G., and Gottesman, S. 2001. Identification of novel small RNAs using comparative genomics and microarrays. Genes Dev. 15: 1637-1651.

Wassarman, K.M., Zhang, A., and Storz, G. 1999. Small RNAs in Escherichia coli. Trends Microbiol. 7: 37-45.

Waterhouse, P.M., Wang, M.B., and Lough, T. 2001. Gene silencing as an adaptive defence against viruses. Nature. 411: 834-842.

Weinberg, R.A., Loening, U., Willems, M., and Penman, S. 1967. Acrylamide gel electrophoresis of HeLa cell nucleolar RNA. Proc. Natl. Acad. Sci. USA. 58: 1088-1095.

Weinberg, R.A., and Penman, S. 1968. Small molecular weight monodisperse nuclear RNA. J. Mol. Biol. 38: 289-304.

Weinstein, L.B., and Steitz, J.A. 1999. Guided tours: from precursor snoRNA to functional snoRNP. Curr. Opin. Cell Biol. 11: 378-384.

Wiemann, S., Weil, B., Wellenreuther, R., Gassenhuber, J., Glassl, S., Ansorge, W., Bocher, M., Blocker, H., Bauersachs, S., Blum, H., Lauber, J., Dusterhoft, A., Beyer, A., Kohrer, K., Strack, N., Mewes, H.W., Ottenwalder, B., Obermaier, B., Tampe, J., Heubner, D., Wambutt, R., Korn, B., Klein, M., and Poustka, A. 2001. Toward a Catalog of Human Genes and Proteins: Sequencing and Analysis of 500 Novel Complete Protein Coding Human cDNAs. Genome Res. 11: 422-435.

Wightman, B., Ha, I., and Ruvkun, G. 1993. Posttranscriptional regulation of the heterochronic gene lin-14 by lin- 4 mediates temporal pattern formation in C. elegans. Cell. 75: 855-862.

Woese, C.R., Magrum, L.J., Gupta, R., Siegel, R.B., Stahl, D.A., Kop, J., Crawford, N., Brosius, J., Gutell, R., Hogan, J.J., and Noller, H.F. 1980. Secondary structure model for bacterial 16S ribosomal RNA: phylogenetic, enzymatic and chemical evidence. Nucleic Acids Res. 8: 2275-2293.

Wu-Scharf, D., Jeong, B., Zhang, C., and Cerutti, H. 2000. Transgene and transposon silencing in Chlamydomonas reinhardtii by a DEAH-box RNA helicase. Science. 290: 1159-1162.

Yang, Z., Zhu, Q., Luo, K., and Zhou, Q. 2001. The 7SK small nuclear RNA inhibits the CDK9/cyclin T1 kinase to control transcription. Nature. 414: 317-322.

Zieve, G., and Penman, S. 1976. Small RNA species of the HeLa cell: metabolism and subcellular localization. Cell. 8: 19-31.

From: *Frontiers in Computational Genomics*

10

Genome-Scale Phylogenetic Trees

Yuri I. Wolf, Igor B. Rogozin, Nick V. Grishin, and Eugene V. Koonin

Abstract

Genome comparisons indicate that horizontal gene transfer and differential gene loss are major evolutionary phenomena that involve a large fraction, if not the majority, of the genes, at least in prokaryotes. The extent of these events casts doubt on the very feasibility of constructing a "tree of life" because the trees for different genes often tell different stories. However, alternative approaches to tree construction that attempt to determine tree topology on the basis of comparisons of complete gene sets seem to reveal a phylogenetic signal that supports the three-domain evolutionary scenario and suggests the possibility of delineation of previously undetected major clades of prokaryotes. If the validity of these whole-genome approaches to tree building is confirmed by analyses of numerous new genomes that are currently sequenced at an increasing rate, it will seem that the concept of a universal, "species" tree still makes sense, but this tree should be reinterpreted as a prevailing trend in the evolution of genome-scale gene sets rather than a complete picture of evolution.

1. A Super-Brief History of Phylogenetic Trees

The idea that the evolution of life can be represented as a tree, with leaves corresponding to extant species and nodes to extinct ancestors, apparently belongs to Charles Darwin himself and is epitomized in the famous single illustration of the *Origin of Species* (Darwin, 1859). The earliest elaborations on this concept are associated with the name of Ernst Haeckel (Haeckel, 1997). Haeckel's and other early trees were based on a general idea of an hierarchy of relationships between species and higher taxa. Gradually, quantitative criteria have been developed to measure the degree of morphological difference that was thought to reflect evolutionary distance. Another major direction in evolutionary tree construction is cladistics, founded by Willy Hennig, a methodology that employs inferred shared derived characters to identify monophyletic lineages, or clades (Scott-Ram, 1990).

The possibility of using molecular sequences for phylogenetic tree construction was first suggested by Francis Crick and realized by Emile Zuckerkandl and Linus Pauling who built their trees using aligned sequences of the only two protein families that has been extensively characterized at the time, the cytochromes c and the globins (Zuckerkandl and Pauling, 1965). This seminal analysis was done under the molecular clock hypothesis, the notion that a given gene evolves at a constant rate as long as the function of the gene product remains unchanged (Zuckerkandl and Pauling, 1965). With the accumulation of numerous sequences from diverse species, a variety of phylogenetic studies showed that, even if molecular clock could be adopted as a useful null hypothesis, it is violated all too often, which may easily result in incorrect tree topologies. This problem that is inherent in phylogenetic tree analysis, together with the necessity for developing measures of evolutionary distance that take into account multiple substitutions in the same site rate variation between sites within the same gene, brought about a plethora of increasingly sophisticated methods for tree construction. The most popular phylogenetic approaches, each existing in a number of flavors, are distance methods, such as neighbor-joining and least-square method (Fitch- Margoliash), maximum parsimony, and maximum likelihood (Hillis *et al.*, 1996; Nei and Kumar, 2000; Page and Holmes, 1998).

In the early days of molecular phylogenetics, a gene tree was, more or less automatically, equated with the species tree. This view was epitomized in the use of ribosomal RNA sequences as the principal molecular phylogenetic markers. Phylogenetic analysis of rRNA transformed our understanding of the history of life once and for all, resulting in the discovery of a previously unrecognized domain of life, the Archaea, and in a tree topology that has been aptly called the "standard model" of evolution (Doolittle and Handy, 1998; Woese, 1987; Woese *et al.*, 1990). This model involves the early descent of the bacterial clade from the last universal common ancestor and a subsequent separation of archaea and eukaryotes. The standard model has been challenged by the so-called eocyte tree, which was based on a distinct tree topology produced by a method of tree construction developed by James Lake. The eocyte tree derived eukaryotes from one of the branches of the archaea, the crenarchaeota, and thus implied that archaea are, in fact, a paraphyletic group (Rivera and Lake, 1992). However, the standard model was supported by an alternative analysis of rRNA trees (Gouy and Li, 1989) and by many features shared by archaea and eukaryotes, particularly in the composition and organization of the molecular machineries for translation, transcription and replication (Brown and Doolittle, 1997; Leipe *et al.*, 1999), and perforce of these combined arguments, seemed to withstand the challenge. In addition to the formulation of the standard model, phylogenetic analysis of rRNAs brought "the winds of (evolutionary) change" onto taxonomy by revealing, supporting or correcting many major clades among bacteria, archaea, and eukaryotes (Olsen *et al.*, 1994). Thus, this appeared to be the approach of choice to produce the true species tree depicting the history of life, or The Tree of Life, should one choose to adopt loftier language. It was recognized that the tree topology substantially depends on the method used for tree reconstruction, but the very validity of the rRNA-based approach to species phylogeny was not seriously questioned in the pre-genomic era.

The disturbing signs, however, appeared soon after the number of different gene families employed for phylogenetic analysis has become substantial. The problem was that different genes often yielded different trees. This incongruence between tree topologies invaded even the sacred of sacred of phylogenetic taxonomy, the three-domain classification of life. In particular, archaeal genes systematically showed different phylogenetic affinities, the components of information-processing systems typically affiliating with eukaryotes, whereas metabolic enzymes and structural proteins displayed bacterial connections (Golding and Gupta, 1995; Gupta and Golding, 1996). All this was just a foreshadow of things to come once multiple complete genome sequences have become available for comparison.

2. Comparative Genomics Threatens the Species Tree Concept

Systematic comparisons of complete gene sets showed beyond reasonable doubt that there is much more to evolution than vertical inheritance. Horizontal gene transfer and lineage-specific gene loss have come to the fore as major evolutionary phenomena, at least in the prokaryotic world (Doolittle, 1999b; Doolittle, 1999c; Koonin *et al.*, 2001; Koonin *et al.*, 1997; Ragan, 2001a; Ragan, 2001b). The prominence of these events is apparent even without detailed phylogenetic tree analysis. Indeed, examination of the phyletic patterns of orthologous gene families reveals remarkable patchiness, with most families including only a few lineages, and many having an odd composition such as, for example, two bacterial and one archaeal species (Koonin *et al.*, 2001; Tatusov *et al.*, 2001). It seems impossible to explain these patterns without invoking massive, lineage-specific gene loss and horizontal transfer. A recent quantitative analysis suggested that these processes contributed to the evolution of a substantial majority of orthologous sets of prokaryotic proteins (Snel *et al.*, 2002).

Phylogenetic tree analysis for multiple gene families sends the same message. A detailed study of 28 protein families from prokaryotes suggested that, after probable horizontal transfers were removed, there was no reliable phylogenetic signal left in the trees (Teichmann and Mitchison, 1999). Similar results were obtained for proteins that comprise the conserved core of archaeal genomes: while they all showed greater conservation within the archaeal domain than outside it, no clear consensus phylogeny within the archaea could be determined (Nesbo *et al.*, 2001).

Thus, comparative genomics appears to potentially undermine the very idea of a Tree of Life, at least as long as the prokaryotic life is concerned (since prokaryotes comprise two of the three primary kingdoms, a Tree of Life without them is out of the question) (Pennisi, 1998; Pennisi, 1999). However, genomics that seems to "uproot" any simple-minded tree of life based on a single gene or a small group of genes, might also have the potential to offer a way to salvage the concept itself, at least in a "weak" form. The determination of multiple, complete genome sequences of bacteria, archaea and eukaryotes created the opportunity for a new level of phylogenetic analysis that is based not on a phylogenetic tree for selected molecules, *e.g.* rRNA, but (ideally) on the entire body of information contained in the genomes or on a rationally selected, substantial part of this information. Below we briefly discuss different genome-based phylogenetic approaches (or, as we call them, "genome-tree" methods) and the first results in large-scale prokaryotic phylogeny that emerge from the application of these methods and bear on the very feasibility of a Tree of Life.

3. Genome-Trees: Can Genomes Help Build a Consensus?

Methods Based on Gene Content

The most obvious criterion for genome comparison is based on the analysis of gene repertoires. Closely related species share a large proportion of genes; in contrast, distantly related species should have lost a significant fraction of the genes inherited from their last common ancestor, rendering the proportion of shared genes low. If this process carries on in a regular fashion *(i.e.* inter-genomic distances based on gene repertoires can be mapped to time scale uniformly across lineages), it could be used for phylogenetic reconstruction.

The latter requirement raises an obvious *a priori* objection. The well-studied plasticity of prokaryotic genomes results in gene content being malleable by selective pressures, both with respect to gene loss *(e.g.* adaptation to parasitism) and gene acquisition via horizontal transfer *(e.g.* adaptation to extreme environments). As first presciently noted by Charles Darwin himself, traits that are subject to strong selection are less suitable for phylogenetic reconstruction than neutral traits because of highly non-uniform rates of changes and the tendency for convergent evolution among the former. This makes gene content comparisons a relatively weak tool to study prokaryote phylogeny *per se*, although, if treated as a mean to study similarities and differences between genomes, rather than evolutionary relationships, this approach can produce interesting results.

To use gene content for phylogeny or similarity dendrogram reconstruction, one needs, first, to establish orthology (or, in simpler analysis schemes, homology) relationships between genes and, second, to choose a method to translate gene presence-absence data into a tree structure. Both goals can be achieved by various means. Snel and co-workers (Snel *et al.*, 1999) (Huynen *et al.*, 1999) and Korbel and co-workers (Korbel *et al.*, 2002) defined orthologs as intergenomic best hits (BeTs) according to Smith-Waterman expectation (e)-values, computed the fraction of shared genes for genome pairs, converted it to intergenomic distance and then applied neighbor-joining or least-squares method to construct trees. Tekaia and co-workers (Tekaia *et al.*, 1999) used BLAST E-values to establish homology and turned the

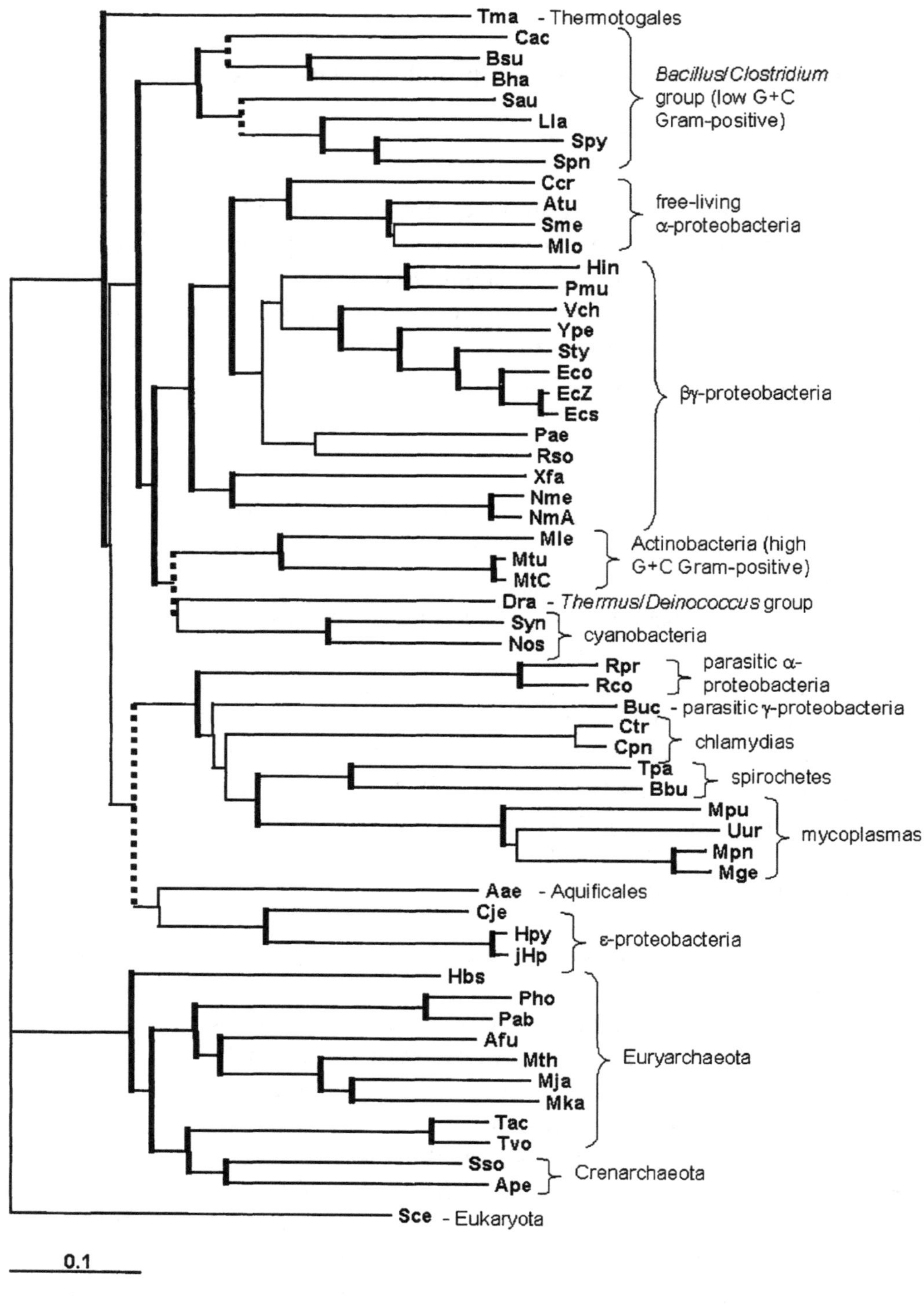
Tma - Thermotogales
Cac
Bsu
Bha
Sau
Lla
Spy
Spn
Bacillus/Clostridium group (low G+C Gram-positive)
Ccr
Atu
Sme
Mlo
free-living α-proteobacteria
Hin
Pmu
Vch
Ype
Sty
Eco
EcZ
Ecs
Pae
Rso
Xfa
Nme
NmA
βγ-proteobacteria
Mle
Mtu
MtC
Actinobacteria (high G+C Gram-positive)
Dra - Thermus/Deinococcus group
Syn
Nos
cyanobacteria
Rpr
Rco
parasitic α-proteobacteria
Buc - parasitic γ-proteobacteria
Ctr
Cpn
chlamydias
Tpa
Bbu
spirochetes
Mpu
Uur
Mpn
Mge
mycoplasmas
Aae - Aquificales
Cje
Hpy
jHp
ε-proteobacteria
Hbs
Pho
Pab
Afu
Mth
Mja
Mka
Tac
Tvo
Euryarchaeota
Sso
Ape
Crenarchaeota
Sce - Eukaryota
0.1

pairwise matrix of shared gene fraction into factorial space, in which they computed intergenomic distances; these were used to build hierarchical classification trees. Fitz-Gibbon and House (Fitz-Gibbon and House, 1999; House and Fitz-Gibbon, 2002) used intergenomic FASTA z-scores followed by single linkage clustering to identify groups of orthologs, then applied parsimony analysis to reconstruct trees. Lin and Gerstein (Lin and Gerstein, 2000) relied on orthology as defined in the Clusters of Orthologous Groups of proteins (COGs) (Tatusov *et al.*, 2001) and built either parsimony or least-square trees. Natale and coworkers (Natale *et al.*, 2000) used the coefficient of co-occurrence (Jaccard coefficient) of genomes in COGs for calculating inter-genomic distances, from which neighbor-joining trees were constructed. Wolf and co-workers (Wolf *et al.*, 2001a) also employed COGs to construct gene presence-absence (in each genome) matrices, which then were used to

Figure 1. An updated gene-content tree of prokaryotes. The inter-genome distances were calculated as follows: i) a set of COGs represented in each genome was determined; ii) the distance between a pair of genomes (A and B) was calculated as $D_{AB}=1-J_{AB}$ where J_{AB} is the Jaccard coefficient which reflects the similarity between the COG sets A and B and is calculated as $J_{AB} = |A \cap B| / |A \cup B|$ ($J_{AB} \subset [0..1]$); iii) a least-squares tree was calculated from the pairwise distance matrix. Bootstrap analysis was performed by resampling the entire set of 4075 COGs. Thick solid lines indicate internal nodes with bootstrap support >90%; thick dotted lines indicate internal nodes with bootstrap support between 70% and 90%, and thin lines show nodes with <70% support.

Species abbreviations. Archaea: Afu, *Archaeoglobus fulgidus*; Hbs, *Halobacterium sp.* NRC-1; Mja, *Methanococcus jannaschii*; Mka, *Methanopyrus kandleri* AV19; Mth, *Methanothermobacter thermautotrophicus*; Tac, *Thermoplasma acidophilum*; Tvo, *Thermoplasma volcanium*; Pho, *Pyrococcus horikoshii*; Pab, *Pyrococcus abyssi*; Ape, *Aeropyrum pernix*; Sso, *Sulfolobus solfataricus*; Sce, *Saccharomyces cerevisiae* (a eukaryotic genome included in addition to the prokaryotes); bacteria: Aae, *Aquifex aeolicus*; Tma: *Thermotoga maritime*; Dra: *Deinococcus radiodurans*; Cac: *Clostridium acetobutylicum*; Mtu, *Mycobacterium tuberculosis* H37Rv; MtC, *Mycobacterium tuberculosis* CDC1551; Mle, *Mycobacterium leprae*; Lla, *Lactococcus lactis*; Spy, *Streptococcus pyogenes* M1 GAS; Spn, *Streptococcus pneumoniae* TIGR4; Sau, *Staphylococcus aureus* N315; Bsu, *Bacillus subtilis*; Bha, *Bacillus halodurans*; Syn, *Synechocystis sp.*; Nos, *Nostoc sp.* PCC 7120; Eco, *Escherichia coli* K12; EcZ, *Escherichia coli* O157:H7 EDL933; Ecs, *Escherichia coli* O157:H7; Ype, *Yersinia pestis*; Sty, *Salmonella typhimurium* LT2; Buc, *Buchnera sp.* APS; Vch, *Vibrio cholerae*; Pae, *Pseudomonas aeruginosa*; Hin, *Haemophilus influenzae*; Pmu, *Pasteurella multocida*; Xfa, *Xylella fastidiosa* 9a5c; Nme, *Neisseria meningitidis* MC58; NmA, *Neisseria meningitidis* Z2491; Rso, *Ralstonia solanacearum*; Hpy, *Helicobacter pylori* 26695; jHp, *Helicobacter pylori* J99; Cje, *Campylobacter jejuni*; Atu, *Agrobacterium tumefaciens* strain C58 (Cereon); Sme, *Sinorhizobium meliloti*; Mlo, *Mesorhizobium loti*; Ccr, *Caulobacter crescentus*; Rpr, *Rickettsia prowazekii*; Rco, *Rickettsia conorii*; Ctr, *Chlamydia trachomatis*; Cpn, *Chlamydophila pneumoniae*; Tpa, *Treponema pallidum*; Bbu, *Borrelia burgdorferi*; Uur, *Ureaplasma urealyticum*; Mpu, *Mycoplasma pulmonis*; Mpn, *Mycoplasma pneumoniae*; Mge, *Mycoplasma genitalium*.

constructed Dollo parsimony trees. Clarke and co-workers (Clarke *et al.*, 2002) computed the ratio of orthologs (identified as reciprocal BeTs) to the number of genes in the smaller genome and constructed least-squares trees. In addition, Wolf and co-workers (Wolf *et al.*, 1999) and Lin and Gerstein (Lin and Gerstein, 2000) built dendrograms on the basis of the predicted protein fold composition of genomes; the former work used linear correlation coefficient between fold abundance lists and the latter relied on presence-absence data.

The trees produced by different applications of this approach are not directly comparable because of the different species sets and different methods for intergenomic distance calculation used by each group. Nevertheless, several major trends are recognizable. Enough phylogenetic information is retained in gene repertoires to provide reliable classification on both ends of the phylogenetic distance scale. Gene content trees show good separation between primary kingdoms and consistently group together closely related species. However, on the intermediate distances, which involve relationships between major lineages, this approach appears to be less suitable for phylogenetic inference. The major factor that determines tree topology appears to be the relative amount of gene loss in different genomes, *e.g.* the major division in the bacterial branch is between the free-living and parasitic forms, which results in well-defined major lineages (*e.g.* Proteobacteria) being broken up (Wolf *et al.*, 2001a). This is readily explained by the above-mentioned genome plasticity of prokaryotic genomes, which strongly manifests in common trends in genome reduction under the selective pressure during the adaptation to parasitism. Attempts to overcome this effect included simple removal of parasites from the analysis (Fitz-Gibbon and House, 1999; House and Fitz-Gibbon, 2002) and normalization of the intergenomic distances by the number of genes of the smaller genome in each pair (Korbel *et al.*, 2002; Snel *et al.*, 1999). The latter method results in reasonable phylogenetic reconstructions, with most of the major prokaryotic lineages recovered. The legitimacy of using the fraction of shared genes for phylogenetic reconstructions has been challenged on the grounds that this is a phenetic rather than phylogenetic character (Doolittle, 1999a). This objection seems to apply to any distance-based method of phylogenetic analysis; however, it is well known that, combined with a proper calculation of evolutionary distances, such methods often result in correct trees. More importantly, however, gene content analysis seems to have less resolution power than some of the other genome-tree approaches (see below).

Figure 1 shows a gene content tree constructed using the latest COG data and the co-occurrence coefficient for calculating intergenomic distances. The separation of free-living and parasitic bacteria is apparent (as discussed above), but, in addition, a potentially phylogenetically meaningful branch unifying cyanobacteria, actinobacteria and Deinococcales appears in this tree (see below).

Methods Based on Gene Order

The same logic as above applies to genome comparisons based on gene order. Rearrangements continuously shuffle the genomes, gradually breaking ancestral gene strings. The operonic organization of a prokaryotic genome complicates makes the kinetics of this process. On the one hand, the selective advantage of physical proximity for co-regulation makes some gene arrays less prone to break-up than others, thus extending the range of evolutionary distances over which gene order comparison is technically possible (Lathe *et al.*, 2000; Wolf *et al.*, 2001b). On the other hand, selective forces acting on operons make them sensitive to the influence of the environmental niche occupied by the organism at any particular time. Furthermore, operons are especially liable to being transferred as a whole, accentuating the effect of lateral transfer on the tree topology (Lawrence, 1997).

Wolf and coworkers (Wolf *et al.*, 2001a) identified pairs of proteins belonging to COGs whose physical proximity is conserved in several genomes. The presence-absence matrices of these pairs were analyzed using Dollo parsimony and neighbor joining, which produced essentially the same topology. Korbel and coworkers (Korbel *et al.*, 2002) counted adjacent pairs of BeT-derived orthologs shared by two genomes, converted the fraction of such pairs to distance and used these distances to construct neighbor-joining or least-squares trees.

Due to the high rate of intragenomic rearrangements, the gene order trees are (at least in theory) especially suitable to resolving the phylogeny of closely related species (Suyama and Bork, 2001). Generally, this approach behaves in a manner similar to the gene content methods, providing a good separation between Archaea and Bacteria and keeping closely related species together, but offering poor resolution on intermediate distances. Both groups described the influence of horizontal gene transfer on these trees.

Methods Based on Evolutionary Distances Between Orthologs

Evolutionary distances measured between different pairs of orthologs in the given two genomes show a broad distribution. In theory, this is due to the genuine variability of mean protein evolution rates caused by the variability in the strength of selective constraints acting on functionally different proteins. In practice, several other factors add to the rate variance, including sampling errors, misidentification of orthologs and lateral gene transfer. Nevertheless, if, for the majority of ortholog pairs, the time of divergence coincides with the divergence of species, it seems reasonable to expect that the distribution of the distances retains enough phylogenetic information to be used for tree construction.

Grishin and coworkers (Grishin *et al.*, 2000), Wolf and coworkers (Wolf *et al.*, 2001a) and Clarke and coworkers (Clarke *et al.*, 2002) used reciprocal BeTs to identify pairs of likely orthologs for a pair of genomes and to obtain a distribution of distances between orthologs. Grishin and coworkers [43] computed the scaling factor for the best-fitting approximation of the distance distribution function, Wolf and coworkers (Wolf *et al.*, 2001a) used the median identity percentage between orthologs transformed into distance, and Clarke and coworkers (Clarke *et al.*, 2002) relied on the median of the relative BLAST score. Least-squares trees were built upon the pairwise distance matrices in all three publications. In addition, Clarke and coworkers (Clarke *et al.*, 2002) employed a statistical test to identify and remove "phylogenetically discordant" sequences (those that displayed an abnormal pattern of similarity to orthologs in several genomes) to reduce the potential effect of lateral gene transfer and misidentification of orthologs. This procedure resulted in an increased bootstrap support but had little effect on tree topology.

The trees constructed by Wolf and coworkers (Wolf *et al.*, 2001a) and Clarke and coworkers (Clarke *et al.*, 2002) include enough genomes to allow a meaningful comparison and are remarkably similar. Both trees confidently separate Archaea from Bacteria. Both recover most of the known major bacterial lineages: proteobacteria (also providing good separation between α-, βγ-and ε-proteobacteria), low G+C gram-positive bacteria (including mycoplasmas), spirochetes and chlamydia. Both trees further suggest a relationship between spirochetes and chlamydia, and both place bacterial hyperthermophyles *(Thermotoga* and *Aquifex*) near the root of the bacterial subtree, followed by *Deinococcus*, *Mycobacterium* and *Synechocystis*. The trees fail to support the monophyly of euryarchaeota, place *Halobacterium* at the root of the archaeal subtree and unify methanogens with *Pyrococcus* species.

Methods Based on Concatenated Alignments of Orthologous Protein Sequences

Traditional sequence-based phylogeny relies on gradual sequence change over time. The three main problems with using single genes to determine the relationships between species are the insufficient number of informative sites, the variability of evolutionary rates in different lineages, and the effect of lateral gene transfer. The former two factors add uncertainty to reconstructions; the latter one leads to protein phylogenies being genuinely different from (the hypothetical) species phylogeny. In an attempt to overcome these problems, one can concatenate many sequence alignments into one and use the combined long sequence for tree reconstruction. If there is no systematic bias in the pattern of horizontal transfers involving the concatenated genes or the likelihood of such transfer is reduced by the careful choice of genes, the trees reconstructed from such an alignment have the potential to provide excellent resolution. There are, however, several limitations to this approach. Firstly, all of the concatenated alignments must include exactly the same set of species represented by one protein each, which limits the choice of families to universally represented ones with no paralogs (or require a selection of paralogs, which might be a source of bias in itself). Secondly, concatenation forces a single sequence change model on all proteins (including branch lengths and intra-protein variability of evolutionary rates), which, in general, is not necessarily true.

Teichmann and Mitchison analyzed a concatenated alignment, mostly comprised of translation-related proteins, using the neighbor-joining algorithm (Teichmann and Mitchison, 1999). They found that horizontal gene transfer in three families likely shaped up their trees; after the elimination of probable transfer candidates, the remaining set failed to produce a resolved tree. In contrast, a concatenated alignment, mostly of ribosomal proteins, constructed by Hansmann and Martin (Hansmann and Martin, 2000) resulted in well-supported maximum likelihood trees, especially after the systematic removal of poorly conserved sites. Brown and coworkers made a concatenated set dominated by translation-related proteins, to produce parsimony and neighbor-joining trees, which had very similar topologies (Brown *et al.*, 2001). Removal of phylogenetically discordant sequences resulted in a well-resolved phylogeny of prokaryotes. Wolf and coworkers used concatenated alignments of universal ribosomal proteins (to minimize the possibility of lateral gene transfer within the data set) to produce maximum likelihood trees (Wolf *et al.*, 2001a). Brochier and coworkers (Brochier *et al.*, 2002) and Forterre and coworkers (Forterre *et al.*, 2002; Matte-Tailliez *et al.*, 2002) used, respectively, a diverse set of translation-related proteins from bacteria and the set of archaeal ribosomal proteins for maximum likelihood tree construction; in these studies, phylogenetic tree construction was preceded by principal component analysis to identify and remove phylogenetically discordant *(i.e.* apparently subject to lateral gene transfer) sequences.

The trees constructed using concatenated alignments share many features with each other and with the ortholog-distance-based trees described above. Especially notable is the similarity of the trees produced by Brown and coworkers (Brown *et al.*, 2001), Wolf and coworkers (Wolf *et al.*, 2001a) and Brochier and coworkers (Brochier *et al.*, 2002). Like the ortholog-distance trees, these trees support most of the well-established bacterial clades. In addition, all three show a common clade for spirochetes and chlamydia. Wolf and coworkers (Wolf *et al.*, 2001a) and Brochier and coworkers (Brochier *et al.*, 2002) place bacterial hyperthermophyles on a common branch at the base of the bacterial subtree, whereas, in the trees of Brown and coworkers (Brown *et al.*, 2001), they form separate deep branches. Wolf and coworkers (Wolf *et al.*, 2001a) and Brochier and coworkers (Brochier *et al.*, 2002) agree on joining *Deinococcus*, *Mycobacterium* and *Synechocystis* in a single clade. The trees of Brown and coworkers (Brown *et al.*, 2001) support a *Deinococcus-Synechocystis* clade, but join actinobacteria with green sulfur bacteria (the latter are absent from the work of Wolf and coworkers (Wolf *et al.*, 2001a); in the work of Brochier and coworkers (Brochier *et al.*, 2002), cyanobacteria belong to a distinct branch together with cytophagales. Both Wolf and coworkers (Wolf *et al.*, 2001a) and Brown and coworkers (Brown *et al.*, 2001) support methanogen-pyrococci and *Thermoplasma-Archaeoglobus* clades within the archaea but disagree on the position of crenarchaeota; the phylogeny of concatenated archaeal ribosomal proteins constructed by Forterre and coworkers (Forterre *et al.*, 2002; Matte-Tailliez *et al.*, 2002) does not reproduce the methanogen-pyrococci clade (nor the monophyly of methanogens themselves) or the *Thermoplasma-Archaeoglobus* clade, but identifies crenarchaeota and euryarchaeota as sister clades.

Approaches Based on Multiple Trees

Combining phylogenetic information contained in multiple, independently reconstructed trees allows, in principle, to achieve the same kind of resolution enhancement as with concatenated alignments without the inherent limitations of the latter. However, these approaches face a methodological problem of their own, namely reconciliation of many trees, possibly including different sets of species and paralogs.

Sicheritz-Ponten and Andersson (Sicheritz-Ponten and Andersson, 2001) analyzed over 8000 individual neighbor-joining trees based on proteins from seven prokaryotic species, systematically counting the nearest neighbors of the proteins in question. The concept of "phylome" (the set of phylogenetic trees including proteins from the given genome) proposed by these researchers, while not producing global trees *per se*, allows one to capture trends in the placement of the given species in the trees. Wolf and coworkers (Wolf *et al.*, 2001a) constructed 132 trees representing (nearly) universal protein families with low propensity for lineage-specific duplications. The nearest neighbor census for particular species and/or higher taxa was used to validate hypotheses formulated during the analysis of other types of genome-trees. Daubin and Gouy (Daubin and Gouy, 2001) employed a "supertree" approach by combining matrix representations (Ragan, 1992a; Ragan, 1992b) (with weights assigned to internal nodes on the basis of bootstrap probabilities) of individual maximum likelihood trees and analyzing the resulting matrix using the neighbor-joining method. Zhaxybaeva and Gogarten (Zhaxybayeva and Gogarten, 2002) employed maximum likelihood mapping of three topologies possible for a quartet of species to poll the trends existing in multiple four-ortholog trees. Two methodologically important studies should be mentioned here, although they did are not directly address prokaryote phylogeny. Page (Page, 2000) used a method that minimizes the total number of "apparent duplications" in the tree to reconcile many individual trees in an attempt to reconstruct the phylogeny of vertebrates. Bapteste and coworkers (Bapteste *et al.*, 2002) analyzed the phylogeny of eukaryotes using the sum of log-likelihoods for competing topologies instead of concatenating alignments, in order to avoid forcing the same parameters of the evolutionary model on different protein families.

The supertree of Daubin and Gouy (Daubin and Gouy, 2001) is generally consistent with the ortholog-distance trees and concatenated-alignment trees described above. It successfully recovers known taxonomic groups, supports a separate clade for bacterial hyperthermophyles (although it is placed as a sister group to proteobacteria instead of in the bacterial root), joins *Deinococcus*, *Mycobacterium* and *Synechocystis* in another clade, and also groups methanogens and pyrococci. Unlike many other approaches, the supertree places spirochetes and chlamydia separately in the bacterial root instead of joining them on a common branch and recovers crenarchaeota and euryarchaeota as sister groups.

The Emerging Consensus

Figure 2 shows a synthetic tree of prokaryotes in which we took into account the results of the genome-tree analyses discussed above (specifically, ortholog distance trees (Wolf *et al.*, 2001a; Clarke *et al.*, 2002), concatenated alignment trees (Brown *et al.*, 2001; Brochier *et al.*, 2002; Forterre *et al.*, 2002; Matte-Tailliez *et al.*, 2002; Wolf *et al.*, 2001a) and the supertree of Daubin and Gouy (Daubin and Gouy, 2001)) and attempted to depict the apparent consensus. At least two major new clades are strongly supported by different types of analysis and appear reliable: i) chlamydia-spirochaetes among bacteria and ii) methanogens-pyrococci among euryarchaeota. In addition, several other major groupings were supported by some but not other approaches and should be considered tentative for the moment, *e.g.* the unification of Cyanobacteria, Deinococcales and Actinobacteria or of Aquificales and Thermotogales.

4. Conclusions and Outlook: Redefining the Concept of the Tree of Life

The results of comparative genomics suggest that the simple notion of a single Tree of Life that would accurately and definitively depict the evolution of all life forms is gone forever. Individual genes, especially those of prokaryotes, follow their unique evolutionary trajectories. This uniqueness stems from the fact that different genes evolved at different points during the history of life and that, in addition to vertical inheritance, evolution of most orthologous families involved multiple instances of lineage-specific gene loss and gene acquisition via horizontal transfer. However, those same comparative-genomic studies that have "uprooted" the Tree of Life provide hope that the concept could be rescued, albeit in a limited sense. Taken together, the results achieved by genome-tree approaches indicate that there is, after all, a phylogenetic signal in the sequences of prokaryotic proteins, but it is weak because of massive gene loss and horizontal transfer and possibly also because of (relatively) rapid divergence of the major lineages

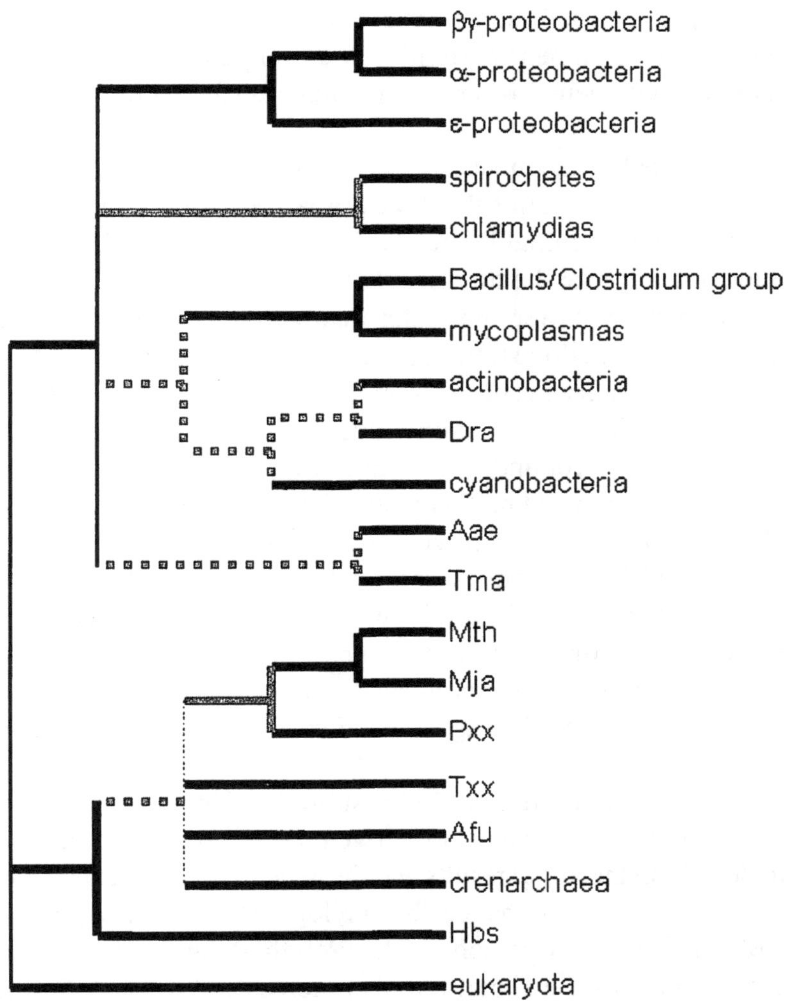

Figure 2. A tentative consensus of genome-trees of prokaryotes. Black branches are conventional clades reproduced by genome-tree approaches; solid red branches are new clades that we consider to be firmly established by genome-trees; broken red branches are tentative clades suggested by genome-trees; thin lines are unresolved multifurcations. Species/clades abbreviation: the same as in Figure 1 and Txx, two *Thermoplasma* species (*T. acidophilum* and *T. volcanium*); Pxx, two *Pyrococcus* species (*P. horikoshii* and *P. abyssi*). ***Colour plate at back of book.***

in the deep past. In order to capture this faint signal, analysis of genome-wide protein sets or of carefully selected subsets is required. With all due caution, these approaches seem to converge on some major new taxa, such as the spirochetes-chlamydia clade, whose existence was not even suspected in the pre-genomic era (Figure 2). However, the concept of the Tree of Life is bound to change in the post-genomic world. It cannot be thought of as a definitive "species tree" (something that does not even exist in reality) anymore but only as a central trend in the rich patchwork of evolutionary history, replete with gene loss and horizontal transfer events (Woese, 2000).

Phylogenies based on sequences of single molecules, including the good old rRNA, have helped define the three primary kingdoms and are extremely useful in resolving terminal branches. For the latter task, they will remain the method of choice for the foreseeable future, given that rRNA sequencing is still incomparably easier and faster than genome sequencing. However, these approaches have done little for our understanding of the fundamentally interesting evolutionary events in between. Moreover, such phylogenies can seriously mislead in cases of anomalous evolutionary modes in particular lineages. The recent conundrum around the phylogenetic position of the hyperthemophilic archaeal methanogen *Methanopyrus kandleri* is a case in point. Ribosomal RNA-based phylogeny placed this organism at the base of the crenarchaeal branch and some analyses suggested an even deeper branching (Burggraf *et al.*, 1991). However, once the genome sequence was determined, genome-trees based on concatenated ribosomal proteins, gene content and gene order all unequivocally pointed to a monophyletic clade of archaeal methanogens including *M. kandleri* (Slesarev *et al.*, 2002).

What's in store for genome-trees? This subfield of molecular phylogenetics is still very young, and the most effective methods remain to be developed. Nevertheless, both the initial results and the general notion that using genome-wide information helps enhance the phylogenetic signal suggest that the future belongs to these approaches. The growth of the number of sequenced genomes accentuates the signal but also increases the noise. Furthermore, powerful tree-building algorithms, such as maximum likelihood, tend to rapidly become prohibitively expensive with the increase in the number of analyzed species. Careful selection of slowly evolving species that are most apt to phylogenetic analysis potentially could allow taking advantage of the progress of genomics while avoiding the above problems. Such analysis should be expected to support or refute some tentative new clades suggested by current genome-trees (Figure 2).

Beyond doubt, major difficulties lie ahead. To mention just one: all alignment-based phylogenetic methods face the problem of site selection, and it becomes all the more pertinent with long concatenated alignments employed in genome-tree analysis. It has been shown that elimination of subsets of aligned residues in concatenated ribosomal protein alignments changes the resulting tree topology, in some cases dramatically (Hansmann and Martin, 2000), but there is no theory that would indicate which subsets tell the right story. This and other problems will occupy molecular phylogeneticists and evolutionary genomicists for years to come. What is already clear is that genomics has brought an extra layer of complexity to molecular evolution but also the information that is required for generating a new, richer picture of evolution.

Acknowledgements

We thank Roman Tatusov for providing updated, pre-release COG data and Mark Ragan and Herve Philippe for providing preprints of their work.

References

Bapteste, E., Brinkmann, H., Lee, J.A., Moore, D.V., Sensen, C.W., Gordon, P., Durufle, L., Gaasterland, T., Lopez, P., Muller, M., and Philippe, H. 2002. The analysis of 100 genes supports the grouping of three highly divergent amoebae: *Dictyostelium*, *Entamoeba*, and *Mastigamoeba*. Proc. Natl. Acad. Sci USA. 99: 1414-1419.

Brochier, C., Bapteste, E., Moreira, D., and Philippe, H. 2002. Eubacterial phylogeny based on translational apparatus proteins. Trends Genet. 18: 1-5.

Brown, J.R. and Doolittle, W.F. 1997. Archaea and the prokaryote-to-eukaryote transition. Microbiol. Mol. Biol. Rev. 61: 456-502.

Brown, J.R., Douady, C.J., Italia, M.J., Marshall, W.E., and Stanhope, M.J. 2001. Universal trees based on large combined protein sequence data sets. Nat. Genet. 28: 281-285.

Burggraf, S., Stetter, K.O., Rouviere, P., and Woese, C.R. 1991. *Methanopyrus kandleri*: an archaeal methanogen unrelated to all other known methanogens. Syst. Appl. Microbiol. 14: 346-351.

Clarke, G.D., Beiko, R.G., Ragan, M.A., and Charlebois, R.L. 2002. Inferring genome trees by using a filter to eliminate phylogenetically discordant sequences and a distance matrix based on mean normalized BLASTP scores. J. Bacteriol. 184: 2072-2080.

Darwin, C. 1859. On the Origin of Species. Murray, London.

Daubin, V. and Gouy, M. 2001. Bacterial molecular phylogeny using supertree approach. Genome Inform. Ser Workshop Genome Inform. 12: 155-164.

Doolittle, R.F. and Handy, J. 1998. Evolutionary anomalies among the aminoacyl-tRNA synthetases. Curr. Opin. Genet. Dev. 8: 630-636.

Doolittle, W.F. 1999a. Lateral gene transfer, genome surveys, and the phylogeny of prokaryotes. Science. 286: 1434a.

Doolittle, W.F. 1999b. Lateral genomics. Trends Cell Biol. 9: M5-M8.

Doolittle, W.F. 1999c. Phylogenetic classification and the universal tree. Science. 284: 2124-2129.

Fitz-Gibbon, S.T. and House, C.H. 1999. Whole genome-based phylogenetic analysis of free-living microorganisms. Nucleic Acids Res. 27: 4218-4222.

Forterre, P., Brochier, C. and Philippe, H. 2002. Evolution of the Archaea. Theoretical Population Biology (in press).

Golding, G.B. and Gupta, R.S. 1995. Protein-based phylogenies support a chimeric origin for the eukaryotic genome. Mol. Biol. Evol. 12: 1-6.

Gouy, M. and Li, W.H. 1989. Phylogenetic analysis based on rRNA sequences supports the archaebacterial rather than the eocyte tree. Nature. 339: 145-147.

Grishin, N.V., Wolf, Y.I., and Koonin, E.V. 2000. From complete genomes to measures of substitution rate variability within and between proteins. Genome Res. 10: 991-1000.

Gupta, R.S., and Golding, G.B. 1996. The origin of the eukaryotic cell. Trends Biochem Sci. 21: 166-71.

Haeckel, E. 1997. The Wonders of Life: A Popular Study of Biological Philosophy. De Young Press, New York.

Hansmann, S., and Martin, W. 2000. Phylogeny of 33 ribosomal and six other proteins encoded in an ancient gene cluster that is conserved across prokaryotic genomes: influence of excluding poorly alignable sites from analysis. Int. J. Syst. Evol. Microbiol. 50: 1655-1663.

Hillis, D.M., Moritz, C., and Mable, B.K. 1996. Molecular Systematics. Sinauer Assoc., New York.

House, C.H., and Fitz-Gibbon, S.T. 2002. Using homolog groups to create a whole-genomic tree of free-living organisms: an update. J. Mol. Evol. 54: 539-47.

Huynen, M.A., Snel, B., and Bork, P. 1999. Lateral gene transfer, genome surveys, and the phylogeny of prokaryotes. Science. 286: 1443a.

Koonin, E.V., Makarova, K.S., and Aravind, L. 2001. Horizontal gene transfer in prokaryotes: quantification and classification. Annu. Rev. Microbiol. 55: 709-742.

Koonin, E.V., Mushegian, A.R., Galperin, M.Y., and Walker, D.R. 1997. Comparison of archaeal and bacterial genomes: computer analysis of protein sequences predicts novel functions and suggests a chimeric origin for the archaea. Mol. Microbiol. 25: 619-637.

Korbel, J.O., Snel, B., Huynen, M.A., and Bork, P. 2002. SHOT: a web server for the construction of genome phylogenies. Trends Genet. 18: 158-

162.
Lathe, W.C., 3rd, Snel, B., and Bork, P. 2000. Gene context conservation of a higher order than operons. Trends Biochem. Sci. 25: 474-479.
Lawrence, J.G. 1997. Selfish operons and speciation by gene transfer. Trends Microbiol. 5: 355-359.
Leipe, D.D., Aravind, L., and Koonin, E.V. 1999. Did DNA replication evolve twice independently? Nucleic Acids Res. 27: 3389-401.
Lin, J., and Gerstein, M. 2000. Whole-genome trees based on the occurrence of folds and orthologs: implications for comparing genomes on different levels. Genome Res. 10: 808-818.
Matte-Tailliez, O., Brochier, C., Forterre, P., and Philippe, H. 2002. Archaeal phylogeny based on ribosomal proteins. Mol. Biol. Evol. 19: 631-639.
Natale, D.A., Shankavaram, U.T., Galperin, M.Y., Wolf, Y.I., Aravind, L., and Koonin, E.V. 2000. Towards understanding the first genome sequence of a crenarchaeon by genome annotation using clusters of orthologous groups of proteins (COGs). Genome Biol. 1: RESEARCH0009.
Nei, M., and Kumar, S. 2000. Molecular Evolution and Phylogenetics. Oxford University Press, Oxford.
Nesbo, C.L., Boucher, Y., and Doolittle, W.F. 2001. Defining the core of nontransferable prokaryotic genes: the euryarchaeal core. J. Mol. Evol. 53: 340-350.
Olsen, G.J., Woese, C.R., and Overbeek, R. 1994. The winds of (evolutionary) change: breathing new life into microbiology. J. Bacteriol. 176: 1-6.
Page, R.D. 2000. Extracting species trees from complex gene trees: reconciled trees and vertebrate phylogeny. Mol. Phylogenet. Evol.14: 89-106.
Page, R.D., and Holmes, E. 1998. Molecular Evolution : A Phylogenetic Approach. Blackwell Science, Inc., London.
Pennisi, E. 1998. Genome data shake tree of life. Science. 280: 672-674.
Pennisi, E. 1999. Is it time to uproot the tree of life? Science. 284: 1305-1307.
Ragan, M.A. 1992a. Matrix representation in reconstructing phylogenetic relationships among the eukaryotes. Biosystems. 28: 47-55.
Ragan, M.A. 1992b. Phylogenetic inference based on matrix representation of trees. Mol. Phylogenet. Evol. 1: 53-58.
Ragan, M.A. 2001a. Detection of lateral gene transfer among microbial genomes. Curr. Opin. Genet. Dev. 11: 620-626.
Ragan, M.A. 2001b. On surrogate methods for detecting lateral gene transfer. FEMS Microbiol. Lett. 201: 187-191.
Rivera, M.C., and Lake, J.A. 1992. Evidence that eukaryotes and eocyte prokaryotes are immediate relatives. Science. 257: 74-76.
Scott-Ram, N.R. 1990. Transformed Cladistics, Taxonomy and Evolution. Cambridge University Press, Cambridge.

Sicheritz-Ponten, T., and Andersson, S.G. 2001. A phylogenomic approach to microbial evolution. Nucleic Acids Res. 29: 545-552.

Slesarev, A.I., Mezhevaya, K.V., Makarova, K.S., Polushin, N.N., Shcherbinina, O.V., Shakhova, V.V., Belova, G.I., Aravind, L., Natale, D.A., Rogozin, I.B., Tatusov, R.L., Wolf, Y.I., Stetter, K.O., Malykh, A.G., Koonin, E.V., and Kozyavkin, S.A. 2002. The complete genome of hyperthermophile Methanopyrus kandleri AV19 and monophyly of archaeal methanogens. Proc. Natl. Acad. Sci. USA. 99: 4644-4649.

Snel, B., Bork, P., and Huynen, M.A. 1999. Genome phylogeny based on gene content. Nat. Genet. 21: 108-110.

Snel, B., Bork, P., and Huynen, M.A. 2002. Genomes in flux: the evolution of archaeal and proteobacterial gene content. Genome Res. 12: 17-25.

Suyama, M., and Bork, P. 2001. Evolution of prokaryotic gene order: genome rearrangements in closely related species. Trends Genet. 17: 10-13.

Tatusov, R.L., Natale, D.A., Garkavtsev, I.V., Tatusova, T.A., Shankavaram, U.T., Rao, B.S., Kiryutin, B., Galperin, M.Y., Fedorova, N.D., and Koonin, E.V. 2001. The COG database: new developments in phylogenetic classification of proteins from complete genomes. Nucleic Acids Res. 29: 22-28.

Teichmann, S.A., and Mitchison, G. 1999. Is there a phylogenetic signal in prokaryote proteins? J. Mol. Evol. 49: 98-107.

Tekaia, F., Lazcano, A., and Dujon, B. 1999. The genomic tree as revealed from whole proteome comparisons. Genome Res. 9: 550-557.

Woese, C.R. 1987. Bacterial evolution. Microbiol. Rev. 51: 221-271.

Woese, C.R. 2000. Interpreting the universal phylogenetic tree. Proc. Natl. Acad. Sci. USA 97: 8392-8396.

Woese, C.R., Kandler, O., and Wheelis, M.L. 1990. Towards a natural system of organisms: proposal for the domains Archaea, Bacteria, and Eucarya. Proc. Natl. Acad. Sci. USA 87: 4576-4579.

Wolf, Y.I., Brenner, S.E., Bash, P.A., and Koonin, E.V. 1999. Distribution of protein folds in the three superkingdoms of life. Genome Res. 9: 17-26.

Wolf, Y.I., Rogozin, I.B., Grishin, N.V., Tatusov, R.L., and Koonin, E.V. 2001a. Genome trees constructed using five different approaches suggest new major bacterial clades. BMC Evol. Biol. 1: 8.

Wolf, Y.I., Rogozin, I.B., Kondrashov, A.S., and Koonin, E.V. 2001b. Genome alignment, evolution of prokaryotic genome organization and prediction of gene function using genomic context. Genome Res. 11: 356-372.

Zhaxybayeva, O. and Gogarten, J.P. 2002. Bootstrap, Bayesian probability and maximum likelihood mapping: exploring new tools for comparative genome analyses. BMC Genomics. 3: 4.

Zuckerkandl, E. and Pauling, L. 1965. Molecules as documents of evolutionary history. J. Theor. Biol. 8: 357-366

From: *Frontiers in Computational Genomics*

11

Mathematical Modeling of the Evolution of Domain Composition of Proteomes: A Birth-and-Death Process with Innovation

Georgy P. Karev, Yuri I. Wolf, Andrey Y. Rzhetsky, Faina S. Berezovskaya, and Eugene V. Koonin

Abstract

A simple model of evolution of the domain composition of proteomes is described, with the following elementary processes: i) domain birth (duplication with divergence), ii) death (inactivation and/or deletion), and iii) innovation (emergence from non-coding or non-globular sequences or acquisition via horizontal gene transfer). This formalism can be described as a birth, death and innovation model (BDIM). The formulas for equilibrium frequencies of domain families of different size and the total number of families at equilibrium were derived for a general BDIM. All asymptotics of equilibrium frequencies of domain families possible for the given type of models are found and their appearance depending on model parameters is investigated. It is proved that the power law asymptotics appears if, and only if, the model is

balanced, *i.e.* domain duplication and deletion rates are asymptotically equal up to the second order. It is further proved that any power asymptotic with the power ≠-1 can appear only if the hypothesis of independence of the duplication/deletion rates on the size of a domain family is rejected. Specific cases of BDIMs, namely simple, linear, polynomial and rational models, are considered in details and the distributions of the equilibrium frequencies of domain families of different size are determined for each case. We apply the developed formalism to the analysis of the domain family size distributions in prokaryotic and eukaryotic proteomes and show a good fit between these empirical data and a particular form of the model, the second-order balanced linear BDIM. The developed approach is oriented at a mathematical description of evolution of domain composition, but a simple reformulation could be applied to models of other evolving networks with preferential attachment.

Introduction

Massive sequencing of genomes from all walks of life, including multiple representatives of diverse lineages of bacteria, archaea and eukaryotes, creates unprecedented opportunities for comparative-genomic studies (Dacks and Doolittle, 2001; Koonin *et al.*, 2000; Lander *et al.*, 2001). One of the mainstream avenues of comparative genomics is the analysis of protein and domain composition of predicted proteomes (Chervitz *et al.*, 1998; Lander *et al.*, 2001; Rubin *et al.*, 2000). These studies often concentrate on domains rather than entire proteins because many proteins have variable multidomain architectures, particularly complex in eukaryotes. As soon as first genome sequences of bacteria became available, it has been shown that a large fraction of the proteome of each species, from approximately one third in bacteria with very small genomes, to a significant majority in species with larger genomes, consists of families of paralogs, proteins that evolved via gene duplication at different stages of evolution (Brenner *et al.*, 1995; Koonin *et al.*, 1995; Labedan and Riley, 1995; Tatusov *et al.*, 1996). Again, a comprehensive analysis of paralogous relationships between proteins is probably best performed at the level of individual domains, firstly, because many proteins share only a subset of common domains, and secondly, because domains can be conveniently and with a reasonable accuracy detected using the available collections of domain-specific sequence profiles (Marchler-Bauer *et al.*, 2002; Ponting *et al.*, 1999). Comparisons of domain repertoires revealed both substantial similarities between different species, particularly with respect to the relative abundance of house-keeping domains, and major differences (Chervitz *et al.*, 1998; Rubin *et al.*, 2000). The latter are typically manifest in lineage-specific expansions of protein/domain families, which probably point to unique adaptations (Jordan *et al.*, 2001; Lespinet *et al.*, 2002). It was also demonstrated that more complex organisms, such as humans, have a greater variety of domains and, in general, more complex domain architectures of proteins than simpler life forms (Koonin *et al.*, 2000; Lander *et al.*, 2001).

Lineage-specific expansions and gene loss events detected as the result of comparative analysis of the domain compositions of different proteomes have been interpreted mostly qualitatively, in terms of the underlying biological phenomena, such as adaptation associated with expansion or coordinated loss of functionally linked sets of genes (Aravind *et al.*, 2000). A complementary approach involves quantitative comparative analysis of the frequency distributions of proteins or domains in different proteomes. In several studies, it has been pointed out that these distributions appear to fit the power law: $P(i) \cong ci^{-\gamma}$ where $P(i)$ is the frequency of domain families including exactly i members, c is a normalization constant and positive γ is a parameter, which typically assumes values between 1 and 3 (Huynen and van Nimwegen, 1998; Qian *et al.*, 2001; Rzhetsky and Gomez, 2001; Wuchty, 2001). Obviously, in double-logarithmic coordinates, the plot of P as a function of i is a straight line with a negative slope. Power distributions appear in numerous biological, physical and other contexts, which seem to be fundamentally different, *e.g.* the number of links between documents in the Internet, the population of towns or the number of species that become extinct within a year (Albert and Barabasi, 2002; Albert *et al.*, 2000; Amaral *et al.*, 2000; Barabasi, 2002; Barabasi and Albert, 1999; Dorogovtsev and Mendes, 2001; Gisiger, 2001; Jeong *et al.*, 2001; Jeong *et al.*, 2000; Krapivsky and Redner, 2001).

These distributions are scale-free, *i.e.* the shape of the distribution remains the same regardless of scaling of the analyzed variable. In particular, scale-free behavior has been observed in networks describing diverse phenomena, *e.g.* the metabolic pathways of an organism or infectious contacts during an epidemic spread (Barabasi, 2002; Gisiger, 2001; Jeong *et al.*, 2001; Jeong *et al.*, 2000). The principal pattern of network evolution that ensures the emergence of power distributions (and, accordingly, scale-free properties) is preferential attraction, whereby the probability of a node acquiring a new connection increases with the number of connections this node already has.

The importance of the analysis of frequency distributions of domain or proteins lies in the fact that specific forms of such distributions are linked to models of evolution. Therefore, by exploring the distributions, inferences can be made on the mode and parameters of genome evolution. In order to achieve this goal, the connections between domain frequency distributions and evolutionary models need to be explored theoretically within as general class of models as possible. In this work, we undertake such a mathematical analysis using a simple model of evolution with duplication (birth), elimination (death) and *de novo* emergence (innovation) of a domain as elementary processes. Depending on the parameters of the model, all asymptotics of equilibrium frequencies of domain families of different size possible for this type of models were identified, and analytical conditions on birth and death rates producing power asymptotics are determined. We prove that the power law asymptotics appears if, and only if, the model is balanced, *i.e.* domain duplication and deletion rates are asymptotically equal up to the second order, and that any power asymptotic with the power $\neq$-1 can appear only if the hypothesis of independence of the duplication/deletion rates on the size of a domain family is rejected. We applied the developed formalism to the analysis of the frequency distributions of domains in individual prokaryotic and eukaryotic genomes and showed a good fit between these data and a particular form of the model, the second-order balanced linear BDIM

Mathematical Theory and Model

1. Fundamental Definitions and Assumptions

We treat a genome as a "bag" of coding sequence for protein domains, which we will simply call ***domains*** for brevity. Domains are treated as independently evolving units disregarding the dependence between domains that tend to belong to the same multidomain protein. Each domain is considered to be a member of a ***family*** (including families with only one member). We consider three types of elementary events: i) domain ***birth,*** which generates a new member in the same family; the principal mechanism of birth is duplication with divergence but additional mechanisms may be considered including acquisition of a family member from a different species via horizontal gene transfer which generates a new domain in the same family (Koonin *et al.*, 2001), ii) domain ***death,*** which results from domain inactivation and/or deletion, and c) domain ***innovation,*** which generates a new family with one member; innovation may occur via domain evolution from a non-coding sequence or a sequence of a non-globular protein, via horizontal gene transfer from another species, or via radical modification of a domain following a duplication. The innovation rate, which we designate ν, is considered constant for a given genome. The rates of elementary events are considered to be independent of time (only homogeneous models are considered) and of the nature (structure, biological function etc.) of individual families.

In a finite genome, the maximal number of domains in a family obviously cannot exceed the total number of domains and, in reality, is probably much smaller; let N be the maximal possible number of domain family members. We consider classes of domain families, which have only one common feature, namely the number of members (obviously, a class may include families that are completely unrelated biologically, Figure 1). Let f_i be the number of domain families in i-th class, *i.e.* families that have exactly i domains in the given genome, i=1,2,…N. Birth of a domain in a family belonging to class i results in relocation of this family from class i to class

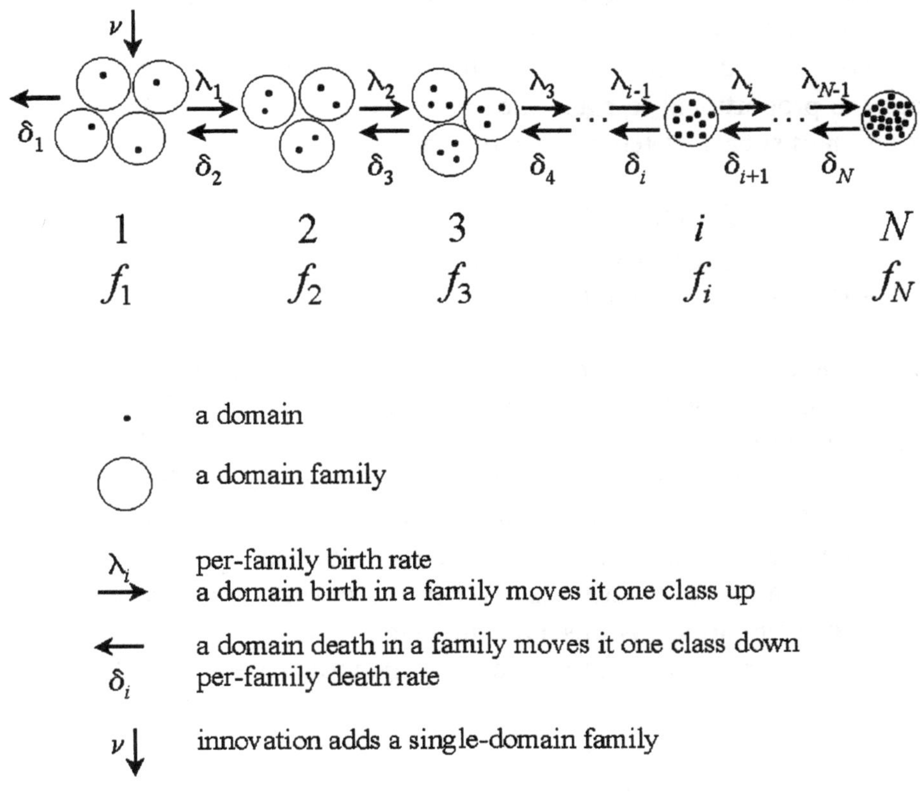

Figure 1. Domain dynamics and evolutionary events under BDIM.

i+1 (decrease of f_i and increase of f_{i+1} by 1). Conversely, death of a domain relocates the corresponding family to class i-1; death of a domain of class 1 results in the elimination of the corresponding family (this being the only mechanism of family death). We consider time to be continuous and suppose it very unlikely that the more than one elementary event occur during a short time interval; formally, the probability that more than one event occurs during an interval Δt is $o(\Delta t)^2$.

In what follows, we describe models of evolution of the domain composition of proteomes, which involves the above elementary processes and, accordingly, can be described as birth, death and innovation models (BDIMs).

2. The Formulation of the Model

2.1. The Simple BDIM

Let us formulate the following **hypothesis of independence**: i) all elementary events are independent of each other; ii) the rates of individual domain birth (λ) and death (δ) do not depend on i (number of domains in a family). Under this hypothesis, the instantaneous rate with which a domain family leaves class i is proportional to i and the following simple BDIM describes the evolution of such a system of domain classes:

$$\begin{aligned}
&\mathrm{d}f_1(t)/\mathrm{d}t = -(\lambda+\delta)f_1(t) + 2\delta f_2(t) + \nu \\
&\mathrm{d}f_i(t)/\mathrm{d}t = (i-1)\lambda f_{i-1}(t) - i(\lambda+\delta)f_i(t) + (i+1)\,\delta f_{i+1}(t) \text{ for } 1<i<N, \qquad (2.1)\\
&\mathrm{d}f_N(t)/\mathrm{d}t = (N-1)\lambda f_{N-1}(t) - N\delta f_N(t).
\end{aligned}$$

Similar models have been considered previously in several different contexts (Feller, 1967-1968 v. 1, ch. 17; Ijuri and Simon, 1977). We will see (in s.3.2) that the solution of model (2.1) evolves to equilibrium, with a specific distribution of the domain family sizes, f_i, namely $f_i \sim (\lambda/\delta)^i/i$; in particular, if $\lambda=\delta$, then $f_i \sim 1/i$. Thus, under the simple BDIM, if the birth rate equals the death rate, the abundance of a domain class is inversely proportional to the size of the families in this class. When the observations do not fit this particular asymptotic (as observed in several studies on distributions of protein family sizes), a different, more general model needs to be developed.

2.2. The Master BDIM

A more general BDIM emerges when the Hypothesis of Independence is rejected. Instead of constructing specific hypotheses regarding the dependence between the elementary events, let us suppose simply that the domain birth and death rates in a family of class i do not show simple proportionality to i. For the general case, we designate these rates, respectively, λ_i and δ_i; in the specific case of the simple BDIM (2.1), $\lambda_i=\lambda i$ and $\delta_i=\delta i$. Then we have the following ***master model***:

$$d f_1(t)/dt = -(\lambda_1 +\delta_1) f_1(t)+ \delta_2 f_2(t)+\nu$$
$$d f_i(t)/dt = \lambda_{i-1} f_{i-1}(t) -(\lambda_i +\delta_i) f_i(t)+ \delta_{i+1} f_{i+1}(t) \text{ for } 1<i<N, \quad (2.2)$$
$$d f_N(t)/dt = \lambda_{N-1} f_{N-1}(t) -\delta_N f_N (t).$$

Let $F(t)=\sum_{i=1}^{N} f_i(t)$ be the total number of domain families at instant t; it follows from (2.2) that

$$dF(t)/dt =\nu -\delta_1 f_1(t) \quad (2.3)$$

The system (2.2) has an equilibrium solution $f_1,\dots f_N$ defined by the equalities $df_i(t)/dt = 0$ for all i; this solution is described below under Proposition 1. Accordingly, there exists an equilibrium solution of equation (2.3), which we will designate F_{eq} (the total number of domain families at equilibrium). At equilibrium, $\nu=\delta_1 f_1$, *i.e.* the processes of innovation and death of *single* domains (more precisely, the death of domain families of class 1, *i.e.* singletons) are balanced.

We can rewrite the model (2.2) in terms of the frequency of a domain family of class i $p_i(t)= f_i(t)/ F(t)$. Let $x(t)=y(t)/Y(t)$; then

$$dx/dt =[dy/dt /y -dY/dt /Y]\, x.$$

Applying this identity to $p_i(t)$ and rewriting equation (2.3) in the form

$$[dF(t)/dt]/ F(t)=\nu /F(t) -\delta_1 p_1(t) \quad (2.3')$$

we get the following model for frequencies of the domain families (*master model for frequencies*), which is equivalent to (2.2):

$$d\,p_1(t)/dt = -(\lambda_1+\delta_1)\,p_1(t)+\delta_2 p_2(t)+\nu/F(t)-(\nu/F(t)-\delta_1 p_1(t))\,p_1(t), \tag{2.4}$$
$$d\,p_i(t)/dt = \lambda_{i-1}p_{i-1}(t)-(\lambda_i+\delta_i)\,p_i(t)+\delta_{i+1}\,p_{i+1}(t)-(\nu/F(t)-\delta_1 p_1(t))\,p_i(t) \text{ for } 1<i<N,$$
$$d\,p_N(t)/dt = \lambda_{N-1}p_{N-1}(t) - \delta_N\,p_N(t) - (\nu/F(t)-\delta_1 p_1(t))]\,p_N(t).$$

System (2.4) should be solved together with equation (2.3).

2.3. The Master BDIM and Markov Processes

Let us note that system (2.4) for frequencies is *non-linear* and so it is not a system of Kolmogorov equations for state probabilities of any homogeneous Markov process. Next, let us suppose that the actual genome had ample time to arrive at equilibrium with respect to the total number of domain families, such that $F(t)= F_{eq}$.

It has to be emphasized that this relation does not imply $dp_i(t)/dt=0$ or $df_i(t)/dt=0$; that is, the system may rearrange the frequencies of individual families, although the total number of families remains stable. If $F(t)= F_{eq}$, the master system (2.4) turns into

$$d\,p_1(t)/dt = -(\lambda_1+\delta_1)\,p_1(t)+\delta_2 p_2(t)+\nu/F_{eq} \tag{2.5}$$
$$d\,p_i(t)/dt = \lambda_{i-1}p_{i-1}(t) - (\lambda_i+\delta_i)\,p_i(t)+\delta_{i+1}\,p_{i+1}(t) \text{ for } 1<i<N,$$
$$d\,p_N(t)/dt = \lambda_{N-1}p_{N-1}(t) - \delta_N\,p_N(t).$$

System (2.5) can be rewritten as a matrix equation

$$d\boldsymbol{p}(t)/dt = \boldsymbol{p}(t)Q,$$

where $\boldsymbol{p}(t)=\{p_1(t),\dots p_N(t)\}$ and the matrix $Q=(q_{ij})$ is defined by equalities $q_{11}=-(\lambda_1+\delta_1)+\nu/F_{eq}$, $q_{21}=\delta_2+\nu/F_{eq}$, $q_{s1}=\nu/F_{eq}$ for all $s>2$;
$q_{i-1,i}=\lambda_{i-1}$, $q_{i,i}=-(\lambda_i+\delta_i)$, $q_{i+1,i}=\delta_{i+1}$, $q_{k,i}=0$ for all $k, |i-k|>1$, $i=2,\dots N-1$,
$q_{N-1,N}=\lambda_{N-1}$, $q_{N,N}=-\delta_N$.

It is easy to see that the sum of elements of each row (except for the first one) of the matrix Q is equal to $\nu/F_{eq}>0$. Therefore the matrix Q cannot be a matrix of transition rates for any Markov process (the sum of elements of each row of a matrix of transition rates for Markov process with continuous time should be non-positive (see Feller, 1967-1968 v. 1, ch.17, s. 8; Gihman and Skorohod, 1975 v. 2, ch. 3, s. 2); in other words, there is no Markov process with continuous time and state space $\{1,2,\dots N\}$ whose state probabilities satisfy the system (2.5).

Thus, neither the initial BDIMs (2.1) or (2.2) nor the equilibrium model (2.5) cannot be described by any Markov process with continuous time.

Remark. If $\nu=0$ in system (2.5), then system (2.5) turns into a system of state probabilities for a Markov process of birth and death with continuous time.

3. Equilibrium in BDIMs

3.1. Equilibrium Sizes and Frequencies of the Domain Family System

Let us suppose that the genome had ample time to arrive at a complete equilibrium state, in which not only $dF(t)/dt=0$, but also $df_i(t)/dt=0$ for all i. Thus, the equilibrium sizes of domain families f_i satisfy the system

$$\begin{aligned} &-(\lambda_1+\delta_1)f_1+\delta_2 f_2+\nu=0,\\ &\lambda_{i-1}f_{i-1}-(\lambda_i+\delta_i)f_i+\delta_{i+1}f_{i+1}=0 \text{ for } 1<i<N,\\ &\lambda_{N-1}f_{N-1}-\delta_N f_N=0. \end{aligned} \tag{3.1}$$

It should be emphasized that the master model does not assign *a priori* the value of F_{eq}; this value has to be computed depending on the model parameters.

The following statement is central for further analysis.

Proposition 1. *The master BDIM (2.2) has the single equilibrium state* $(f_1, \ldots f_N)$, *which is the sole solution* of *system (3.1):*

$$f_1=\nu/\delta_1$$

$$f_i=\nu\prod_{j=1}^{i-1}\lambda_j/\prod_{j=1}^{i}\delta_j \text{ for all } i=2,\ldots N. \tag{3.2}$$

The single equilibrium state (3.2) is globally asymptotically stable.

In addition (formally assuming $\prod_{j=1}^{i-1}\lambda_j=1$ for $i=1$),

$$F_{eq}=\nu\Big(\sum_{i=1}^{N}\prod_{j=1}^{i-1}\lambda_j/\prod_{j=1}^{i}\delta_j\Big) \tag{3.3}$$

This proposition ascertains that all evolutionary trajectories of system (2.2) exponentially tend to the single equilibrium state (3.2). The proof is given in the Mathematical Appendix.

Remark. Let us denote the ratio of the birth rate to the innovation rate

$$G(N) \equiv \sum_{i=1}^{N-1} \lambda_i f_i /\nu,$$

and the ratio of the death rate to the innovation rate

$$I(N) \equiv \sum_{i=1}^{N} \delta_i f_i /\nu.$$

Then, according to Proposition 1, for any BDIM in its equilibrium state

$$G(N)- I(N) = \sum_{i=1}^{N-1} \prod_{j=1}^{i} \lambda_j /\delta_j - \sum_{i=2}^{N-1} \prod_{j=1}^{i} \lambda_j /\delta_j -1= -1.$$

The principal goal of the subsequent treatment is the analysis of asymptotic behaviors of equilibrium frequencies and sizes of domain families $(f_1, \ldots f_N)$ at large N. We will differentiate two cases of asymptotic behavior according to the following:

Definition. Let $\{q_i\}$, $\{s_i\}$ be sequences of real numbers; let us denote $q_i \cong s_i$ if $\lim q_i/s_i=1$ and $q_i \sim s_i$ if $\lim q_i/s_i=c=const$ and $0<c<\infty$. We will also use this notation for finite but long enough sequences.

3.2. Equilibrium Frequencies for the Simple BDIM

Let us apply Proposition 1 to simple BDIM (2.1) with $\lambda_i = \lambda i$, $\delta_i = \delta i$.

Definition. A simple BDIM is *balanced* if $\theta=\lambda/\delta=1$, that is, if the rates of individual domain birth and death are equal.

A random discrete variable ξ has the *logarithmic* distribution with parameter $\theta<1$, if

$$P(\xi = i) = \theta^{\,i}/i\,[-\ln(1-\theta)]^{-1},\ i=1,2,\ldots$$

A random variable ξ has the *truncated logarithmic* distribution with parameter θ, if

$$P(\xi = i) = C_n \theta^{\,i}/i,\ i=1,2,\ldots n,\ C_n=1/\sum_{j=1}^{n} \theta^{\,j}/j.$$

Then we have:

Proposition 2.

1) *For any simple BDIM (2.1)*

$$f_i = (\nu/\delta)\theta^{i-1}/i = (\nu/\lambda)\theta^i/i\,, \tag{3.4}$$

$$F_{eq} = \sum_{i=1}^{N} f_i = \nu/\delta \sum_{j=1}^{N} \theta^{j-1}/j\,, \tag{3.5}$$

and

$$p_i = (1/F_{eq})(\nu/\delta)\theta^{i-1}/i = (\theta^i/i) \Big/ \sum_{j=1}^{N} \theta^j/j \tag{3.6}$$

that is, the equilibrium frequencies have the truncated logarithmic distribution if $\theta<1$.

2) *If a simple BDIM is balanced, then*

$$F_{eq} = \nu/\delta \sum_{j=1}^{N} 1/j\,, \tag{3.7}$$

and for all $i=1,2,\ldots N$

$$p_i = \nu/\delta\, F_{eq}/i = \Big(\sum_{j=1}^{N} 1/j\Big)^{-1} / \, i. \tag{3.8}$$

The proof is given in the Mathematical Appendix.

Thus, a simple BDIM can have equilibrium frequencies only of the form $p_i=C\theta^i/i$, where $C=const$ and θ is the distribution parameter. In particular, the equilibrium frequencies for a balanced simple BDIM have the power distribution with the power equal to –1.

Simple methods exist for preliminary graphical estimation of the single distribution parameter θ (Johnson *et al.*, 1992 ch. 7, s. 7). We will prove in the following section that, if we observe a power asymptotic of empirically observed equilibrium frequencies, then (assuming that the system can be described by a BDIM), the rates λ_i and δ_i should be asymptotically equal at large *i*. If, additionally, the power of the asymptotic is *not equal* to –1, then the system dynamics *cannot* be described by a *simple* BDIM. Thus, it is necessary to consider more general models, such as the Master BDIM (2.2).

3.2. Asymptotic Behavior of Equilibrium Frequencies of a Master BDIM: Main Theorems

Let us consider the master BDIM (2.2); we showed in s.3.1 that its equilibrium frequencies solve the system

$$-(\lambda_1+\delta_1)p_1+\delta_2 p_2+\nu/F_{eq}=0,$$
$$\lambda_{i-1}p_{i-1}-(\lambda_i+\delta_i)p_i+\delta_{i+1}p_{i+1}=0 \text{ for } 1<i<N,$$
$$\lambda_{N-1}p_{N-1}-\delta_N p_N=0. \quad (3.9)$$

The following theorem describes all possible types of asymptotic behaviors of the equilibrium frequencies and defines the connections between these asymptotics depending on model parameters. In particular, if we have no information on the exact form of dependence of the rates of birth and death of domains on the size of a domain family, the theorem can be used to describe the qualitative type of dynamics of asymptotic behaviors of the equilibrium frequencies.

We will prove that asymptotic behavior of a solution of system (3.9) is completely defined by the asymptotic relation between λ_i and δ_i.

More precisely, let us define a function $\chi(i)=\lambda_{i-1}/\delta_i$; we consider only functions of a power growth, *i.e.* $\chi(i)\sim i^s$ at $i\to\infty$ for a real s. We will see that this is not a serious restriction because the most realistic situations correspond to the case of s=0. So, let us suppose that, for large i, the following expansion is valid:

$$\chi(i)\equiv\lambda_{i-1}/\delta_i=i^s\theta(1+a/i+O(1/i^2)) \quad (3.10)$$

where s, a are real numbers and θ is positive. Evidently, if $s\neq0$, $\chi(i)$ tends either to 0 ($s<0$) or to ∞ ($s>0$) at large i.

Definition. Let us refer to a BDIM (2.2), (3.10) as

i. *non-balanced*, if $s\neq0$;
ii. *first-order balanced*, if $s=0$ and $\theta\neq1$, *i.e.*
$\lambda_{i-1}/\delta_i=\theta\,(1+a/i+O(1/i^2))$ at large i; (3.11)
iii. *second-order balanced*, if $s=0$, $\theta=1$ and $a\neq0$, *i.e.*
$\lambda_{i-1}/\delta_i=1+a/i+O(1/i^2))$ for large i; (3.12)
iv. *high-order balanced*, if $s=0$, $\theta=1$ and $a=0$, *i.e.*
$\lambda_{i-1}/\delta_i=1+O(1/i^2))$ for large i.

We will show that the first three coefficients, s, θ and a of asymptotic expansion (3.10) for $\chi(i)=\lambda_{i-1}/\delta_i$ exactly specify all possible asymptotic behaviors of equilibrium frequencies of BDIM.

Theorem 1. *The equilibrium frequencies p_i of BDIM (2.2) have the following asymptotics*

i. *if the model is non-balanced, then*
   $p_i \sim \Gamma(i)^s \theta^i i^a$, where $\Gamma(i)$ is the Γ-function;
ii. *if the model is first-order balanced, then*
   $p_i \sim \theta^i i^a$;
iii. *if the model is second-order balanced, then*
   $p_i \sim i^a$;
iv. *if the model is high-order balanced, then*
   $p_i \sim 1$

(Figure 2).

The proof is given in the Mathematical Appendix.

It follows from this theorem that, if a BDIM is non-balanced, then its equilibrium frequencies p_i (and stable family sizes f_i) increase or decrease extremely fast (hyper-exponentially) with the growth of i. In contrast, if a BDIM has a non-zero order of balance, asymptotic behavior is observed.

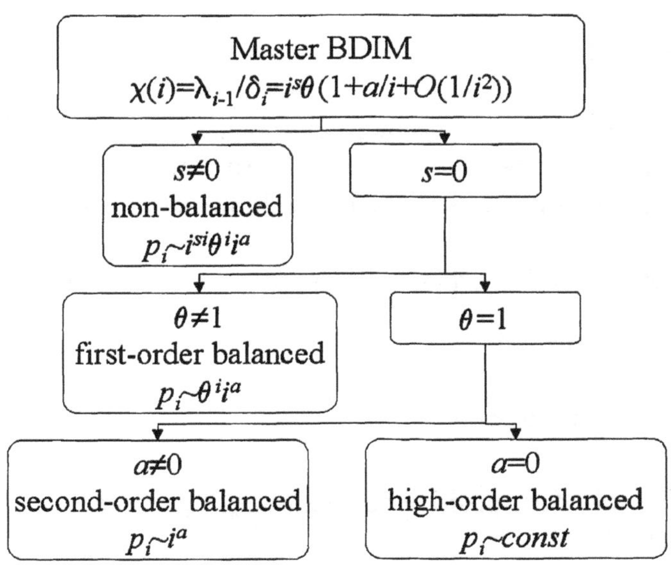

Figure 2. Different orders of balance in BDIMs.

Let us recall that a random discrete variable ξ has the *Pascal* (or *negative binomial*) *distribution* with parameters (r,q), $r>0$, $0<q<1$, if $P(\xi=k)=\Gamma(r+k)/[\Gamma(r)\,\Gamma(1+k)]\,(1-q)^r q^k$ (see Johnson *et al.*, 1992).
We will say that sequence $\{p_i\}$ follows (or asymptotically has) a discrete probabilistic distribution $\{q_i\}$, if $p_i \sim q_i$ for large enough i.

Corollary 1. *For a first-order balanced BDIM with* $\theta<1$,

a) *if* $a>-1$, *the equilibrium frequencies* p_i *follow Pascal distribution with parameters* $(a+1,\theta)$;
b) *if* $a=-1$, *the equilibrium frequencies follow truncated logarithmic distribution with parameter* θ;
c) *if* $a=0$, *the equilibrium frequencies follow geometric distribution with parameter* θ.

The following implication of Theorem 1 is of principal interest.

Corollary 2. *Equilibrium frequencies of a BDIM have a power asymptotic behavior if and only if the BDIM is second-order balanced.*

Corollary 3. *For high-order balanced BDIM, if* $\lambda_{i-1}/\delta_i=1$ *for all i, then the only possible distribution of equilibrium frequencies is uniform,* $p_i=const$ *for all i. Moreover, even if* $\lambda_{i-1}/\delta_i=1+O(1/i^2)$, *the equilibrium frequencies asymptotically have the uniform distribution.*

4. Rational BDIM

Rational models compose a rather general class of BDIMs (Figure 3), for which the asymptotic behavior of the equilibrium frequencies and equilibrium sizes of domain families can be completely investigated.

Let us suppose that the birth and death rates are of the form

$$\lambda_i=\lambda P(i)=\lambda\prod_{k=1}^{n}(i+a_k)^{\alpha_k}, \qquad (4.1)$$

$$\delta_i=\delta Q(i)=\delta\prod_{k=1}^{m}(i+b_k)^{\beta_k}$$

for $i>0$, where λ, δ are positive constants, α_k, β_k are real and a_k, b_k are non-negative for all $k=1,\dots N$.

We will refer to BDIM (2.2), (4.1) as *rational* BDIM.

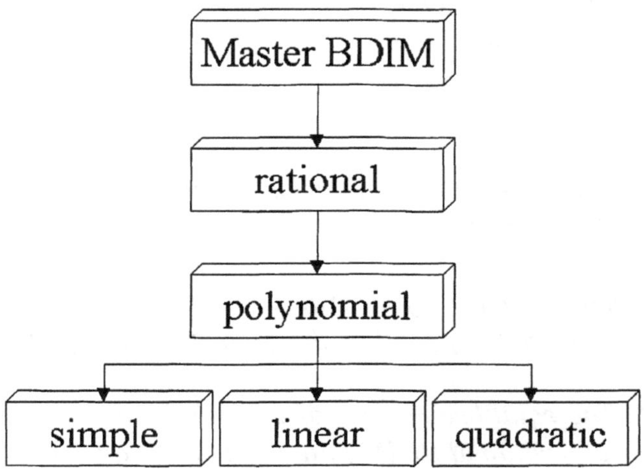

Figure 3. Hierarchy of BDIM types.

It is known that a very wide class of mathematical functions can be well approximated by rational functions of the form (4.1) (see, *e.g.* Henrici, 1986). Particular cases of a rational BDIM are *simple* BDIM with $P(i)=i$, $Q(i)=i$, *linear* BDIM with $P(i)=i+a_1$, $Q(i)=i+b_1$, where a_1, b_1 are constants, and polynomial BDIM, if $P(i)$ and $Q(i)$ are polynomials on i.

The following theorem describes all possible asymptotic behaviors of the equilibrium frequencies of a rational BDIM. Let us denote

$$\theta=\lambda/\delta,$$

$$\eta=\sum_{k=1}^{n}\alpha_k-\sum_{k=1}^{m}\beta_k,$$

$$\rho=\sum_{k=1}^{n}a_k\alpha_k-\sum_{k=1}^{m}b_k\beta_k,$$

$$\beta=\sum_{k=1}^{m}\beta_k.$$

Theorem 2. *The equilibrium sizes of domain families of a rational BDIM have the following asymptotics*

$$f_i \cong C\nu/\lambda\,\Gamma(i)^{\eta}\theta^{i}\, i^{\rho-\beta} \qquad (4.2)$$

where the constant $C=\prod_{k=1}^{m}(\Gamma(1+b_k))^{\beta_k}/\prod_{k=1}^{n}\Gamma(1+a_k)^{\alpha_k}$. (4.3)

The proof is given in the Mathematical Appendix.

Corollary 1. *If* $\eta=0$, *then the rational BDIM is first-order balanced and the* *sequence of equilibrium numbers of domain families* $\{f_i\}$ *has a power-exponential asymptotics*

$$f_i \cong C\nu/\lambda\, \theta^{\,i}\, i^{\rho-\beta}. \qquad (4.4)$$

In particular, if $\rho-\beta > -1$, *the equilibrium frequencies* p_i *follow the Pascal distribution with parameters* $(\rho-\beta+1,\theta)$;

if $\rho-\beta = -1$, *then frequencies* p_i *follow the truncated logarithmic distribution; if* $\rho-\beta=0$, *then frequencies* p_i *follow the geometric distribution.*

Corollary 2. *The equilibrium sizes of domain families* f_i *and equilibrium frequencies* p_i *of a rational BDIM follow the power distribution if and only if* $\eta=0$ *and* $\lambda=\delta$, *i.e. the BDIM is second-order balanced, then*

$$f_i \cong C\nu/\lambda\, i^{\rho-\beta}. \qquad (4.5)$$

Formula (4.4) gives the asymptotics for the equilibrium sizes of domain families f_i and, accordingly, for the total number of families F_{eq}. The exact expressions for these quantities can be extracted from the proofs of Theorem 2 and Lemma (see Mathematical Appendix).

Proposition 5.

i. The equilibrium sizes of domain families f_i *of a balanced (first or higher order) rational BDIM are*

$$f_i = C\nu/\delta\theta^{\,i-1} \prod_{k=1}^{n}[(\Gamma(i+a_k))^{\alpha_k}] / \prod_{k=1}^{m}[(\Gamma(i+1+b_k))^{\beta_k}] \text{ for all } i=1,2,\ldots$$

where

$$C=\prod_{k=1}^{m}[(\Gamma(1+b_k))^{\beta_k}] / \prod_{k=1}^{n}[\Gamma(1+a_k)^{\alpha_k}].$$

ii. The total number of domains at equilibrium is

$$F_{eq} = C\nu/\delta\Big(\sum_{j=1}^{N}\theta^{\,j-1}\prod_{k=1}^{n}(\Gamma(j+a_k))^{\alpha_k} / \prod_{k=1}^{m}(\Gamma(j+1+b_k))^{\beta_k}\Big).$$

For the rational 2nd order balanced BDIM, the ratio of the birth rate to the innovation rate is

$$G(N)= \sum_{i=1}^{N-1}\theta^{\,i}\prod_{k=1}^{m}[\Gamma(i+1+a_k)/\Gamma(1+a_k)]^{\alpha_k} / [\Gamma(i+1+b_k)/\Gamma(1+b_k)]^{\beta_k}.$$

The asymptotic formulas for equilibrium frequencies of rational BDIM could be considered as particular cases of the corresponding formulas of general theorem 1. Proposition 5 allows calculating the constants in the corresponding asymptotic formulas for the sizes of domain families for a rational BDIM. If only equilibrium frequencies are analyzed, the values of these constants become irrelevant because they contract. However, if the actual values of f_i and F_{eq} are of interest, the values of the constants are required.

5. Properties of the Main Types of Rational BDIMs

5.1. Simple BDIM

As shown above, a simple BDIM can have equilibrium frequencies only of the form $p_i = C\theta^i/i$, $C = const$; in particular, if the distribution parameter $\theta<1$, we get the (truncated) logarithmic distribution. Logarithmic distribution fit many observed distribution in biology, *e.g.*, the distribution of species by the number of individuals in populations or, what is more relevant, the distribution of protein folds by the number of families per fold (Wolf *et al.*, 2000). Thus, a simple BDIM could be potentially used for modeling the dynamics of biological systems that have logarithmic distributed equilibrium densities. We examine this possibility in greater detail starting with the case $\lambda=\delta$ (second-order balanced simple BDIM).

We can extract from Proposition 2 some additional information, which could be helpful for estimating the model parameters. It is known that

$\sum_{i=1}^{N} 1/i = \ln N + C_E + O(1/N)$, where C_E is the *Euler* constant, $C_E=0.5772157\ldots$

More precisely, the approximation $\sum_{i=1}^{N} 1/i = \ln N + C_E + N^{-1}/2 - N^{-2}/12$ has an error less then 10^{-6} for $N>10$. Thus, from (3.7), we obtain an interesting formula

$$F_{eq} \cong (\nu/\delta)[\ln N + C_E] \tag{5.1}$$

This means that, in the equilibrium state of the system, the total number of domain families grows only slowly ($\sim\ln N$) with the increase of the maximal number (N) of domains in a family (which is equal to the maximal possible number of domain families).

Furthermore, according to equation (2.3), in the equilibrium state of a simple BDIM $\nu/\delta = f_1$, so we have

$$F_{eq} / f_1 \cong \ln N + C_E \quad (5.2)$$

Formula (5.1) can be used for estimating the model parameters on the basis of empirical data.

In the more general case $\lambda \neq \delta$, we can also obtain an estimate of the rate of innovation ν. If $\lambda < \delta$ (θ<1), then the series in the righthand part of (3.5) quickly converges,

$$\sum_{i=1}^{N} \theta^{i-1}/i \rightarrow -\ln(1-\theta)/\theta,$$

and so $-\ln(1-\theta)/\theta$ is a good approximation for the sum $\sum_{i=1}^{N} \theta^{i-1}/i$ for large N.

Then

$$F_{eq} = (\nu/\delta) \sum_{i=1}^{N} \theta^{i-1}/i = (\nu/\lambda) \sum_{i=1}^{N} \theta^{i}/i \cong \nu/\lambda\,(-\ln(1-\theta)),$$

and

$$\nu/\delta = F_{eq}\theta/\,(-\ln(1-\theta))\,. \quad (5.3)$$

Taking into account that $\nu/\delta = f_1$ (2.3), we have a relation

$$F_{eq}/f_1 \cong -\ln(1-\theta)/\theta\,, \quad (5.4)$$

which allows the parameter θ to be estimated on the basis of empirical data. If the N can be estimated independently and is not very large, we can use more exact relations:

$$\sum_{i=1}^{N} \theta^{i}/i \cong -\ln(1-\theta) + Ei(-N(1-\theta)) - N^{-1}/2 + N^{-2}/12,$$

where the function $Ei(u) = \int_{-\infty}^{u} e^{x}/x\,dx$.

Further, if $(1-\theta)N$ is small (*i.e.*, θ is very close to 1), then the approximation

$$\sum_{i=1}^{N} \theta^{i}/i \cong C_E - N(1-\theta)$$

has an error less then $[N(1-\theta)]^2/4$ and in this case

$$F_{eq}/f_1 \cong (C_E - N(1-\theta))\,/\theta\,. \quad (5.5)$$

If $(1-\theta)N$ is large, then the following inequalities provide simple bounds for $F_{eq}/f_1=\sum_{i=1}^{N}\theta^{i-1}/i$:

$$-\ln(1-\theta)/\theta-\theta^N/[(N+1)(1-\theta)]<\sum_{i=1}^{N}\theta^{i-1}/i<-\ln(1-\theta)/\theta-\theta^N[1/(N+1)-\theta/(N+2)]. \quad (5.6)$$

For the simple BDIM, the ratio of the rate of duplications to the innovation rate is

$$G(N)=\sum_{i=1}^{N-1}\lambda_i f_i/v=\sum_{i=1}^{N-1}\theta^i=\theta(1-\theta^{N-1})/(1-\theta),$$

so $G(N)\to\infty$ if $\theta>1$ and $G(N)\to 1/(1-\theta)$ if $\theta<1$ at $N\to\infty$.

If the simple BDIM is the 2nd order balanced, $\theta=1$, then $G(N)=N-1$.

Thus, for the simple 2nd order balanced BDIM the number of duplications is in N-1 times more then the number on innovations at time unit.

The total number of domains in the equilibrium state for the simple BDIM is

$$M(N)=\sum_{i=1}^{N} if_i= v/\lambda\theta(1-\theta^N)/(1-\theta).$$

If the simple BDIM is the 2nd order balanced, then $G(N)= v/\lambda N$.

5.2. Linear BDIM

We saw that the Hypothesis of Independence of birth and death rates of individual domains on each other and on the size of domain families is incompatible with any power distribution with the power $\neq$-1 of the equilibrium frequencies. The simplest case of a BDIM, which can have, depending on the parameters, three types of asymptotic behavior described by Theorem 1 (excluding the first one, hyper-exponential, which corresponds to a non-balanced BDIM; all linear BDIMs are balanced) and, in particular, any power asymptotics, is a model with *linear birth and death* rates of the form:

$$\lambda_i=\lambda(i+a_1),\ \delta_i=\delta(i+b_1), \text{ where } a_1, b_1 \text{ are constants.} \quad (5.7)$$

The parameters a_1 and b_1 account, in the simplest possible form, for the deviation of the domain birth and death rates from those under the Hypothesis

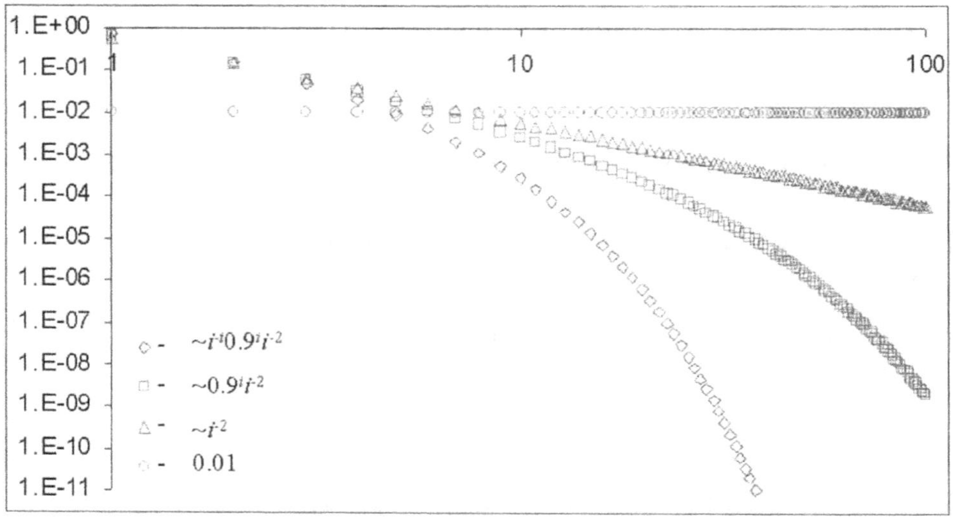

Figure 4. Asymptotics of equilibrium distributions for balanced BDIMs of different orders. ***Colour plate at back of book.***

of Independence. More precisely, according to (5.7), the average birth rate per domain in a family of the size i is $\lambda_i/i = \lambda + \lambda a_1/i$. So, for small i, the average birth rate is close to $\lambda + \lambda a_1$, whereas, for large i, it tends to λ. Similarly, the average death rate changes from $\delta + \delta b_1$ in a small family to the limit value δ in a large family. Thus, if a_1 and b_1 are positive (which seems to be the case for the available data; see below), both the birth rate and the death rate per domain decrease with the increase of the class number (size of the respective domain families); conversely, if a_1 and b_1 are negative, these rates increase with the class number (Figure 17).

Corollary 1 of Theorem 2 implies that:
equilibrium frequencies p_i of a linear BDIM have asymptotics

$$p_i \sim \theta^{\,i}\, i^{(a_1 - b_1 - 1)}, \text{ where } \theta = \lambda/\delta. \tag{5.8}$$

In particular, if $\lambda \neq \delta$ and $a_1 = b_1$, then a linear BDIM is first-order balanced and the equilibrium frequencies p_i follow the logarithmic distribution (in this case, the linear BDIM is asymptotically equivalent to the simple BDIM); if $\lambda = \delta$, the linear BDIM is second-order balanced and the equilibrium frequencies p_i follow the power distribution

$$p_i \sim i^{(a_1 - b_1 - 1)}. \tag{5.9}$$

Thus, the dependence of the domain frequency on the family size is actually

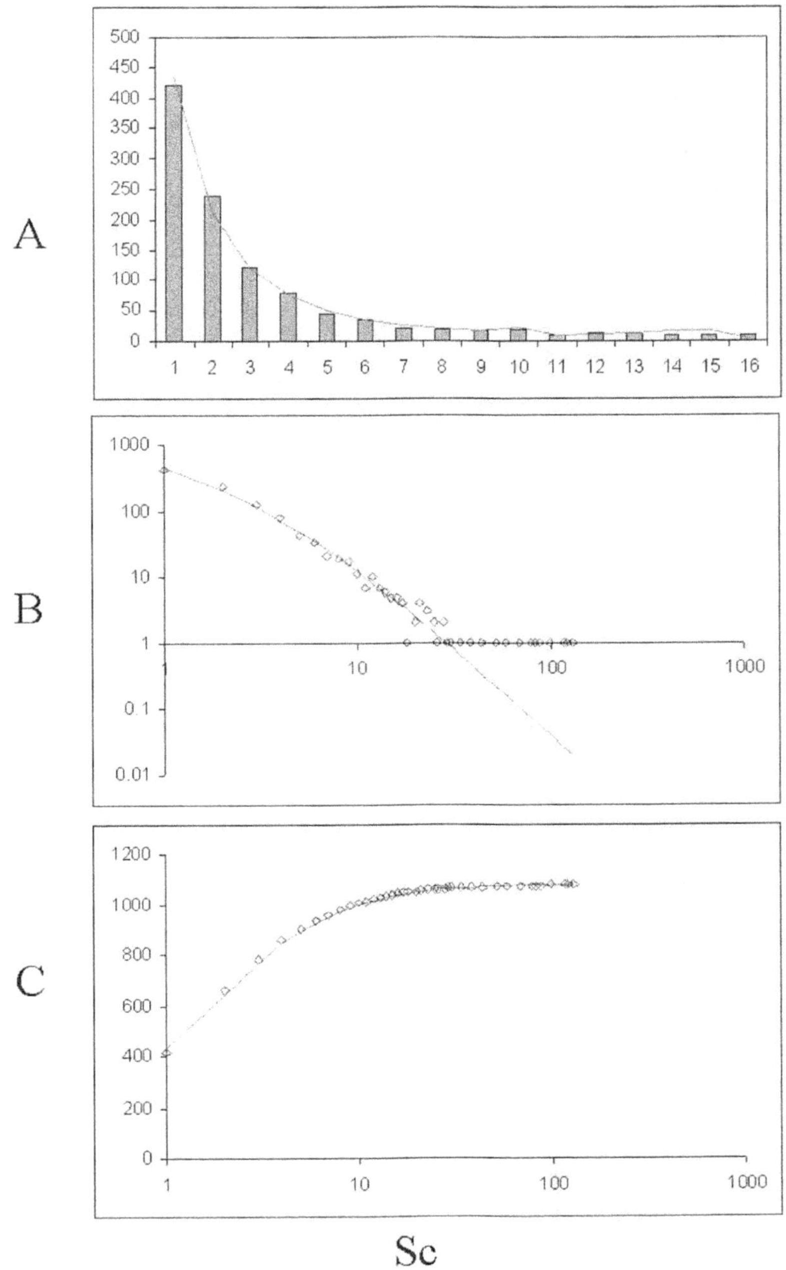

Figure 5. Fit of empirical domain family size distributions to the second-order balanced linear BDIM. *Saccharomyces cerevisiae*. A. Distribution of the size of domain families grouped into bins

B. Domain family size distribution in double logarithmic coordinates.

C. Cumulative distribution function of domain family size.

The line shows the prediction of the second-order balanced linear BDIM.

Colour plate at back of book.

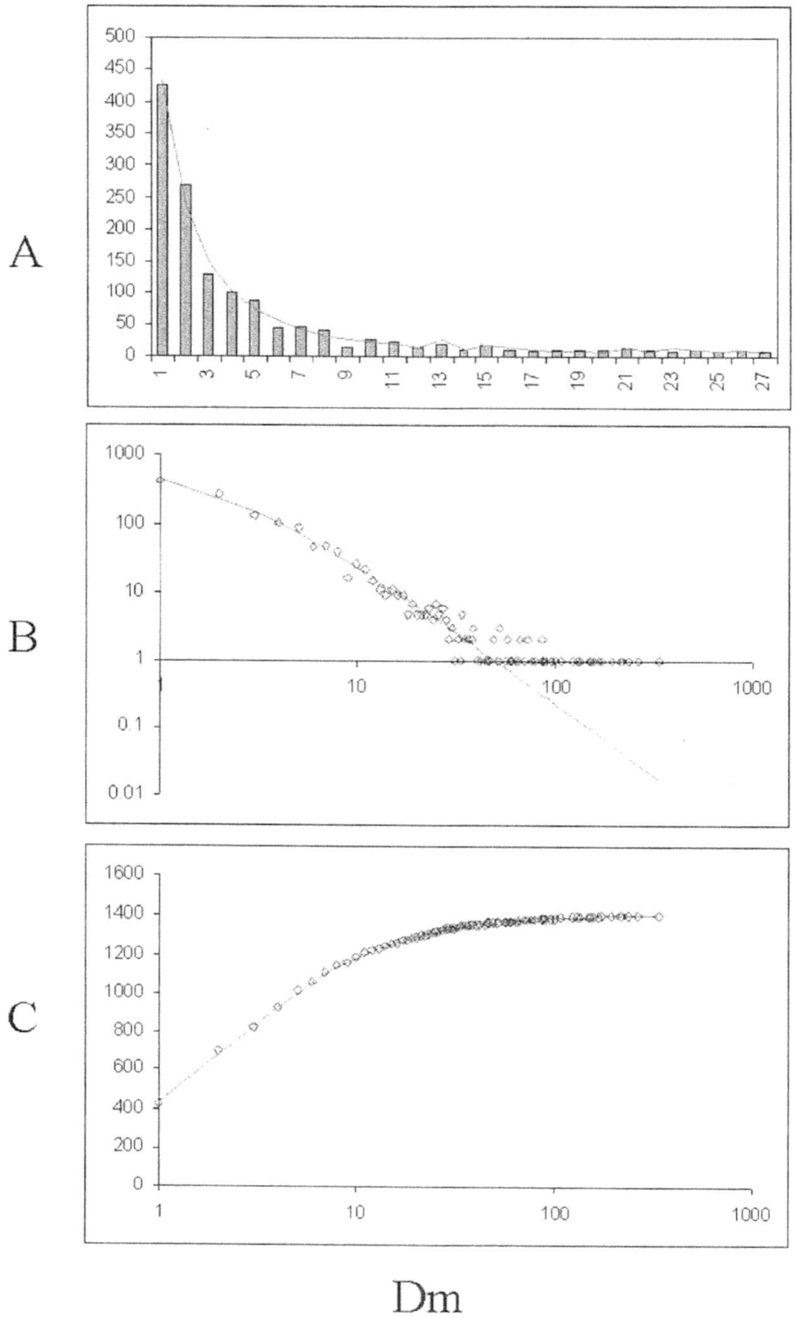

Figure 6. Fit of empirical domain family size distributions to the second-order balanced linear BDIM. *Drosophila melanogaster*. ***Colour plate at back of book.***

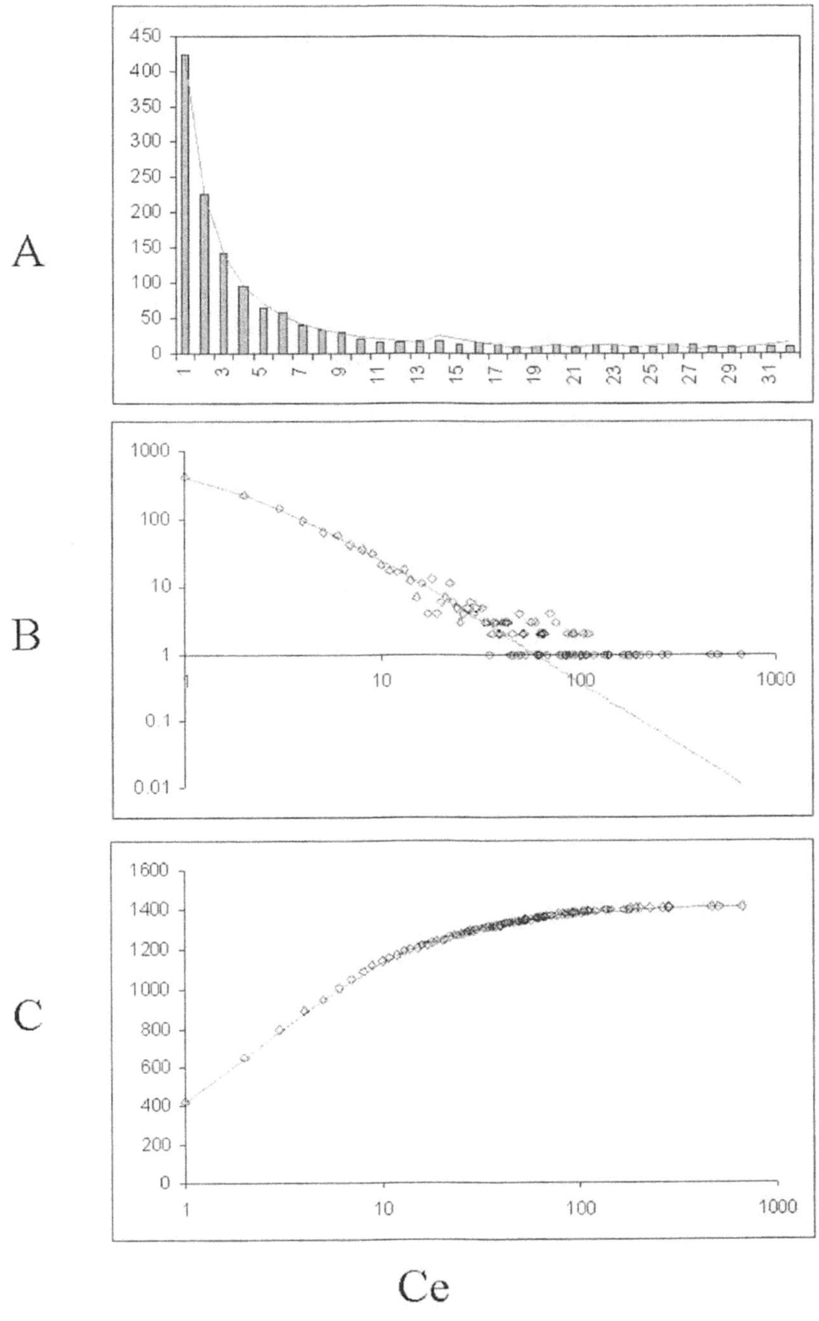

Figure 7. Fit of empirical domain family size distributions to the second-order balanced linear BDIM. *Caenorhabditis elegans*. ***Colour plate at back of book.***

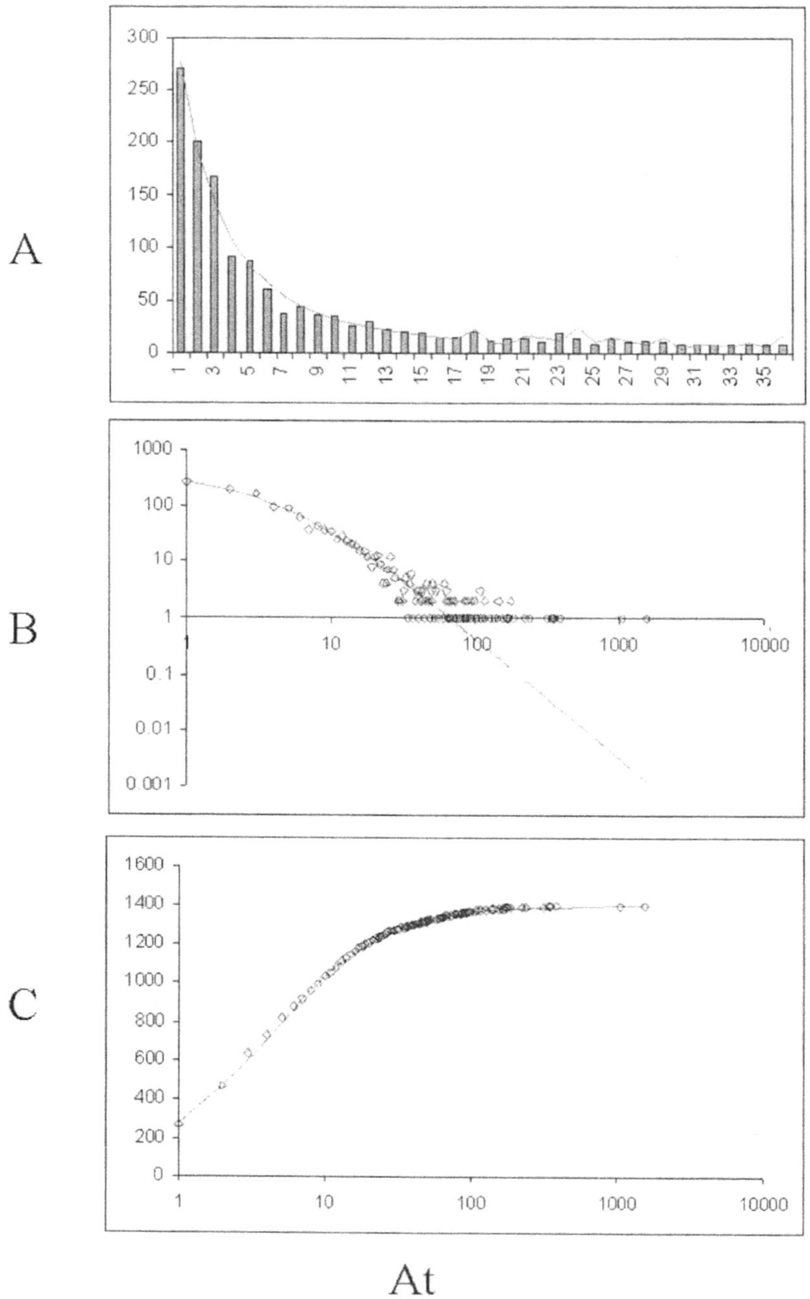

Figure 8. Fit of empirical domain family size distributions to the second-order balanced linear BDIM. *Arabidopsis thaliana*. ***Colour plate at back of book.***

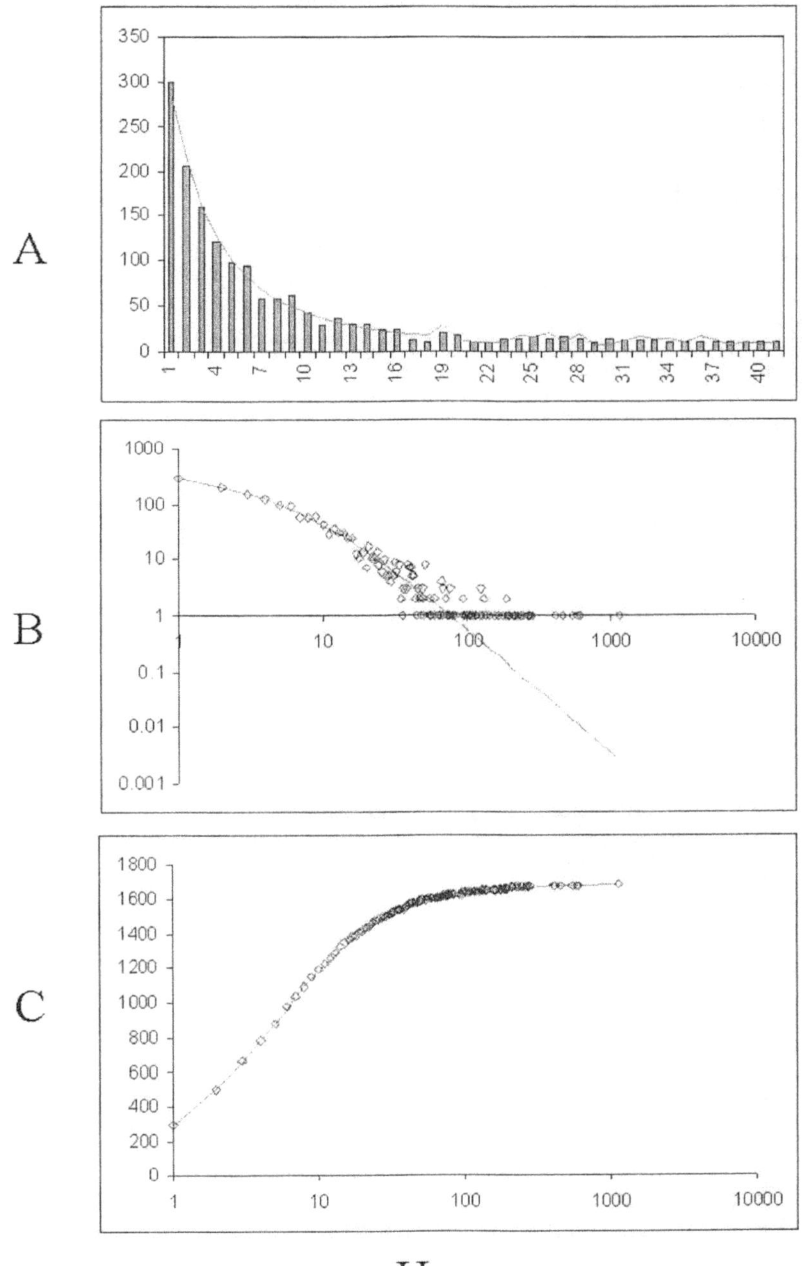

Figure 9. Fit of empirical domain family size distributions to the second-order balanced linear BDIM. *Homo sapiens*. ***Colour plate at back of book.***

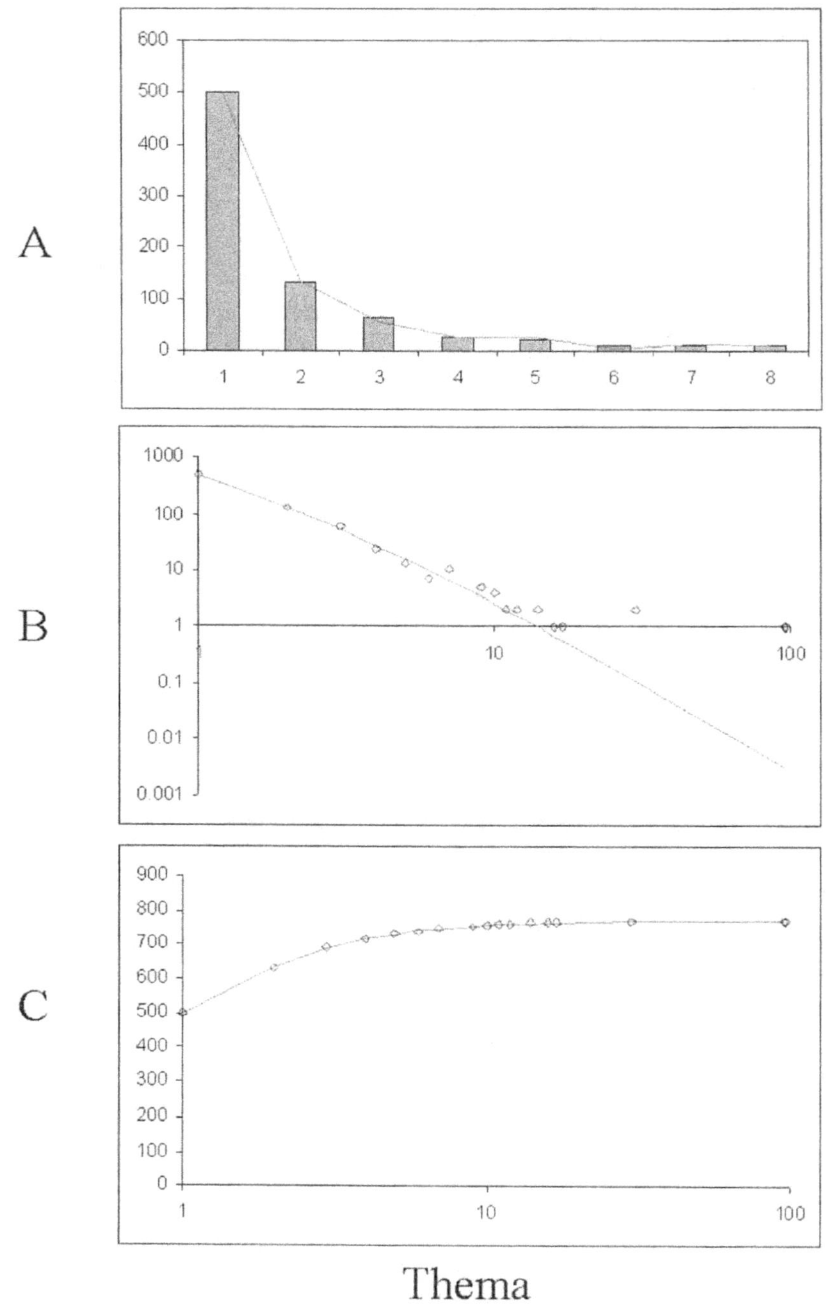

Figure 10. Fit of empirical domain family size distributions to the second-order balanced linear BDIM. *Thermotoga maritima*. ***Colour plate at back of book.***

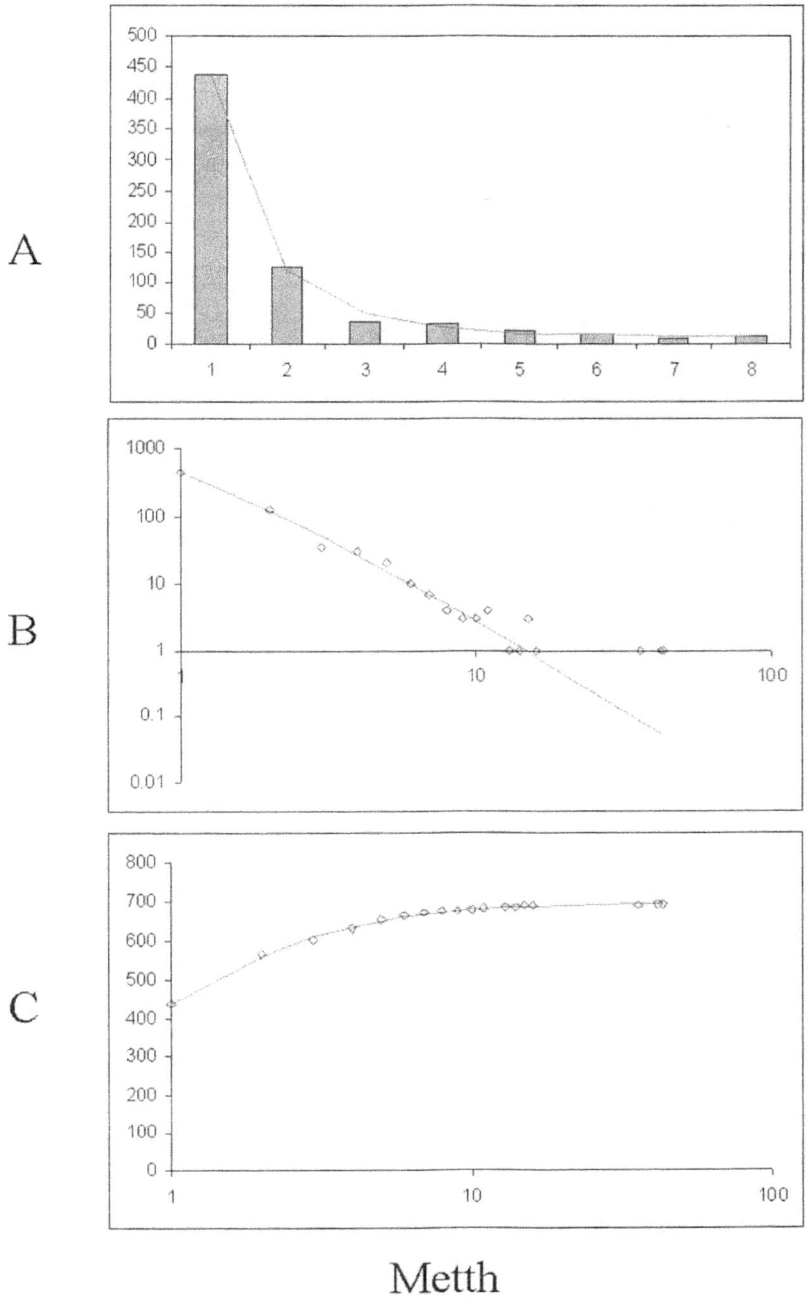

Figure 11. Fit of empirical domain family size distributions to the second-order balanced linear BDIM. *Methanothermobacter thermautotrophicus*. ***Colour plate at back of book.***

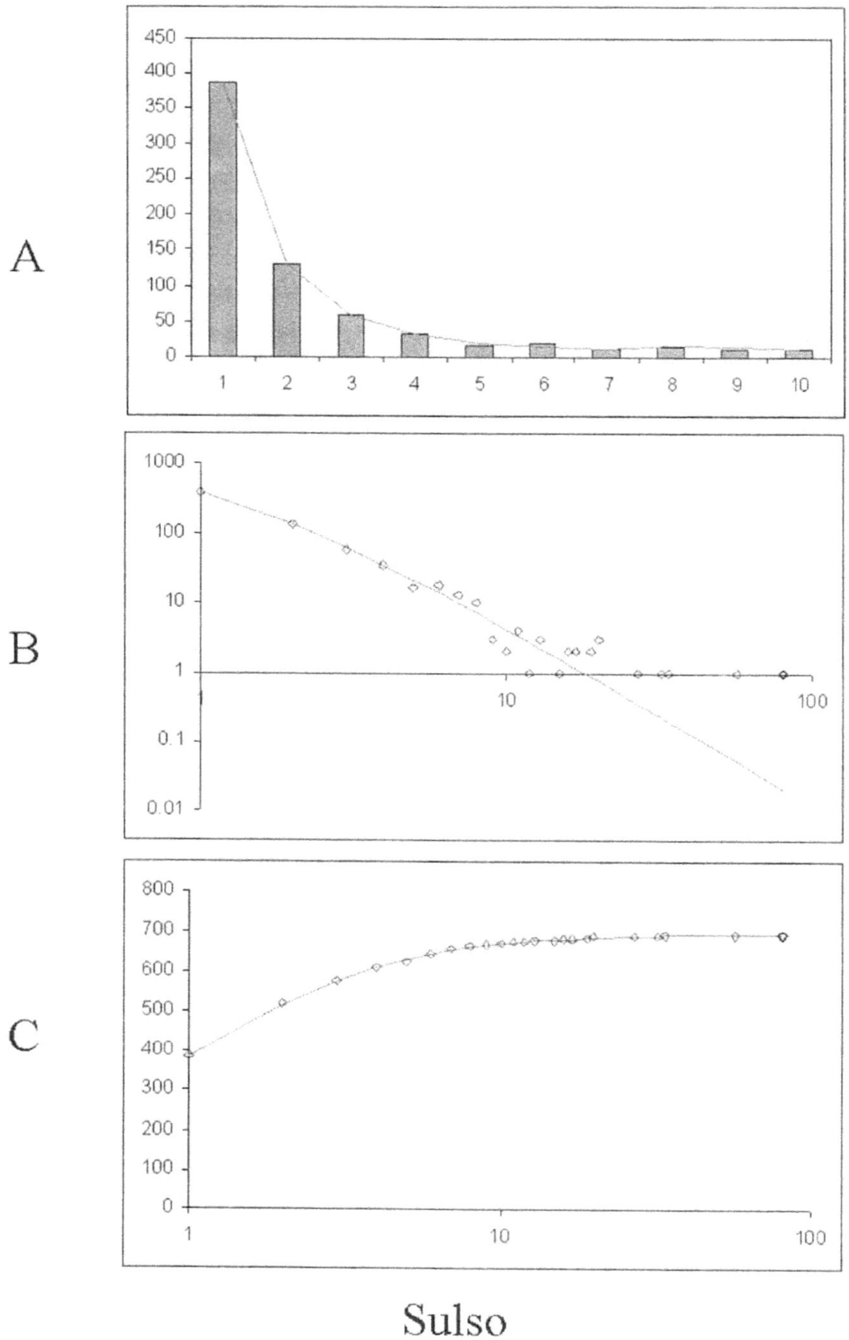

Figure 12. Fit of empirical domain family size distributions to the second-order balanced linear BDIM. *Sulfolobus solfataricus*. ***Colour plate at back of book.***

A
B
C

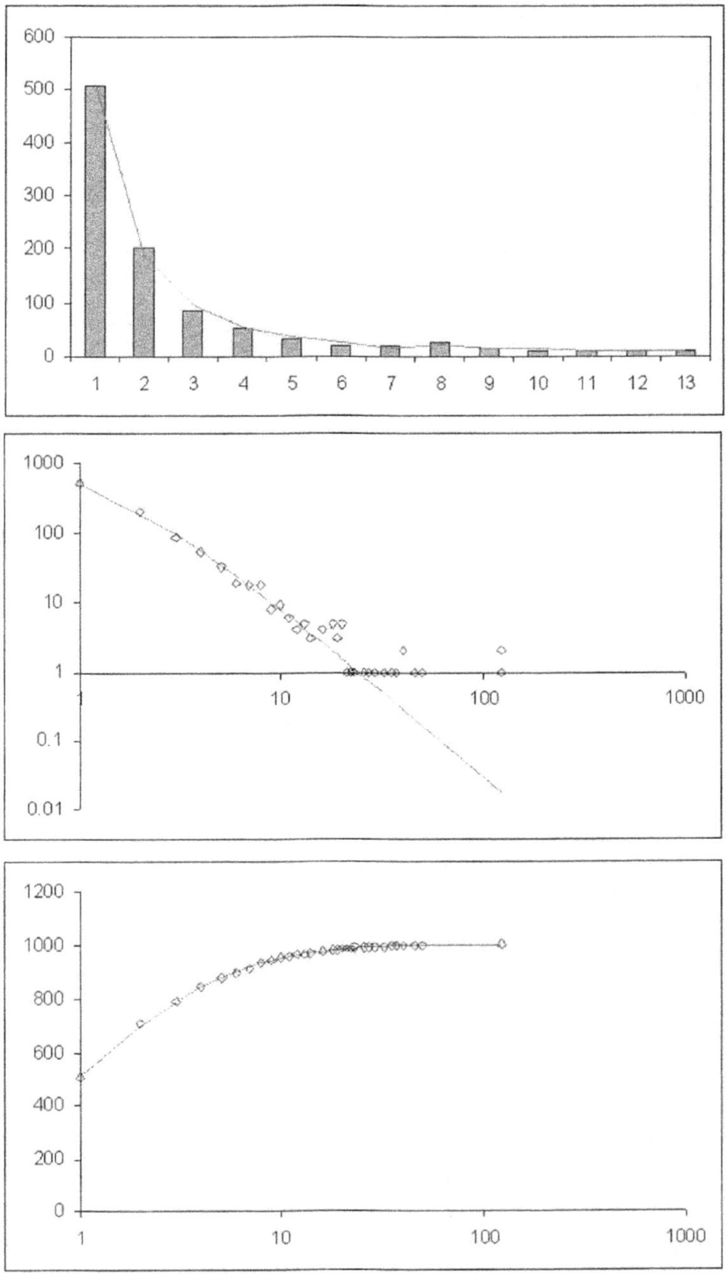

Bacsu

Figure 13. Fit of empirical domain family size distributions to the second-order balanced linear BDIM. *Bacillus subtilis*. ***Colour plate at back of book.***

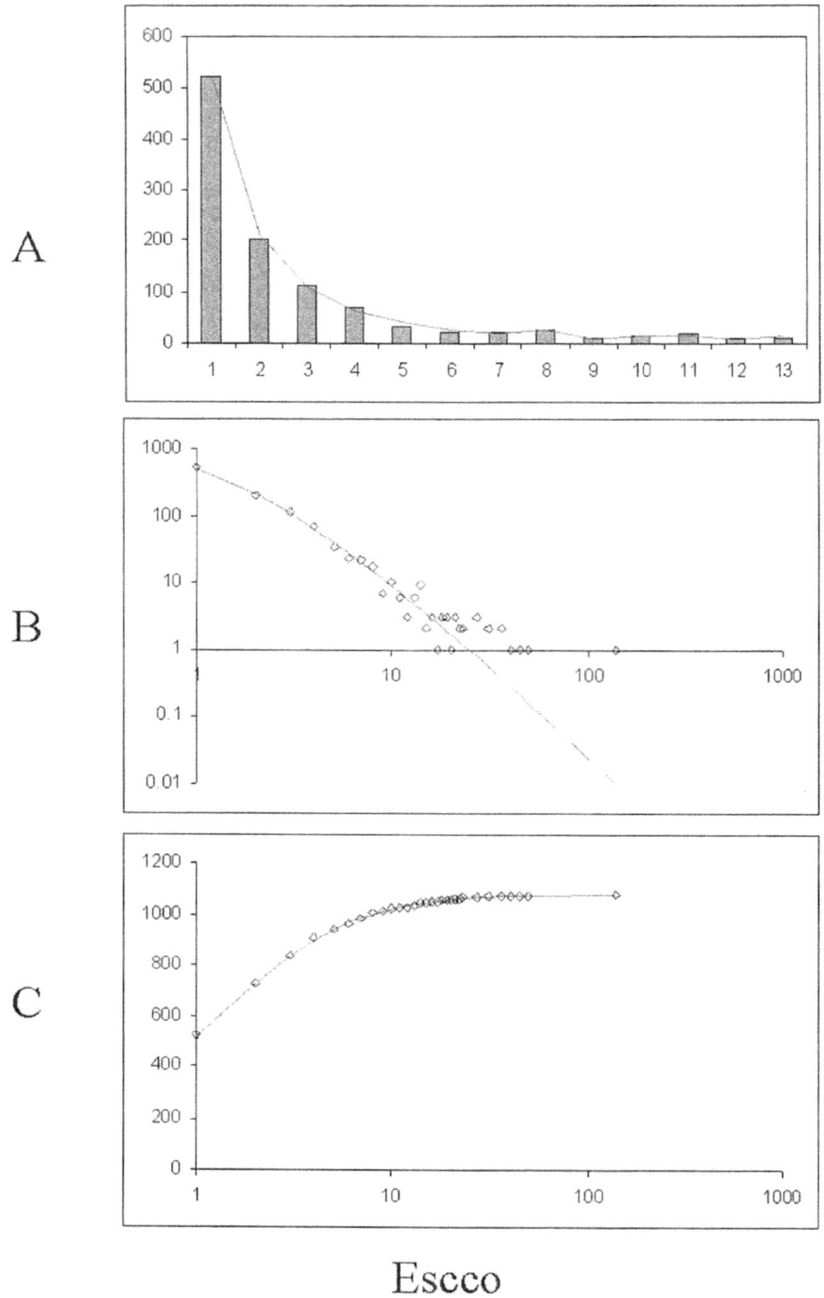

Figure 14. Fit of empirical domain family size distributions to the second-order balanced linear BDIM. *Escherichia coli*. ***Colour plate at back of book.***

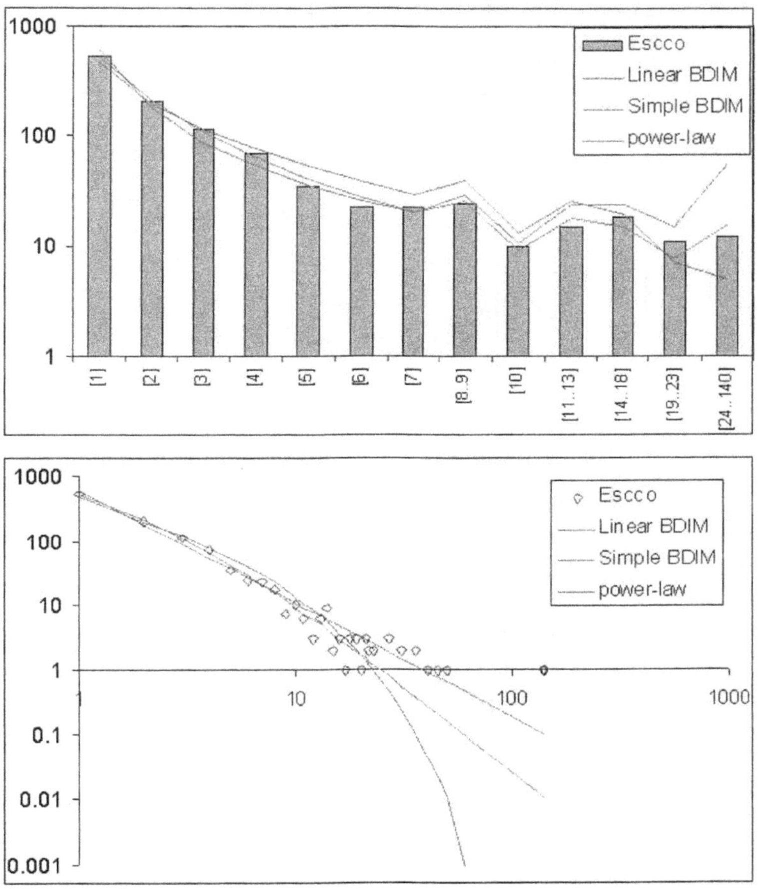

Figure 15. Comparison of different approximations of the empirical domain family size distribution. *Escherichia coli.*
Magenta line: second-order balanced linear BDIM
Red line: simple BDIM
Cyan line: power law
Colour plate at back of book.

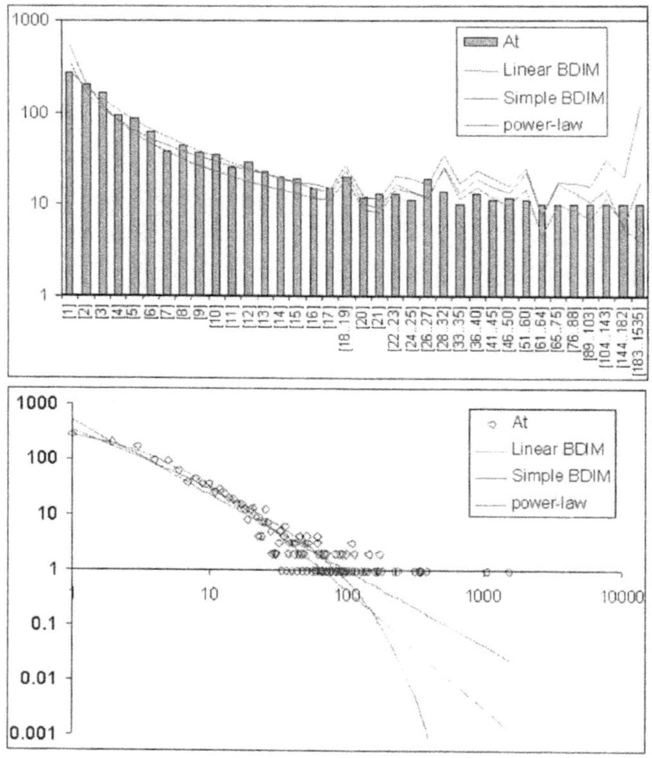

Figure 16. Comparison of different approximations of the empirical domain family size distribution. *Arabidopsis thaliana*. ***Colour plate at back of book.***
Magenta line: second-order balanced linear BDIM
Red line: simple BDIM
Cyan line: power law

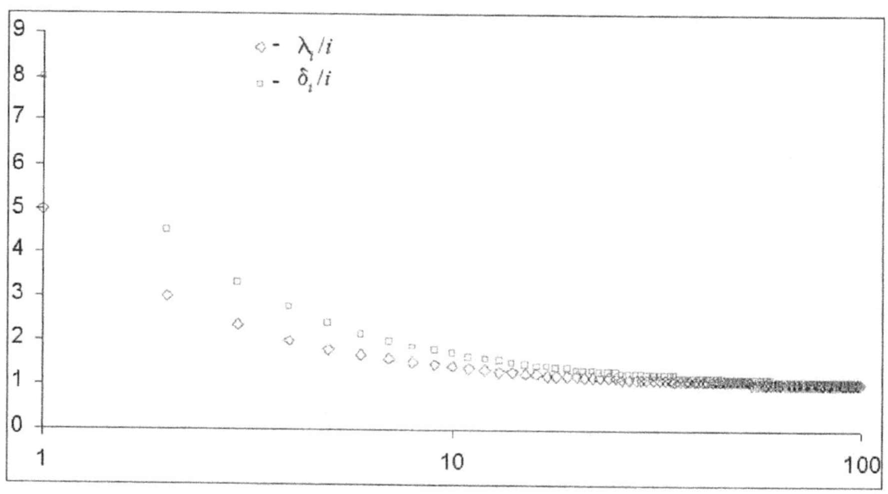

Figure 17. Per domain birth and death rates for the second-order balanced BDIM. ***Colour plate at back of book.***

determined by the difference a_1-b_1. If $a_1 > b_1$, the birth rate decreases faster than the death rate with the increase of family size, i. e. there seems to be a "competition" between domains in a family; in contrast, if $a_1 < b_1$, the death rate drops faster, *i.e.* "synergy" between domains appears to exist (Figure 17).

More detailed information can be obtained using Proposition 4:

i) for a first-order balanced linear BDIM, the equilibrium sizes f_i of domain families are

$f_i = c_1\nu/\delta\theta^{i-1}\Gamma(i+a_1)/(\Gamma(i+1+b_1))$ for all i

where

$c_1 = \Gamma(1+b_1)/\Gamma(1+a_1)$;

and the equilibrium total number of domain families is

$$F_{eq} = c_1\nu/\delta[\sum_{j=1}^{N}\theta^{j-1}\Gamma(j+a_1)/(\Gamma(j+1+b_1)]. \quad (5.10)$$

ii) for a second-order balanced linear BDIM (θ=1),

$f_i = c_1\nu/\delta\,\Gamma(i+a_1)/\Gamma(i+1+b_1)$

and

$$F_{eq} = c_1\nu/\delta(\sum_{j=1}^{N}\Gamma(j+a_1)/(\Gamma(j+1+b_1))=$$

$$\nu/\delta(1-\frac{\Gamma(1+b_1)\Gamma(1+a_1+N)}{\Gamma(1+a_1)\Gamma(1+b_1+N)})/(b_1-a_1) \quad (5.11)$$

According to (2.3), in the equilibrium state of a linear BDIM, $d_1 = \nu/\delta_1 = \nu/(\delta(1+b_1))$ and so for a *second-order balanced* linear BDIM we have the formula

$$F_{eq}/f_1 = (1+b_1)\,(1-\frac{\Gamma(1+b_1)\Gamma(1+a_1+N)}{\Gamma(1+a_1)\Gamma(1+b_1+N)})/(b_1-a_1)$$

Suppose that equilibrium frequencies obtained from empirical data follow the power distribution $p_i \sim i^{-\gamma}$; in this case, $-\gamma$ is the slope of the empirical curve ($\ln f_i$ versus $\ln i$) and can be estimated from the data. Assuming that the system is well described by a linear BDIM, (5.9) shows that $a_1 - b_1 = 1 - \gamma$ and $\lambda = \delta$. Thus,

$$f_i = c_1 \nu/\delta\, \Gamma(i + a_1)/\Gamma(i + a_1 + \gamma), \text{ where } c_1 = \Gamma(\gamma + a_1)/\Gamma(1 + a_1), \tag{5.12}$$

$$F_{eq} = c_1 \nu/\delta \sum_{j=1}^{N} \Gamma(j + a_1)/\Gamma(j + a_1 + \gamma) =$$

$$\nu/\delta\left(1 - \frac{\Gamma(1 + a_1 + N)\Gamma(a_1 + \gamma)}{\Gamma(a_1 + \gamma + N)\Gamma(1 + a_1)}\right)/(\gamma - 1)$$

and

$$F_{eq}/f_1 = (a_1 + k)\left(1 - \frac{\Gamma(1 + a_1 + N)\Gamma(a_1 + \gamma)}{\Gamma(a_1 + \gamma + N)\Gamma(1 + a_1)}\right)/(\gamma - 1)$$

where a_1 is the single free parameter.

For the linear 2nd order balanced BDIM, the ratio of the birth rate to the innovation rate is

$$G(N) = \sum_{i=1}^{N-1} \lambda_i f_i/\nu = \sum_{i=1}^{N-1} \Gamma(1+b_1)/\Gamma(1+a_1)\,(i + a_1)\Gamma(i + a_1)/(\Gamma(i+1+b_1)) =$$

$$\sum_{i=1}^{N-1} \Gamma(1+b_1)/\Gamma(1+a_1)\,\Gamma(i+1+a_1)/(\Gamma(i+1+b_1)) =$$

$$\frac{\Gamma(1 + a_1 + N)\Gamma(1 + b_1)}{(1 + a_1 - b_1)\Gamma(b_1 + N)\Gamma(1 + a_1)} + \frac{1 + a_1}{b_1 - a_1 - 1}$$

if $1 + a_1 - b_1 \neq 0$. As

$$\frac{\Gamma(1 + a_1 + N)\Gamma(1 + b_1)}{(1 + a_1 - b_1)\Gamma(b_1 + N)\Gamma(1 + a_1)} \cong \frac{\Gamma(1 + b_1)}{(1 + a_1 - b_1)\Gamma(1 + a_1)} N^{1 + a_1 - b_1},$$

$$G(N) \rightarrow \frac{1 + a_1}{b_1 - a_1 - 1}$$

if $1 + a_1 - b_1 < 0$ and $G(N) \rightarrow \infty$ if $1 + a_1 - b_1 > 0$ at $N \rightarrow \infty$.

The case $1 + a_1 - b_1 = 0$ (slope of the asymptote in double logarithmic coordinates equal to $a_1 - b_1 - 1 = -2$) is a critical one.
In this case

$$G(N) = \sum_{i=1}^{N-1} \Gamma(1+b_1)/\Gamma(b_1)\,\Gamma(i + b_1)/(\Gamma(i+1+b_1)) =$$

$$\sum_{i=1}^{N-1} b_1/(i+b_1) = b_1 \text{ [PolyGamma}(0, b_1+N)\text{- PolyGamma}(0, b_1+1)].$$

Accordingly, $G(N)\to\infty$ at $N\to\infty$.

The total number of domains in the equilibrium state for second-order balanced linear BDIM is

$$M(N)= \sum_{i=1}^{N} if_i=$$

$$v/\delta \sum_{i=1}^{N} i\Gamma(1+b_1)/\Gamma(1+a_1)\,\Gamma(i+a_1)/(\Gamma(i+1+b_1) =$$

$$v/\delta[\frac{a_1}{a_1-b_1}+\frac{1+a_1}{b_1-a_1-1}+\frac{\Gamma(1+b_1)}{\Gamma(1+b_1+N)}(\frac{\Gamma(2+N+a_1)}{(1+a_1-b_1)\Gamma(1+a_1)}-\frac{\Gamma(1+N+a_1)}{(a_1-b_1)\Gamma(a_1)})].$$

If the asymptote slope γ=-1, the linear second-ordered BDIM has the same asymptotic behavior as a simple BDIM (2.1), but a different left part of the graph. If $\gamma\neq$-1, the system cannot be described by a *simple* BDIM even asymptotically, but can be described by a linear BDIM. As indicated above, in this case, the average per-domain birth and death rates depend on the size of the domain family and the difference a_1-b_1 characterizes this dependence.

5.3. Quadratic BDIM

The linear BDIM takes into account a possible dependence of average birth and death rates of individual domains on the size of domain family, but does not imply a specific form of interaction between domains. Let us consider the simplest, pairwise interaction, which leads to $\lambda_i \sim i^2$ and/or $\delta_i \sim i^2$, *i.e.* one or both rates are polynomials on *i* of the second degree. If these degrees are different (*i.e.*, $\lambda_i \sim i$ and $\delta_i \sim i^2$), then the corresponding BDIM is non-balanced and equilibrium frequencies have hyper-exponential asymptotics. Thus, let

$$\lambda_i = \lambda(i^2+r_1 i+r_2),\ \delta_i = \delta(i^2+q_1 i+q_2), \tag{5.13}$$

where r_k, q_k, k=1,2 are constants (such that λ_i, δ_i are positive for all *i*) or

$$\lambda_i = \lambda(i+a_1)(i+a_2),$$
$$\delta_i = \delta(i+b_1)(i+b_2)$$

Then, $r_1=a_1+a_2$, $q_1=b_1+b_2$, and

$\chi(i)=\lambda_{i-1}/\delta_i = \theta(1+(r_1-q_1-2)/i+O(1/i^2))$,

where $\theta=\lambda/\delta$.

According to theorem 3 and Proposition 5, the quadratic BDIM with rates (5.13) has equilibrium sizes of domain families

$$f_i = c_2 v/\delta\, \theta^{\,i-1}\Gamma(i+a_1)\Gamma(i+a_2) / (\Gamma(i+1+b_1)(\Gamma(i+1+b_2)) \cong c_2 v/\delta\, \theta^{\,i-1} i^{\,\rho-2} \quad (5.14)$$

where $\rho=r_1-q_1$ and the constant $c_2=[(\Gamma(1+b_1)\,\Gamma(1+b_2)]/[\Gamma(1+a_1)\,\Gamma(1+a_2)]$, and the total number of domain families at equilibrium

$$F_{eq} = c_2 v/\delta\, (\sum_{j=1}^{N}\theta^{\,j-1}\, \Gamma(j+a_1)\, \Gamma(j+a_2) / (\Gamma(j+1+b_1)\, (\Gamma(j+1+b_2)). \quad (5.15)$$

Note that the asymptotics of frequencies p_i does not depend on free coefficients r_2, q_2 in (5.13), but only on θ and r_1-q_1 (as follows from (5.14)), although the values of f_i are proportional to the constant c_2, which could depend on the free coefficients r_2, q_2. Let us consider the case $r_2=q_2=0$ in more detail.

If only square terms are present in the expressions for the birth and death rates, $\lambda_i=\lambda i^2$, $\delta_i=\delta i^2$, then $a_k=b_k=0$, $k=1,2$ and so $c_2=1$, $f_i = v/\delta\, \theta^{\,i-1}/i^2$ and $F_{eq} = v/\delta \sum_{j=1}^{N}\theta^{\,j-1}/j^2$. So at $N\to\infty$

$$F_{eq} \cong v/\delta \sum_{j}\theta^{\,j-1}/j^2 = v/\lambda\ \text{Polylog}(2,\theta) \quad (5.16)$$

where Polylog is a special function, $\text{Polylog}(k,x)= \sum_{j=1}^{\infty} x^{j}/j^{k}$.

According to (3.2), $f_1= v/\delta_1$; for this particular case of quadratic BDIM, $f_1= v/\delta$ and

$$F_{eq}/f_1 \cong \text{Polylog}(2,\theta). \quad (5.17)$$

Formula (5.17) allows estimating parameter θ from empirical data if N is large enough.

More precisely, $F_{eq} = v/\lambda \sum_{j=1}^{N}\theta^{j}/j^2 = v/\lambda(\text{Polylog}(2,\theta) - \theta^{1+N}\text{LerchPhi}(\theta,2,1+N))$, where LerchPhi is a special function (these and other special functions used below can be computed using program packages Mathematika or Maple). If, additionally, $\theta=1$ (the BDIM is second-order balanced), then

$$f_i = (\nu/\delta)/i^2 = f_1/i^2 \tag{5.18}$$

and, at large N

$$F_{eq} \cong \nu/\delta\, \pi^2/6 \cong 1.645\,\nu/\delta = 1.645 f_1. \tag{5.19}$$

From formulas (5.8), (5.15), we can extract some additional information, which could be helpful for estimating the model parameters at relatively small N. Let us recall definitions of some special functions.

The digamma function $\phi(z)$ is logarithmic derivative of $\Gamma(z)$, $\phi(z) = \Gamma'(z)/\Gamma(z)$.

The function PolyGamma(n,z) is n^{th} derivative of $\phi(z)$, PolyGamma(n,z)= $d^n\phi(z)/dz^n$, such that $\phi(z)$= PolyGamma(0,z).

It is known that

$$\sum_{i=1}^{N} 1/i^2 = \pi^2/6\text{-PolyGamma}(1,1+N),$$

Thus we have

$$F_{eq} = \nu/\delta \sum_{j=1}^{N} 1/j^2 = \nu/\delta\,[\pi^2/6\text{-PolyGamma}(1,1+N)] \tag{5.20}$$

$$F_{eq}/f_1 = \pi^2/6\text{-PolyGamma}(1,1+N),$$

which can be used for estimating unknown parameters of the model.

The values of PolyGamma(1,x) are tabulated and can be computed using standard program packages; for a rough preliminary estimate, PolyGamma(1,x) = $1/x+1/2x^2+O(1/x^4)$).

If linear terms also present in the quadratic BDIM, $\lambda_i = \lambda(i^2+a_1 i)$, $\delta_i = \delta(i^2+b_1 i)$, then

$$f_i = c_2 \nu/\delta\, \theta^{\,i-1}/i\; \Gamma(i+a_1)/\Gamma(i+1+b_1)$$

where $c_2 = \Gamma(1+b_1)/\Gamma(1+a_1)$; $F_{eq} = \Sigma f_i$ could be computed with the help of special functions. In particular, if the BDIM is second-order balanced, $\theta = 1$, then

$$f_i = c_2 \nu/\delta\, \Gamma(i+a_1) / (i\, \Gamma(i+1+b_1)).$$

For this variant of the model, $f_1 = v/\delta_1 = v/(\delta(1+b_1))$, and so

$$f_i = f_1 \frac{\Gamma(2+b_1)\Gamma(i+a_1)}{i\Gamma(1+a_1)\Gamma(i+1+b_1)}.$$

5.4. Polynomial BDIMs

The quadratic models take into account the possible dependence of birth and death rates of individual domains on the simplest, pairwise interactions. If interactions of higher orders are postulated, $\lambda_i \sim P_n(i)$ and/or $\delta_i \sim Q_m(i)$, where $P_n(i)$, $Q_m(i)$ are polynomials on i of the n-th and m-th degrees. Again, if the degrees n and m are different, then the BDIM is non-balanced and equilibrium frequencies have hyper-exponential asymptotics. Thus, let $n=m$,

$$\lambda_i = \lambda R(i) = \lambda \sum_{k=0}^{m} r_k\, i^{m-k},\ \delta_i = \delta Q(i) = \delta \sum_{k=0}^{m} q_k\, i^{m-k} \qquad (5.21)$$

where r_k, q_k are constants and $r_0 = q_0 = 1$. We suppose, of course, that $R(i)$, $Q(i)$ are positive for all natural i. Note that, in this case,
$\chi(i) \equiv \lambda_{i-1}/\delta_i = \theta(1+(r_1 - q_1 - m)/i + O(1/i^2))$, where $\theta = \lambda/\delta$. We will suppose that $\theta \leq 1$.

According to Theorem 3, the polynomial BDIM with rates (5.21) has equilibrium sizes of domain families with *power-exponential asymptotics*

$$f_i \sim \theta^i\, i^{\rho - m} \qquad (5.22)$$

where $\rho = r_1 - q_1$.

In particular, if $\rho - m > -1$, the equilibrium frequencies p_i follow the Pascal distribution with parameters $(\rho - m + 1, \theta)$;

if $\rho - m = -1$, the equilibrium frequencies p_i follow the (truncated) logarithmic distribution;

if $\rho - m = 0$, the equilibrium frequencies p_i follow the geometric distribution; if $\lambda = \delta$, the polynomial BDIM is second-order balanced and the equilibrium frequencies p_i follow the power distribution

$$p_i \sim i^{\rho - m}. \qquad (5.23)$$

Note that the value of power for the power distribution (5.23) is determined only by m, the common degree of the polynomials (5.21), and by ρ, the difference of the coefficients r_1 and q_1, and does not depend on other coefficients. In particular, if $r_1 = q_1$, then $p_i \sim i^{-m}$. This relation could be interpreted as following: if the first two coefficients of polynomial rates λ_i and δ_i are equal, then the power value of the power distribution (5.19) is equal to the "order of interactions" of domains.

Formula (5.22) can be refined. Let $R(i) = \prod_{k=1}^{m}(i + a_k)$, $Q(i) = \prod_{k=1}^{m}(i + b_k)$.

Then (see Proposition 5) the equilibrium numbers of domain families f_i of the polynomial BDIM (5.18) are

$$f_i = Cv/\delta\theta^{i-1}\prod_{k=1}^{m}[\Gamma(i + a_k)/\Gamma(i+1+b_k)]$$

where $C = \prod_{k=1}^{m}[\Gamma(1+b_k)/\Gamma(1+a_k)]$, and the equilibrium total number of domain families

$$F_{eq} = Cv/\delta\sum_{j=1}^{N}\theta^{j-1}\prod_{k=1}^{m}[\Gamma(j+a_k)/\Gamma(j+1+b_k)].$$

For the polynomial model $f_1 = v/\delta_l = v/(\delta\sum_{k=0}^{m} q_k)$, so

$$F_{eq}/f_1 = C\sum_{j=1}^{N}\theta^{j-1}\prod_{k=1}^{m}(\Gamma(j+a_k)/\Gamma(j+1+b_k))/\sum_{k=0}^{m} q_k .$$

This formula could be used for estimating the model parameters.

For the polynomial second-order balanced BDIM, the ratio of the death rate to the innovation rate is

$$G(N) = \sum_{i=1}^{N-1}\lambda_i f_i/v = \left(\prod_{k=1}^{m}\Gamma(1+b_k)/\Gamma(1+a_k)\right)\sum_{i=1}^{N-1}\prod_{k=1}^{m}\Gamma(i+1+a_k)/\Gamma(i+1+b_k) =$$

$$\sum_{i=1}^{N-1}\prod_{k=1}^{m}[\Gamma(i+1+a_k)/\Gamma(1+a_k)] / [\Gamma(i+1+b_k)/\Gamma(1+b_k)].$$

Approximation of the Observed Domain Family Size Distributions in Prokaryotic and Eukaryotic Genomes with Different BDIMs

In order to identify genome-specific domain sets, the CDD library of Position-specific scoring matrices (PSSMs), which includes the domains from the Pfam and SMART databases, was used in RPS-BLAST searches (Marchler-Bauer *et al.*, 2002) against the protein sequences from a set of completely sequenced eukaryotic and prokaryotic genomes (http://www.ncbi.nlm.nih.gov/entrez/query.fcgi?db=Genome). The CDD domain library is partially redundant, so when the results obtained from individual PSSMs showed significant overlap (more than 50% of hits overlapped for more than 50% of their length), the corresponding domains were examined case-by-case for redundancy. PSSMs representing structurally similar domains and producing overlapping lists of hits were joined into "synonymy clusters".

The results of RPS-BLAST searches against the sets of protein sequences from individual genomes were interpreted as follows: non-overlapping hits to the same protein were treated independently; among overlapping hits, only the strongest one (lowest E-value) was recorded; all hits from a synonymy cluster were assigned to one representative domain family. The number of hits that a domain family had in a genome, with the cut-off of E=0.001 was recorded as the number of domains of the given family encoded in the given genome. The CDD domain library certainly does not include all existing domains. In practice, domains from this collection were detected in >50% in each of the analyzed species, with the only exception of human, for which the analyzed protein set is likely to contain a substantial fraction of false predictions (Table 1).

To enable statistical analysis using the χ^2-method for the whole spectrum of the data, including the sparsely distributed classes corresponding to large families, the data needed to be combined. For each genome, the observed domain family frequencies were grouped into bins, each containing at least 10 families; typically, bins corresponding to families with small number of members included a single class (*e.g.* all single-member families, two-member families etc), whereas bins corresponding to large families may span hundreds of classes, most of them empty. Theoretical distribution values for a bin combining observed data from m-th to n-th class were computed as $\sum_{i=m}^{n} f_i$, where f_i is the predicted number of families in the i-th class and depends on the model parameters. Since the model displays only a weak dependence on the maximum number of domains in a family (N), instead of including N as a model parameter, the sum $\sum_{i=1}^{i_{max}} f_i$ (where i_{max} is the number of domains in the most abundant of the observed families), was normalized to equal the total number of families detected in the given genome (a requirement for the χ^2 analysis). χ^2 values were computed to measure the quality of fit between the observed and the theoretical distributions. The distribution parameters

Table 1. Domain families in sequence genomes and parameters of the best-fit evolutionary model[a]

	No. of ORFs in genome	No. of detected domain families	No. of detected domains	No. of ORFs with RPS-BLAST hits	Maximum size of a family	f_1 (f'_1)	a_1	b_1	k	$\nu/\delta = \nu/\lambda$	$G = \sum_{i=1}^{N-1} \lambda_i f_i / \nu$
Sce[b]	6340	1080	4575	3331	130	420 (436)	1.55	3.27	-2.72	1861.8	[3.28..3.53]
Dme	13605	1405	11734	7262	335	426 (435)	1.62	2.79	-2.17	1648.2	[8.44..15.50]
Cel	20524	1418	17054	11090	662	423 (421)	1.13	2.03	-1.89	1273.0	[16.18..∞]
Ath	25854	1405	21238	15006	1535	270 (277)	3.80	4.98	-2.18	1657.7	[17.09..26.89]
Hsa	39883	1681	27844	16755	1151	298 (288)	5.16	6.43	-2.27	2136.2	[17.14..22.88]
Tma	1846	772	1683	1268	97	501 (499)	0.14	2.22	-3.08	1606.4	[1.04..1.06]
Mth	1869	693	1480	1150	43	438 (436)	0.12	2.00	-2.88	1305.3	[1.18..1.27]
Sso	2977	695	1950	1614	81	386 (385)	0.36	2.04	-2.68	1167.8	[1.83..2.00]
Bsu	4100	1002	3413	2502	124	507 (510)	0.48	2.01	-2.53	1534.6	[2.46..2.79]
Eco	4289	1078	3624	2765	140	523 (519)	0.84	2.54	-2.70	1837.0	[2.45..2.61]

a:
f_1(f'_1), the observed (predicted) number of domains in class 1 (represented only once in the genome);
a_1, b_1, parameters of the second-order balanced linear BDIM; k, slope of the power asymptotic;
$\nu/\delta = \nu/\lambda$, ratio of the innovation rate to the per domain death (birth) rate; G, ratio of the innovation rate to the total (per genome) birth rate.

b:
Species abbreviations: Sce, *Saccharomyces cerevisiae*, Dme, *Drosophila melanogaster*, Cel, *Caenorhabditis elegans*, Ath, *Arabidopsis thaliana*, Hsa, *Homo sapiens*, Tma, *Thermotoga maritima*, Mth, *Methanothermobacter thermoautotrophicum*, Sso, *Sulfolobus solfataricus*, Bsu, *Bacillus subtilis*, Eco, *Escherichia coli*.

(θ for the simple BDIM, a_1 and b_1 for the second-order balanced linear BDIM) were adjusted to minimize the χ^2 value.

The simplest model that resulted in a good fit to the observed domain family size distributions was the second-order balanced linear BDIM (Figure 5-14). For all analyzed genomes, $P(\chi^2)$ for this model was >0.05, *i.e.* no significant difference between the model predictions and the observed data was detected. Considering the first-order balanced linear BDIM, which involves varying the parameter θ, did not result in a significant improvement of fit for any of the analyzed genomes (data not shown). In contrast, a fit to a truncated logarithmic distribution (prediction of a simple BDIM) failed for all genomes ($P(\chi^2)<10^{-5}$; Figure 15, 16 and data not shown). An exact power-law distribution, which is often used to approximate protein family frequency distributions, similarly failed to adequately fit the observed data, even when the most deviant class 1 families were excluded ($P(\chi^2)=0.0013$ for *T. maritima*; $P(\chi^2)<10^{-5}$ for the rest of the genomes; Figure 15, 16 and data not shown). Notably, second-order balanced linear BDIM results in a correct prediction of the number of very large families, whereas simple BDIM systematically underestimates the number of families in the highest bins. Conversely, the power-law fit underestimates the slope of the best-fit line (in double logarithmic coordinates) compared to the asymptote of the linear BDIM prediction and, accordingly, significantly overestimates the number of the families in the highest bins.

Fitting the observed domain family size distribution to the second-order balanced linear BDIM resulted in positive values of the parameters a_1 and b_1, with $a_1 < b_1$, for all analyzed genomes (Table 1). Accordingly, domain family size distributions in all cases asymptotically tend to the power law with the power $k<-1$, and for all species except for C. elegans, $k<-2$ (Table 1 and Figure 7). As discussed above, this seems to indicate the existence of "synergy" between domains in a family whereby the likelihood of survival is greater for a domain that belongs to a large family than for a domain from a small family. For all species, we find that the innovation rate is approximately three orders of magnitude greater than the per domain birth (death) rate. Accordingly, the total per genome birth (duplication) rate is comparable to but, typically, several times greater than the innovation rate (Table 1). The ratio of the per genome birth rate to the innovation rate increases with the number of genes or the number of detected domains, with nearly identical rates seen for small prokaryotic genomes and values as high as 20 for the largest plant and animal genomes (Table 1).

General Discussion and Conclusions

We developed a complete mathematical description of the frequency distribution of protein domains encoded in genomes for a simple but not unrealistic models of evolution, which include three types of events: domain duplication (birth), domain elimination (death), and domain innovation. In biological terms, innovation may correspond either to gene acquisition via horizontal gene transfer or to emergence of a new domain from a non-coding sequence or a non-globular protein sequence. We showed that birth-death-innovation models (BDIMs) with different levels of complexity lead to readily distinguishable predictions regarding the distribution of domain family sizes in genomes. In particular, we defined the exact analytic conditions that lead, exactly or asymptotically, to power law distributions, which have recently received ample attention, as they were uncovered in various biological and social contexts (Barabasi, 2002; Gisiger, 2001). Importantly, and in contrast to previous analyses (Huynen and van Nimwegen, 1998; Qian *et al.*, 2001; Rzhetsky and Gomez, 2001; Wuchty, 2001), we showed that power law only asymptotically approximates the domain family size distributions.

Three groups of observations made in this work seem to have the greatest potential for enhancing our understanding of genome evolution and, perhaps, evolution of other complex systems. Firstly, we proved that, within the BDIM framework, there is a unique equilibrium state of the system, which is approached exponentially from any initial state. In this equilibrium state, the number of domain families in each size class remains constant and follows a unique distribution depending on the type and parameters of the BDIM. In particular, power asymptotics emerges when and only when a BDIM is second-order balanced, *i.e.* the rates of domain birth and death are asymptotically equal. Since we showed that the observed distribution of domains in all analyzed genomes is indeed well approximated by a particular form of a power law asymptotics, the results are compatible with the notion that the genomes are close to a steady state with respect to the domain diversity (using the BDIM formalism, F_{eq}, the number of distinct domain families at equilibrium) and distribution (f_i). Taking a broader biological perspective, this result can be interpreted to indicate that evolving lineages go through lengthy periods of relative stasis when the level of genomic complexity remains more or less the same. Under this view, the stasis epochs are punctuated by relatively short periods of dramatic changes when the complexity either greatly increases (the emergence of eukaryotes is the most obvious case in point) or decreases (*e.g.* evolution of parasites). These bursts of evolution might be described as transitions between different BDIMs in the parameter space, with some of trajectories potentially involving non-balanced BDIMs. The analogy between this emerging picture of genome evolution and the punctuated equilibrium concept of species evolution, which has been developed through analysis of the paleontological record (Gould, 2002), is obvious.

Secondly, we showed that the simplest model that adequately describes the observed domain family size distributions is the second-order balanced linear BDIM; in contrast, simple BDIMs do not show a good fit to the observations. This has important implications for the mode of domain family evolution. Simple BDIMs are based on the notion that the likelihood of duplication (birth) or elimination (death) of a domain is uniform across the genome and does not depend on the size or other characteristics of domain families (the *independence hypothesis* as we designate this assumption in this work). Clearly, under the independence hypothesis, a duplication (birth) as well as elimination (death) of a domain is more likely to occur in a large family than in a small one, but only because the overall probability of such an event is proportional to the number of family members, whereas the birth (death) rate *per domain* remains the same. The key observation of this work, that the actual domain frequency distributions are well described by a linear but not by a simple BDIM, shows that the independence hypothesis is an oversimplification. Instead, the linear BDIM includes parameters that describe the dependence of the per domain birth (death) rate on the family size. The asymptotics of the theoretical distribution that is the best fit for the actual data is a power low, with the power equal to a_1-b_1-1 (a_1 and b_1 are the parameters of a linear BDIM). We observed that, for all analyzed genomes, a_1-b_1-1 < -1 ($a_1 < b_1$), which corresponds to "synergy" between domains in a family. In other words, small families are more dynamic during evolution than large families. In particular, under the BDIM formalism, innovation contributes only to single-member families (class 1), which have the greatest evolutionary mobility, and either quickly proliferate and are stabilized or perish. A corollary of this finding is that, in general, large families are older than small ones. Exceptions to this generalization probably point to selection for a specific family size; for example, it seems likely that selection acts against proliferation of certain essential proteins, such as ribosomal proteins, which typically form single-member families (Tatusov *et al.*, 2000). Another pertinent observation is that the linear BDIM adequately accommodates even the largest of the identified domain families. Lineage-specific expansion of paralogous families appears to be one of the principal modes of adaptation (Jordan *et al.*, 2001; Lespinet *et al.*, 2002). Thus, it appears that, quantitatively, adaptive family expansion is within the framework of the BDIMs, although these models do not explicitly incorporate the notion of selection.

Thirdly, the BDIM equilibrium condition with respect to the total number of domain families, $\nu=\delta_1 f_1$ (ν is the innovation rate, δ_1 is the domain death rate for class 1 families, and f_1 is the number of domain families in class 1) allows us to estimate the ratio between domain innovation rate and the domain death and birth rates. Indeed, according to the above and the definition of a second-order linear BDIM, which is the best fit for the actual data, $\lambda=\delta=\nu/f_1(1+b_1)$. Since the number of domain families in class 1 (families with only one member) is in the hundreds for each genome, this translates into an innovation rate that is much greater than the duplication or elimination rate *per domain* (Table 1). Such high innovation rates might appear counter-intuitive, but let us note that the duplication rate over all domain families is a number, which tends to be only several-fold greater than ν or even nearly identical to ν for small prokaryotic genomes (Table 1). Thus, under the second-order balanced linear BDIM, the likelihood of appearance of a new domain in a genome is probably several times less than the likelihood of a duplication or elimination of an existing domain. Nevertheless, the finding that the innovation rate is comparable to the overall duplication/elimination rate seems surprising. If the linear BDIM is indeed a realistic evolutionary model, this emphasizes the critical role of innovation in maintaining the balance (steady state) in genome evolution. In prokaryotes, innovation is likely to involve mainly horizontal gene transfer, whereas, in eukaryotes, it might depend largely on emergence of actual new domains,

This work is an initial step toward modeling genome evolution using the observable characteristics of extant genomes to test theoretical predictions. There is little doubt that BDIMs will be eventually replaced by more sophisticated – and more realistic – formalisms. Nevertheless, even this crude modeling seems to reveal several potentially interesting and non-trivial aspects of genome evolution.

References

Albert, R., and Barabasi, A.L. 2002. Statistical mechanics of complex networks. Reviews of Modern Physics. 74: 47-97.

Albert, R., Jeong, H., and Barabasi, A.L. 2000. Error and attack tolerance of complex networks. Nature. 406: 378-382.

Amaral, L.A., Scala, A., Barthelemy, M., and Stanley, H.E. 2000. Classes of small-world networks. Proc. Natl. Acad. Sci. U S A. 97: 11149-11152.

Aravind, L., Watanabe, H., Lipman, D.J., and Koonin, E.V. 2000. Lineage-specific loss and divergence of functionally linked genes in eukaryotes. Proc. Natl. Acad. Sci USA. 97: 11319-11324.

Barabasi, A.L. 2002. Linked: The New Science of Networks. Perseus Pr, New York.

Barabasi, A.L., and Albert, R. 1999. Emergence of scaling in random networks. Science. 286: 509-512.

Brenner, S.E., Hubbard, T., Murzin, A., and Chothia, C. 1995. Gene duplications in H. influenzae. Nature. 378: 140.

Chervitz, S.A., Aravind, L., Sherlock, G., Ball, C.A., Koonin, E.V., Dwight, S.S., Harris, M.A., Dolinski, K., Mohr, S., Smith, T., Weng, S., Cherry, J.M., and Botstein, D. 1998. Comparison of the complete protein sets of worm and yeast: orthology and divergence. Science. 282: 2022-2028.

Dacks, J.B., and Doolittle, W.F. 2001. Reconstructing/deconstructing the earliest eukaryotes: how comparative genomics can help. Cell 107: 419-425.

Dorogovtsev, S.N., and Mendes, J.F. 2001. Scaling properties of scale-free evolving networks: Continuous approach. Phys. Rev. E Stat. Phys. Plasmas Fluids Relat. Interdiscip. Topics 63: 056125.

Feller, W. 1967-1968. An introduction to probability theory and its application. Wiley, New York.

Gantmacher, F.R. 1989. The theory of matrices. Chelsea Publishing Company, New York.

Gihman, I.I. and Skorohod, A.V. 1975. The theory of stochastic processes. Springer-Verlag, New-York, Heidelberg, Berlin.

Gisiger, T. 2001. Scale invariance in biology: coincidence or footprint of a universal mechanism? Biol. Rev. Camb. Philos. Soc. 76: 161-209.

Gould, S.J. 2002. The Structure of Evolutionary Theory. Harvard Univ. Press, Cambrdige, MA.

Henrici, P. 1986. Applied and computational complex analysis. Wiley, New York.

Huynen, M.A. and van Nimwegen, E. 1998. The frequency distribution of gene family sizes in complete genomes. Mol. Biol. Evol. 15: 583-589.

Ijuri, Y. and Simon, H.A. 1977. Skew distributions and the sizes of business firms. North-Holland Publishing Company, Amsterdam, New York, Oxford.

Jeong, H., Mason, S.P., Barabasi, A.L., and Oltvai, Z.N. 2001. Lethality and centrality in protein networks. Nature. 411: 41-42.

Jeong, H., Tombor, B., Albert, R., Oltvai, Z.N., and Barabasi, A.L. 2000. The large-scale organization of metabolic networks. Nature. 407: 651-654.

Johnson, N.L., Kotz, S., and Kemp, A.W. 1992. Univariate discrete distributions. Wiley, New York.

Jordan, I.K., Makarova, K.S., Spouge, J.L., Wolf, Y.I., and Koonin, E.V. 2001. Lineage-specific gene expansions in bacterial and archaeal genomes. Genome Res. 11: 555-565.

Koonin, E.V., Aravind, L., and Kondrashov, A.S. 2000. The impact of comparative genomics on our understanding of evolution. Cell. 101: 573-576.

Koonin, E.V., Makarova, K.S., and Aravind, L. 2001. Horizontal gene transfer in prokaryotes: quantification and classification. Annu. Rev. Microbiol. 55: 709-742.

Koonin, E.V., Tatusov, R.L., and Rudd, K.E. 1995. Sequence similarity analysis of Escherichia coli proteins: functional and evolutionary implications. Proc. Natl. Acad. Sci. USA. 92: 11921-11925.

Krapivsky, P.L. and Redner, S. 2001. Organization of growing random networks. Phys. Rev. E Stat. Phys. Plasmas Fluids Relat. Interdiscip. Topics 63: 066123.

Labedan, B. and Riley, M. 1995. Widespread protein sequence similarities: origins of Escherichia coli genes. J. Bacteriol. 177: 1585-1588.

Lander, E.S., Linton, L.M., Birren, B., Nusbaum, C., Zody, M.C., Baldwin, J., Devon, K., Dewar, K., Doyle, M., FitzHugh, W., Funke, R., Gage, D., Harris, K., Heaford, A., Howland, J., Kann, L., Lehoczky, J., LeVine, R., McEwan, P., McKernan, K., Meldrim, J., Mesirov, J.P., Miranda, C., Morris, W., Naylor, J., Raymond, C., Rosetti, M., Santos, R., Sheridan, A., Sougnez, C., Stange-Thomann, N., Stojanovic, N., Subramanian, A., Wyman, D., Rogers, J., Sulston, J., Ainscough, R., Beck, S., Bentley, D., Burton, J., Clee, C., Carter, N., Coulson, A., Deadman, R., Deloukas, P., Dunham, A., Dunham, I., Durbin, R., French, L., Grafham, D., Gregory, S., Hubbard, T., Humphray, S., Hunt, A., Jones, M., Lloyd, C., McMurray, A., Matthews, L., Mercer, S., Milne, S., Mullikin, J.C., Mungall, A., Plumb, R., Ross, M., Shownkeen, R., Sims, S., Waterston, R.H., Wilson, R.K., Hillier, L.W., McPherson, J.D., Marra, M.A., Mardis, E.R., Fulton, L.A., Chinwalla, A.T., Pepin, K.H., Gish, W.R., Chissoe, S.L., Wendl, M.C., Delehaunty, K.D., Miner, T.L., Delehaunty, A., Kramer, J.B., Cook, L.L., Fulton, R.S., Johnson, D.L., Minx, P.J., Clifton, S.W., Hawkins, T., Branscomb, E., Predki, P., Richardson, P., Wenning, S., Slezak, T., Doggett, N., Cheng, J.F., Olsen, A., Lucas, S., Elkin, C., Uberbacher, E., Frazier, M., *et al.* 2001. Initial sequencing and analysis of the human genome. Nature. 409: 860-921.

Lespinet, O., Wolf, Y.I., Koonin, E.V., and Aravind, L. 2002. The role of lineage-specific gene family expansion in the evolution of eukaryotes. Genome Res. in press.

Marchler-Bauer, A., Panchenko, A.R., Shoemaker, B.A., Thiessen, P.A., Geer,

L.Y., and Bryant, S.H. 2002. CDD: a database of conserved domain alignments with links to domain three-dimensional structure. Nucleic Acids Res. 30: 281-283.

Ponting, C.P., Schultz, J., Milpetz, F., and Bork, P. 1999. SMART: identification and annotation of domains from signalling and extracellular protein sequences. Nucleic Acids Res. 27: 229-232.

Qian, J., Luscombe, N.M., and Gerstein, M. 2001. Protein family and fold occurrence in genomes: power-law behaviour and evolutionary model. J. Mol. Biol. 313: 673-681.

Rubin, G.M., Yandell, M.D., Wortman, J.R., Gabor Miklos, G.L., Nelson, C.R., Hariharan, I.K., Fortini, M.E., Li, P.W., Apweiler, R., Fleischmann, W., Cherry, J.M., Henikoff, S., Skupski, M.P., Misra, S., Ashburner, M., Birney, E., Boguski, M.S., Brody, T., Brokstein, P., Celniker, S.E., Chervitz, S.A., Coates, D., Cravchik, A., Gabrielian, A., Galle, R.F., Gelbart, W.M., George, R.A., Goldstein, L.S., Gong, F., Guan, P., Harris, N.L., Hay, B.A., Hoskins, R.A., Li, J., Li, Z., Hynes, R.O., Jones, S.J., Kuehl, P.M., Lemaitre, B., Littleton, J.T., Morrison, D.K., Mungall, C., O'Farrell, P.H., Pickeral, O.K., Shue, C., Vosshall, L.B., Zhang, J., Zhao, Q., Zheng, X.H., Zhong, F., Zhong, W., Gibbs, R., Venter, J.C., Adams, M.D., and Lewis, S. 2000. Comparative genomics of the eukaryotes. Science. 287: 2204-2215.

Rzhetsky, A. and Gomez, S.M. 2001. Birth of scale-free molecular networks and the number of distinct DNA and protein domains per genome. Bioinformatics. 17: 988-996.

Tatusov, R.L., Galperin, M.Y., Natale, D.A., and Koonin, E.V. 2000. The COG database: a tool for genome-scale analysis of protein functions and evolution. Nucleic Acids Res. 28: 33-36.

Tatusov, R.L., Mushegian, A.R., Bork, P., Brown, N.P., Hayes, W.S., Borodovsky, M., Rudd, K.E., and Koonin, E.V. 1996. Metabolism and evolution of Haemophilus influenzae deduced from a whole-genome comparison with Escherichia coli. Curr. Biol. 6: 279-291.

Wolf, Y.I., Grishin, N.V., and Koonin, E.V. 2000. Estimating the number of protein folds and families from complete genome data. J. Mol. Biol. 299: 897-905.

Wuchty, S. 2001. Scale-free behavior in protein domain networks. Mol. Biol. Evol. 18: 1694-1702.

Mathematical Appendix. Proofs of Some Statements
Proof of Proposition 1

When system (3.1) is solved consecutively from the last equation to the second one, it becomes obvious that the solution is unique up to a constant multiplier.

Next, if $f_i = f_{i-1}\lambda_{i-1}/\delta_i$, $f_{i+1} = f_{i-1}\lambda_{i-1}\lambda_i/(\delta_i\delta_{i+1})$, then the substitution shows that (f_{i-1}, f_i, f_{i+1}) satisfy the i-th equation of system (3.2). Substituting $f_2 = f_1\lambda_1/\delta_2$ in the first equation, we get $f_1 = \nu/\delta_1$ and, consequently, $f_i = \nu \prod_{k=1}^{i-1}\lambda_k / \prod_{k=1}^{i}\delta_k$ for all $i=2,\ldots N$. By definition, $F_{eq} = \sum_{i=1}^{N} f_i$, so we have (3.3).

Since system (2.2) is linear, the equilibrium state $(f_1, \ldots f_N)$ is asymptotically stable if the real parts of all characteristic values of the matrix

$$\mathbf{A}_N = \begin{matrix} -(\lambda_1+\delta_1) & \delta_2 & 0 & \cdots & \cdots & \cdots & \cdots & \cdots & 0 \\ \lambda_1 & -(\lambda_2+\delta_2) & \delta_3 & 0 & \cdots & \cdots & \cdots & \cdots & 0 \\ \cdots & \cdots & \cdots & \cdots & \cdots & \cdots & \cdots & \cdots & \cdots \\ 0 & \cdots & 0 & \lambda_{i-1} & -(\lambda_i+\delta_i) & \delta_{i+1} & 0 & \cdots & 0 \\ \cdots & \cdots & \cdots & \cdots & \cdots & \cdots & \cdots & \cdots & \cdots \\ 0 & \cdots & \cdots & \cdots & \cdots & 0 & \lambda_{N-2} & -(\lambda_{N-1}+\delta_{N-1}) & \delta_N \\ 0 & \cdots & \cdots & \cdots & \cdots & \cdots & 0 & \lambda_{N-1} & -\delta_N \end{matrix}$$

are negative.

The following theorem (see Gantmacher, 1989) gives the desired criterion: the real part of all the characteristic values of a real matrix $C=|c_{ij}|$, $i,j=1,..n$ with non-negative non-diagonal elements are negative if and only if $(-1)^k D_k > 0$ for all $k=1,..n$, where D_k is the main minor of the matrix C of the k-th order. To apply this theorem, let us consider the n x n matrix, $n \le N$

$$\mathbf{B}_n = \begin{matrix} -(\lambda_1+\delta_1) & \delta_2 & 0 & \cdots & \cdots & \cdots & \cdots & \cdots & 0 \\ \lambda_1 & -(\lambda_2+\delta_2) & \delta_3 & 0 & \cdots & \cdots & \cdots & \cdots & 0 \\ \cdots & \cdots & \cdots & \cdots & \cdots & \cdots & \cdots & \cdots & \cdots \\ 0 & \cdots & 0 & \lambda_{i-1} & -(\lambda_i+\delta_i) & \delta_{i+1} & 0 & \cdots & 0 \\ \cdots & \cdots & \cdots & \cdots & \cdots & \cdots & \cdots & \cdots & \cdots \\ 0 & \cdots & \cdots & \cdots & \cdots & 0 & \lambda_{n-2} & -(\lambda_{n-1}+\delta_{n-1}) & \delta_n \\ 0 & \cdots & \cdots & \cdots & \cdots & \cdots & 0 & \lambda_{n-1} & -(\lambda_n+\delta_n) \end{matrix}$$

It is easy to see that

$$\det \boldsymbol{B}_n = -(\lambda_n + \delta_n)\det \boldsymbol{B}_{n-1} - \lambda_{n-1}\delta_n \det \boldsymbol{B}_{n-2},$$
$$\det \boldsymbol{A}_n = -\delta_n \det \boldsymbol{B}_{n-1} - \lambda_{n-1}\delta_n \det \boldsymbol{B}_{n-2}. \quad \text{(A1)}$$

Using this equality, we can prove that for any n

$$\det \boldsymbol{A}_n = (-1)^n \delta_n \delta_{n-1} \ldots \delta_2 \delta_1.$$

Indeed,

$$\det \boldsymbol{A}_n = -\delta_n \det \boldsymbol{B}_{n-1} - \lambda_{n-1}\delta_n \det \boldsymbol{B}_{n-2} =$$
$$\delta_n((\lambda_{n-1}+\delta_{n-1}) \det \boldsymbol{B}_{n-2} + \lambda_{n-2}\delta_{n-1} \det \boldsymbol{B}_{n-3}) - \lambda_{n-1}\delta_n \det \boldsymbol{B}_{n-2} =$$
$$\delta_n \delta_{n-1} (\det \boldsymbol{B}_{n-2} + \lambda_{n-2} \det \boldsymbol{B}_{n-3}) = \text{(subsequently using (A1))} =$$
$$(-1)^{n-2}\delta_n\delta_{n-1}\ldots \delta_3(\det \boldsymbol{B}_2 + \lambda_2 \det \boldsymbol{B}_1) = (-1)^n \delta_n \delta_{n-1} \ldots \delta_2 \delta_1.$$

Further, it is easy to see that for any n

$$\det \boldsymbol{B}_n = \det \boldsymbol{A}_n - \lambda_n \det \boldsymbol{B}_{n-1}.$$

Taking into account that $B_1 = -(\lambda_1 + \delta_1) < 0$ and that the sign of $\det \boldsymbol{A}_n$ coincides with $(-1)^n$, it is easy to prove that

$$\det \boldsymbol{J}_n > \det \boldsymbol{A}_n \text{ if } \det \boldsymbol{A}_n > 0 \text{ and } \det \boldsymbol{J}_n < \det \boldsymbol{A}_n \text{ if } \det \boldsymbol{A}_n < 0.$$

Thus, the sign of $\det \boldsymbol{B}_n$ coincides with the sign of $\det \boldsymbol{A}_n$ and so $(-1)^n \boldsymbol{B}_n > 0$ for all $n=1,..N$. According to the aforementioned theorem, the real part of all the characteristic values of a real matrix A_N are negative and so the single equilibrium is asymptotically stable, as desired.

Proof of Proposition 2

For simple BDIM (2.1) $f_i = \nu \prod_{k=1}^{i-1} \lambda_k / \prod_{k=1}^{i} \delta_k = (\nu/\delta)\theta^{i-1}/i = (\nu/\lambda)\theta^i/i$, so

$F_{eq} = \sum_{i=1}^{N} f_i = \nu/\lambda \sum_{i=1}^{N} \theta^i/i$. Thus

$$p_i = f_i / F_{eq} = (\theta^i/i) / \sum_{j=1}^{N} \theta^j/j.$$

If a simple BDIM is balanced, then $\theta=1$ and so

$$F_{eq} = \nu/\delta \sum_{j=1}^{N} \theta^j/j,$$

$$p_i = \nu/\delta\, F_{eq}/i = 1/i \left(\sum_{j=1}^{N} 1/j\right)^{-1}.$$

Proof of Theorem 1
The condition (3.10) can be rewritten as $\lambda_{i-1}/\delta_i = i^s\theta(1+a/i+O(1/i^2)) = i^s\theta\,(1+a/i)(1+O(1/i^2))$. Thus we can choose S in such a way that $\prod_{s=S}^{\infty}(1+O(1/s^2))$ converge, $0<\prod_{s=S}^{\infty}(1+O(1/s^2))<\infty$. It follows that

$$\prod_{s=1}^{j}(\lambda_{s-1}/\delta_s) \sim \Gamma(j)^s\theta^j\prod_{s=1}^{j}(1+a/s).$$

According to Proposition 1, $p_i = f_i/F_{eq} \sim \prod_{k=1}^{i-1}\lambda_k/\prod_{k=1}^{i}\delta_k$. So

$$p_i \sim \prod_{s=2}^{i}(\lambda_{s-1}/\delta_s) \sim \Gamma(i)^s\theta^i\prod_{s=2}^{i}(1+a/s) = \Gamma(i)^s\theta^i\Gamma(i+a+1)/\Gamma(i+1).$$

Applying the main asymptotic property of Γ-function, *i.e.* $\Gamma(i+c)/\Gamma(i) \sim i^c$ at large i for any c, we have

$\Gamma(i+a+1)/\Gamma(i+1) \sim i^a$, and so $p_i \sim \Gamma(i)^s\theta^i i^a$.

Proofs of Corollaries 1-3
If a discrete random variable ξ has the Pascal distribution, then

$P(\xi = i) \cong 1/\Gamma(r)\,(1-q)^r i^{r-1}q^i \sim q^i i^{r-1}$ for large i,

and it becomes evident that, for $a>-1$, equilibrium frequencies p_i of the first-order balanced BDIM follows the Pascal distribution with parameters $(a+1,\theta)$. If $a=-1$, then $p_i \sim \theta^i/i$ and so p_i follows the truncated logarithmic distribution with parameter θ. If $a=0$, then $p_i \sim \theta^i$ and p_i follows the geometric distribution. Further, $p_i \sim i^a$, that is the sequence p_i follows the power distribution with the power a, if and only if $\theta=1$, that is, if the BDIM is second-order balanced. Finally, if $\lambda_{i-1}/\delta_i = 1+O(1/i^2)$, that is, if $\theta=1$ and $a=0$, then $p_i \sim$const; in particular, if $\lambda_{i-1} = \delta_i$ for all i, then, according to Proposition 1, $f_i = \nu$ for all i and $p_i = 1/N$.

Reformulation of the Main theorem
We proved that any first-order balanced BDIM has asymptotically a specific form of the distribution of its equilibrium frequencies, which depends on two coefficients of expansion (3.10), θ and a. To find these coefficients for given functions $\lambda(i)=\lambda_i$ and $\delta(i)=\delta_i$, let us note that the main asymptotic relation (3.10) corresponds to the Taylor formula for $\chi(i)$ at the infinitely distant point.

Definition. Let us designate a function of real variable x, $V(\mathrm{x})$, defined by relations:

$V(x)=\mathrm{v}(1/x)$, where $\mathrm{v}(y)=\lambda(y-1)/\delta(y)$

the characteristic function of BDIM.

We can reformulate Theorem 1 in an analytical form, which might be more convenient for calculations.

Theorem 1'. Let the characteristic function of a BDIM, $V(x)$, be defined and finite at $x=0$ together with its two first derivatives, and

$V(0)=\theta;\ 0<\theta<\infty;$
$\mathrm{d}V(0)/\mathrm{d}x=-a.$

Then the equilibrium frequencies of the BDIM have asymptotics
$p_i \sim \theta^i i^a.$

In particular, $\{p_i\}$ follows the power distribution, $p_i \sim i^a$, if and only if $V(0)=1$ and $\mathrm{d}V(0)/\mathrm{d}x=-a$.

Indeed, it is easy to see that relation (3.10) with $s=0$ could be written in the form $g(y)=\theta(1+a/y+O(1/y^2))$ at large y; changing variable $y=1/x$, we see that the last relation is equivalent to the Taylor formula for the function $\mathrm{V}(x)$ at $x=0$: $\mathrm{V}(x)= V(0)+ [\mathrm{dV}(0)/\mathrm{d}x]x +O(x^2)$.

Proof of Theorem 2
According to Proposition 1, system (3.1) has the unique solution:

$f_1=\nu\delta_1, f_i=\nu\prod_{s=1}^{i-1}\lambda_s/\prod_{s=1}^{i}\delta_s$ for all $i=2,\ldots N$. So

$f_i=\nu/\lambda\theta^i\prod_{s=1}^{i-1}P(s)/\prod_{s=1}^{i}Q(s),\ i>1.$

Applying the Lemma (see below), we get

$f_i\cong C\nu/\lambda\,\theta^i\,\Gamma(i)^{\eta}\,i^{\rho-\beta}$, as $i\to\infty$,

where the constant $C=\prod_{k=1}^{m}[(\Gamma(1+b_k))^{\beta_k}]/\prod_{k=1}^{n}[\Gamma(1+a_k)^{\alpha_k}]$.

Lemma. Let $P(j)=\prod_{k=1}^{n}(j+a_k)^{\alpha_k}$, $Q(j)=\prod_{k=1}^{m}(j+b_k)^{\beta_k}$, where a_k, b_k are positive.

Let us denote

$$\eta=\sum_{k=1}^{n}\alpha_k-\sum_{k=1}^{m}\beta_k,\ \rho=\sum_{k=1}^{n}a_k\,\alpha_k-\sum_{k=1}^{m}b_k\beta_k,\ \beta=\sum_{k=1}^{m}\beta_k.$$

Then with fixed j

$$N(j)=\prod_{s=1}^{j-1}P(s)/\prod_{s=1}^{j}Q(s)\cong C\,\Gamma(j)^{\eta}\,j^{(\rho-\beta)}$$

as $j\to\infty$, where

$$C=\prod_{k=1}^{m}[(\Gamma(1+b_k))^{\beta_k}]/\prod_{k=1}^{n}[\Gamma(1+a_k)^{\alpha_k}].$$

Proof.

$$\prod_{s=1}^{j-1}(s+a_k)^{\alpha_k}=[\Gamma(j+a_k)/\Gamma(1+a_k)]^{\alpha_k},$$

$$\prod_{s=1}^{j}(s+b_k)^{\beta_k}=[\Gamma(j+1+b_k)/\Gamma(1+b_k)]^{\beta_k},\text{ so}$$

$$N(j)=\{\prod_{k=1}^{n}[\Gamma(j+a_k)/\Gamma(1+a_k)]^{\alpha_k}\}/\{\prod_{k=1}^{m}[\Gamma(j+1+b_k)/\Gamma(1+b_k)]^{\beta_k}\}=$$

$$C\prod_{k=1}^{n}[(\Gamma(j+a_k))^{\alpha_k}]/\prod_{k=1}^{m}[(\Gamma(j+1+b_k))^{\beta_k}]$$

where

$$C=\prod_{k=1}^{m}[(\Gamma(1+b_k))^{\beta_k}]/\prod_{k=1}^{n}[\Gamma(1+a_k)^{\alpha_k}].$$

Let us use the known asymptotic relation
$\Gamma(t+a)/\Gamma(t)\cong t^a$ with $t\to\infty$.

Thus we have

$$\prod_{k=1}^{n}[(\Gamma(j+a_k))^{\alpha_k}]/\prod_{k=1}^{m}[(\Gamma(j+1+b_k))^{\beta_k}]=$$

$$(\Gamma(j))^{\eta}\prod_{k=1}^{n}[(\Gamma(j+a_k)/\Gamma(j))^{\alpha_k}]/\prod_{k=1}^{m}[(\Gamma(j+1+b_k)/\Gamma(j))^{\beta_k}]\cong$$

$$(\Gamma(j))^{\eta}\,j^{[\sum_{k=1}^{n}a_k\,\alpha_k]}/j^{[\sum_{k=1}^{m}(b_k+1)\beta_k]}=$$

$$(\Gamma(j))^{\eta}\,j^{(\rho-\beta)},$$

and Lemma is completely proved.

Proof of Proposition 5.

It follows from the proof of the Lemma that

$$f_i = Cv/\lambda\ \theta^i \prod_{k=1}^{n}[(\Gamma(j+a_k))^{\alpha_k}]/\prod_{k=1}^{m}[(\Gamma(j+1+b_k))^{\beta_k}] \text{ for } i>1,$$

where $C=\prod_{k=1}^{m}[(\Gamma(1+b_k))^{\beta_k}]/\prod_{k=1}^{n}[\Gamma(1+a_k)^{\alpha_k}]$.

Let us show that f_1 can be expressed by the same formula if i=1. Indeed,

$$Cv/\delta \prod_{k=1}^{n}(\Gamma(1+a_k))^{\alpha_k}/\prod_{k=1}^{m}(\Gamma(1+1+b_k))^{\beta_k}=$$

$$v/\delta(\prod_{k=1}^{m}(\Gamma(1+b_k))^{\beta_k}/\prod_{k=1}^{n}\Gamma(1+a_k)^{\alpha_k}))\ (\prod_{k=1}^{n}(\Gamma(1+a_k))^{\alpha_k}/\prod_{k=1}^{m}(\Gamma(2+b_k))^{\beta_k})=$$

$$v/\delta(\prod_{k=1}^{m}(\Gamma(1+b_k))^{\beta_k}/\prod_{k=1}^{m}(\Gamma(2+b_k))^{\beta_k})= v/\delta(\prod_{k=1}^{m}(1+b_k)^{\beta_k}=f_1$$

as desired. Thus

$$F_{eq} = Cv/\delta(\sum_{j=1}^{N}\theta^{j-1}\prod_{k=1}^{n}(\Gamma(j+a_k))^{\alpha_k}/\prod_{k=1}^{m}(\Gamma(j+1+b_k))^{\beta_k}).$$

Index

A

B

C

D

E

K

L

M

O

P

R

S

Current Issues in Molecular Biology

ISSN 1467-3037 (print)
ISSN 1467-3035 (online)

Current Issues in Molecular Biology (CIMB) publishes high quality critical review papers, in English, in all areas of molecular biology, including plant and animal systems, human genetics, molecular microbiology, genomics, biomedicine, and biotechnology. In addition CIMB publishes articles describing innovative technology of general interest to molecular biologists and biomedical scientists.

Reviews are selected on the basis of topicality, originality, objectivity, and relevance to a broad readership encompassing research scientists, biomedical scientists, and health professionals. All manuscripts undergo a strict peer-review process and are reviewed by a member of the editorial board and at least one other appropriate individual.

CIMB is abstracted in Index Medicus, Medline, Pubmed, BIOSIS, CAB Abstracts, CAB Health, Cambridge Scientific Abstracts, EMBASE (Excerpta Medica), and BIOBASE (Current Awareness in Biological Sciences).

Other Titles of Interest

HLA class II alpha chain

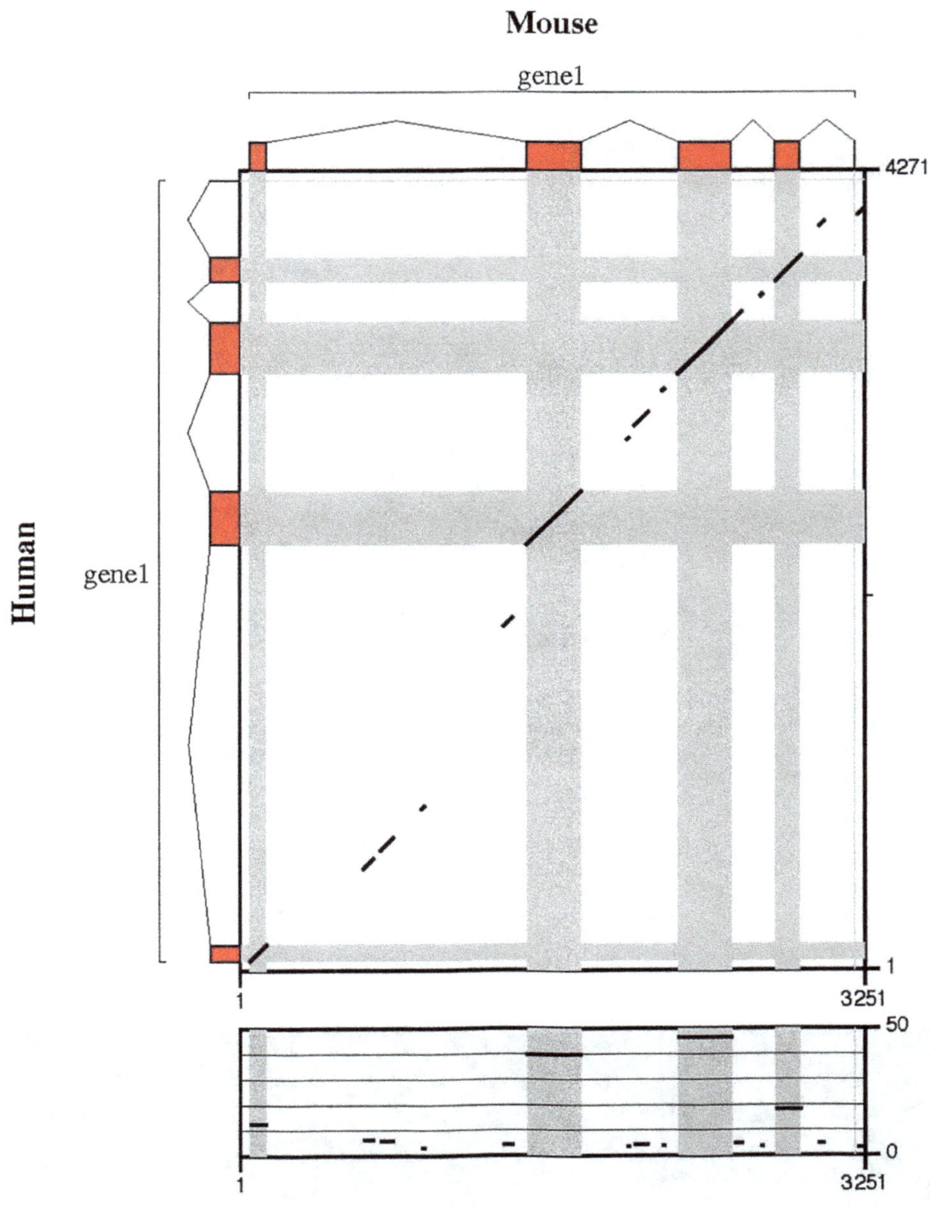

Chapter 1: Figure 1. Pairwise comparison using TBLASTX of the human and mouse genomic sequences coding for the HLA class II alpha chain. Red boxes indicate the coding exons, while black diagonals indicate the conserved alignments. The score of the conserved alignments (divided by 10) is given in the lower panel. While conserved regions between human and mouse fully include the coding exons, a substantial fraction of intronic regions is also conserved. TBLASTX ouptut has been post-processed to produce a maximal-scoring, monotonic path. The figure has been obtained using the program GFF2APLOT (http://www1.imim.es/software/gfftools/GFF2APLOT.html).

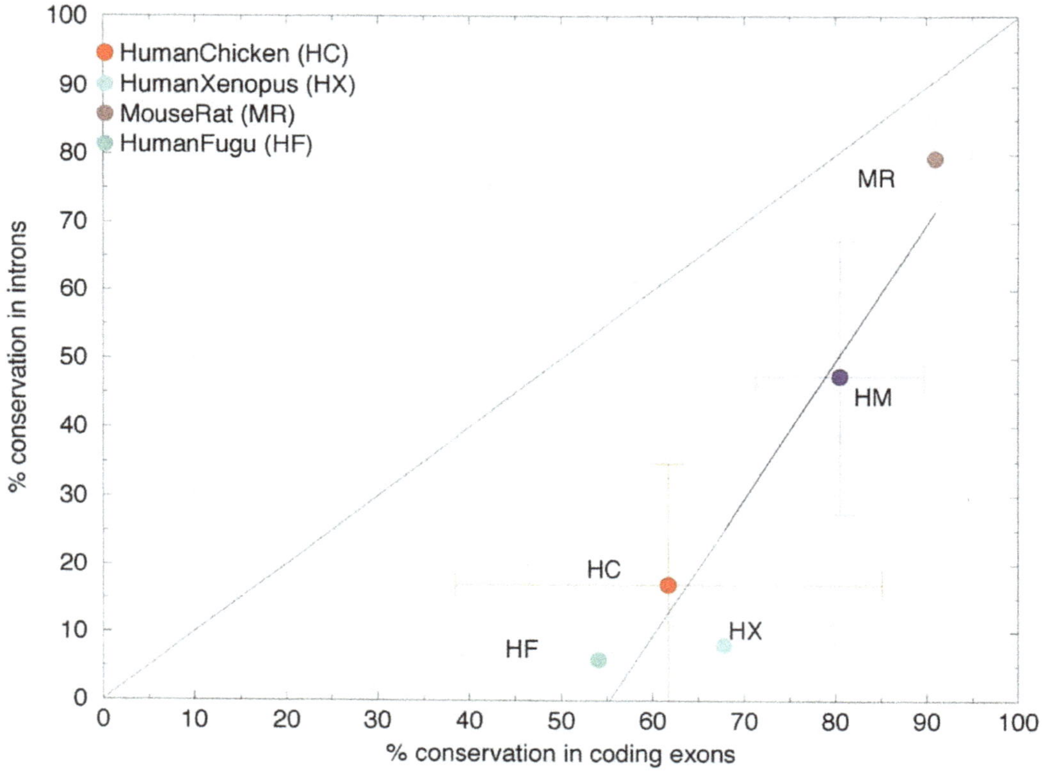

Chapter 1: Figure 2. Conservation in coding exons versus introns. The reported values are averages obtained from pairwise comparisons of a set of orthologous genes (kindly provided by L. Duret). Pairs of some vertebrate species (see the inlet) at varying evolutionary distances have been compared.

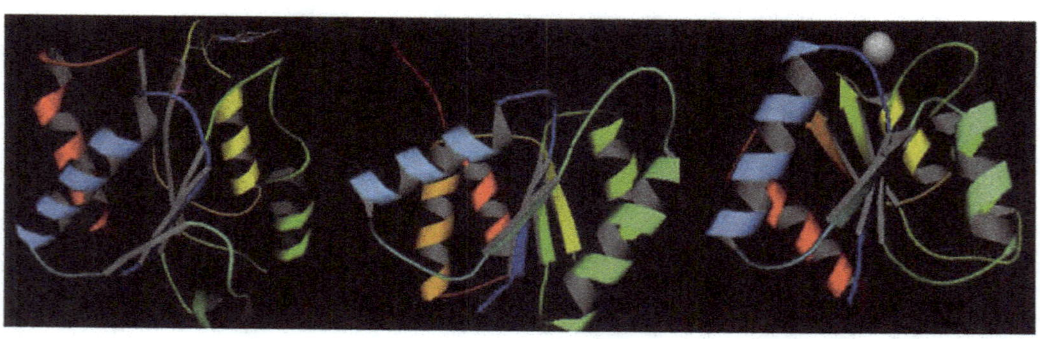

Chapter 2: Figure 1. Distant homologs share significant structural similarity with barely detectable sequence similarity. Fold families: flavodoxin, mannose permease and structures. Sequence similarity is smaller than 15% of identical residues for each pair, structural similarity better than 3Å RMSD over more than 50% of the structure.

A

- phosphoribosyltransferase and viral coat protein, identity: 42%, different folds, different functions

```
 99 IRLKSYCNDQSTGDIKVIGGDDLSTLTGKNVLIVEDIIDTGKTMQTLLSLVRQY.NPKMVKVASLLVKRTPRSVGY 173
    : ||. ||| ||              |.    || |  : |  |  | | || |  || |:|         | ||.| |
214 VPLKTDANDQ.IGDSLY....SAMTVDDFGVLAVRVVNDHNPTKVT..SKVRIYMKPKHVRV...WCPRPPRAVPY 279
```

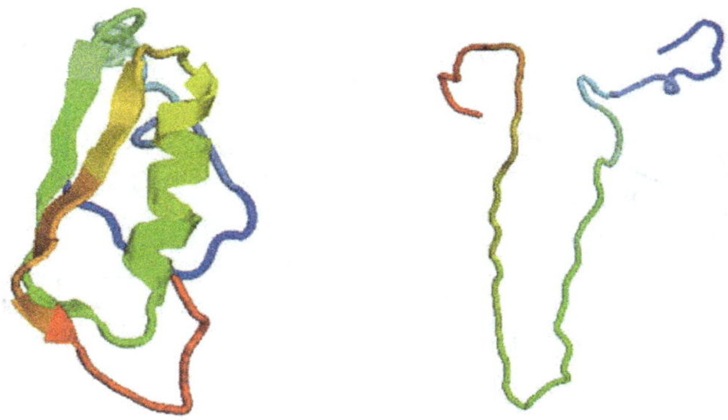

B

- Histon H5 and transcription factor E2F4, identity 7%, similar fold, similar function (DNA binding)

```
• PTYSEMIAAAIRAEKSRGGSSRQSIQKYIKSHYKVGHNADLQIKLSIRRLLAAGVLKQTKGVGASGSFRL
                 |  |                        |         |           |
• GLLTTKFVSLLQEAKD-GVLDLKLAADTLA------VRQKRRIYDITNVLEGIGLIEKKS----KNSIQW
```

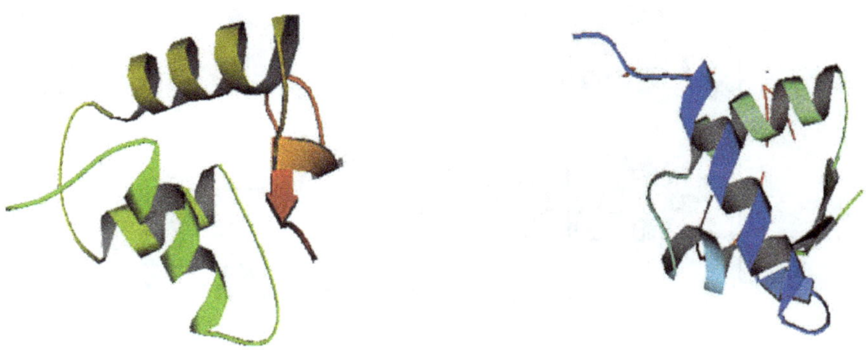

Chapter 2: Figure 2. Sequence similarity is not equivalent to homology. A) Two proteins similar on the sequence level, but not homologous, as shown by their function and structure differences. B). Two proteins, homologous as shown by their function and fold similarity, but with no apparent sequence similarity (7% sequence similarity).

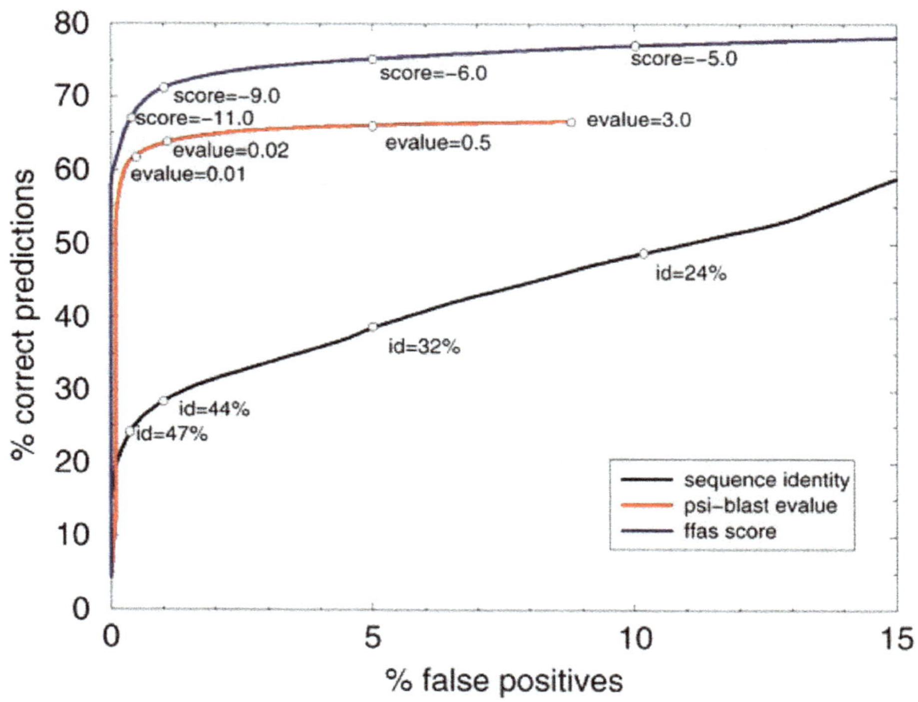

Chapter 2: Figure 3. Comparison of homology recognition programs on a fold recognition benchmark, consisting of several hundred pairs of proteins with similar structures but varying sequence identity. Percentage of correct prediction is given as a function of incorrect ones for every level of statistical significance. (Both numbers do not add to 100% as there are cases where there is no prediction with a given significance level).

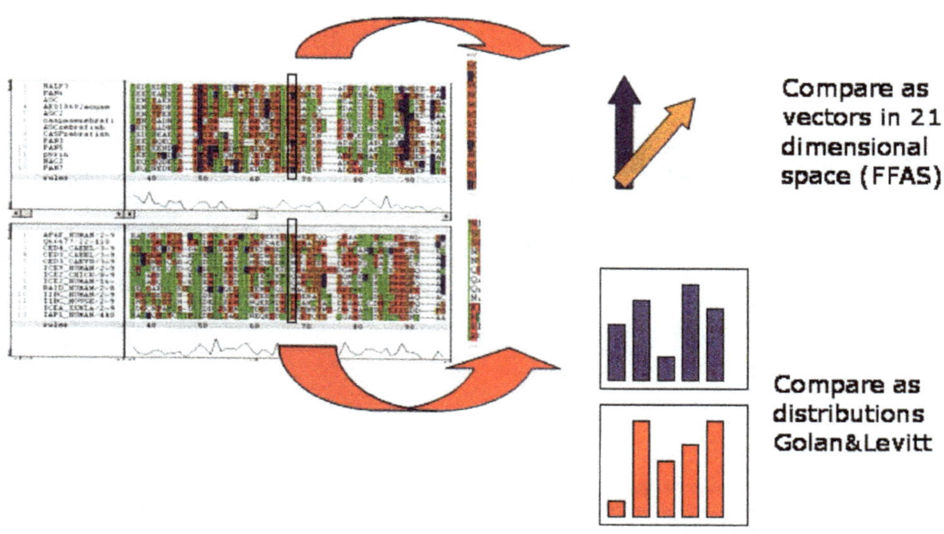

Chapter 2: Figure 5. Two possibilities of implementing a profile-profile scoring systems.

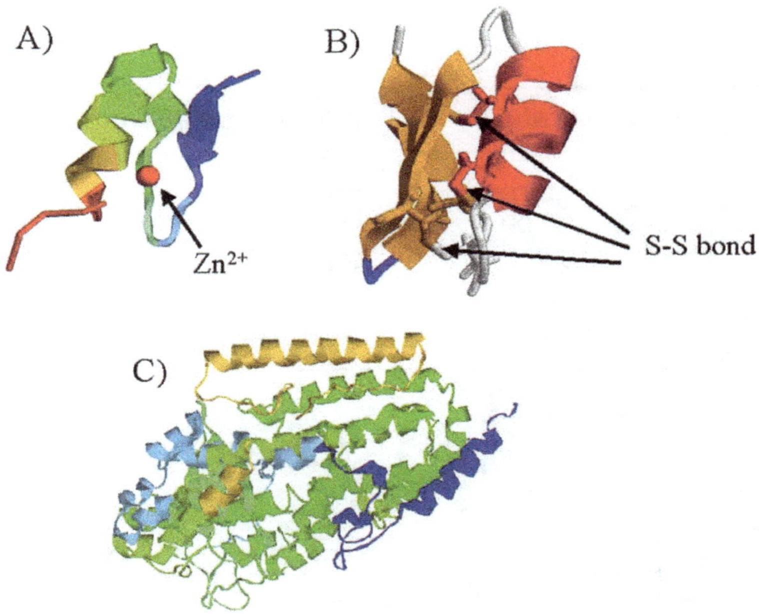

Chapter 5: Figure 1. Protein domains, large and small. All structure figures have been produced using Rasmol (Sayle and Milner-White, 1995). A) The 31 residue long classical C2H2 zinc finger domain. The zinc ion is shown by a dark sphere. B) 37 amino acid long snake toxin domain that is stabilised by three disulphide bridges. C) This α-helical domain from lipoxygenase is over 500 amino acids in length.

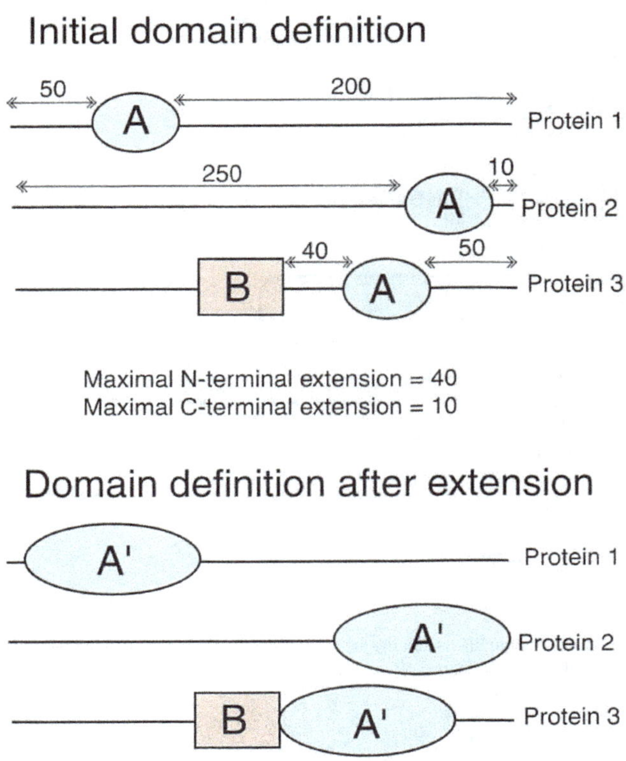

Chapter 5: Figure 2. Definition of domains from sequence alone. Sequence comparison shows a region of similarity (a domain), called A, shared between three proteins. A second domain shown is called B. The extended region of similarity is called A' and is the maximal extent of the predicted domain.

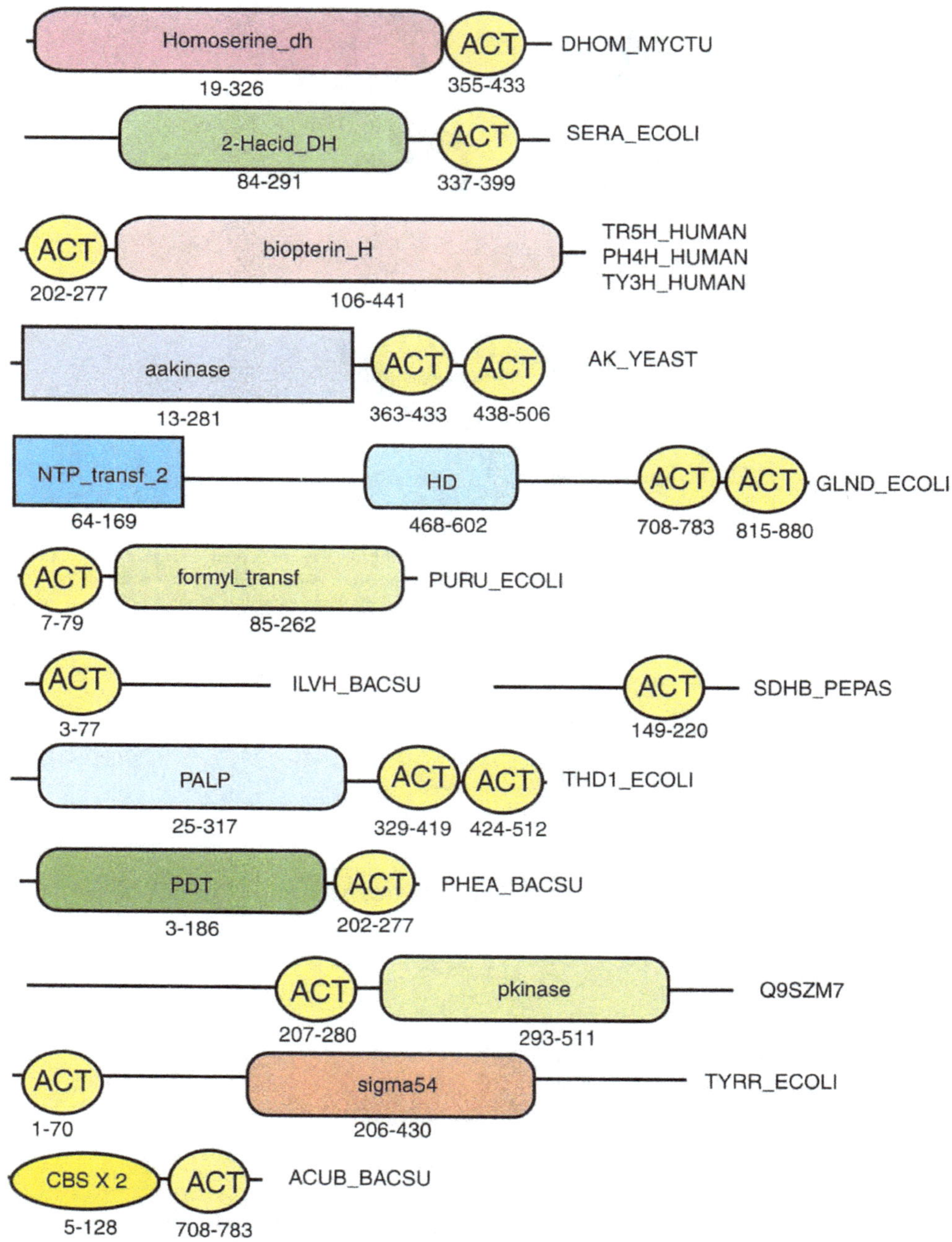

Chapter 5: Figure 3. Domain organisations of ACT domain-containing proteins. Boxes and circles represent domains; within each domain is its Pfam identifier and start and end points of the domain are given below. The Swiss-Prot sequence identifier is given for a representative of each protein.

```
ACUB_BACSU     138 SQIEIKVNDI...TKSLAEISSLCQDLQVKILSVLVYPHD....DPGVKVLVFRVKT...MNPLPFLQALQRNGH..HVVWPSEQRD 212
AK1H_ECOLI     316 AMFSVSGPGMKGMVGMAARVFAAMSRARISVVLITQSSSE........YSISFCVP......QSDCVRAERAM....LEEFYLELKE 384
AK1H_ECOLI     397 AIISVVGDGLRTLRGISAKFFAALARANINIVAIAQGSSE........RSISVVVNN...DDATTGVRVTHQM....LFNTDQVIEV 468
AK_YEAST       363 FVINIHSNKKTLSHGFLAQIFTILDKYKLVVDLISTSEVH..........VSMALPI...PDADSLKSLRQAEE...KLRILGSVDI 433
AK_YEAST       438 SIVSLVGKHMKQYIGIAGTMFTTLAEEGINIEMISQGANE........INISCVIN......ESDSIKALQCI....HAKLLSERTN 506
DHOM_MYCTU     355 YYVSMNVADK...PGVLSAVAAEFAKREVSIAEVRQEGVVDEGGRRVGARIVVVTHL...ATDAALSETVDALDD..LDVVQGVSSV 433
GCVR_ECOLI       8 LVITALGADR...PGIVNTITRHVSSCGCNIEDSRLAMLG......EEFTFIMLLSGS.WNAITLIESTLPLKGA..ELDLLIVMKR 82
GCVR_ECOLI      95 VWVQVDVADS...PHLIERFTALFDAHHMNIAELVSRTQPAENERAAQLHIQITAHSPASADAANIEQAFKALCT..ELNAQGSINV 176
GLND_ECOLI     708 TEIFIWSPDR...PYLFAAVCAELDRRNLSVHDAQIFTTR....DGMAMDTFIVLE....PDGNPLSADRHEVIRFGLEQVLTQSSW 783
GLND_ECOLI     815 SFLELIALDQ...PGLLARVGKIFADLGISLHGARITTIG.....ERVEDLFIIAT............ADRRALNN..ELQQEVHQRL 880
ILVH_BACSU       3 RIITLTVVNR...SGVLNRITGLFTKRHYNIESITVGHTE....TAGVSRITFVVHV...EGENDVEQLTKQLNK..QIDVLKVTDI 77
MAO1_BACSU       8 RTLMIETPSV...PGNLGRVATAIGLLGGDIGEVETVKVG...PNYTMRNITVQVE.....NEEQLQEVIAAVQA..LGEGIRLHTV 81
PH4H_HUMAN      35 ISLIFSLKEE...VGALAKVLRLFEENDVNLTHIESRPSR.....LKKDEYEFFTHLD.KRSLPALTNIIKILRH..DIGATVHELS 110
PHEA_BACSU     202 TLMVMLPQDD..QSGALHRVLSAFSWRNLNLSKIESRPTK.....TGLGHYFFIIDIEKAFDDVLIPGAMQEL....EALGCKVRLL 277
PHEB_BACSU      69 ITLFFHLEDR...SGALSQLLQAVADSGSNVLSIHQTIPL.....QGRANVTLSIST..SAMEEDIHTLMNKLR...KFDFVEKVEI 142
PURU_ECOLI       7 KVLRTICPDQ...KGLIARITNICYKHELNIVQNNEFVDH......RTGRFFMRTEL...EGIFNDSTLLADLDS..ALPEGSVREL 79
RELA_ECOLI     667 LVVRVVANDR...SGLLRDITTILANEKVNVLGVASRSDT....KQQLATIDMTIEI...YNLQVLGRVLGKLNQ..VPDVIDARRL 741
SDHB_PEPAS     149 STILLEYPEQ...RGVISYVSSLLTGSEYNIESLNTKK......NKLTNIVTLTVE....IDKPLTESLKSAILG..VERFTTAKYV 220
SERA_ECOLI     337 RRLMHIHENR...PGVLTALNKIFAEQGVNIAAQYLQTSA....QMGYVVIDIEAD......EDVAEKALQAMKA..IPGTIRARLL 408
SERA_ECOLI SS      .EEEEEEE.......HHHHHHHHHH....EEEEEEEEE.......EEEEEEEEE........HHHHHHHHHHHH......EEEEEE.
SERB_ARCFU      47 YVVTILGKDR...VGIVRDITRAFLDFGINIERTSLTARE......ELISIEFLVD....LGQRDAAEVRKRLRR..EAERLGLDIV 118
SERX_YEAST     398 VRVLYIHQNV...PGVLKTVNDILS..NHNIEKQFSDSNG........EIAYLMADIS.SVDQSDIKDIYEQLNQ..TSAKISIRLL 468
SERY_YEAST     398 VRVLYIHRNV...PGVLKTVNDILS..DHNIEKQFSDSHG........EIAYLMADIS.SVNQSEIKDIYEKLNQ..TSAKVSIRLL 468
SPOT_ECOLI     627 TEIKVEMFNH...QGALANLTAAINTTTSNIQSLNTEEKD.....GRVYSAFIRLT......ARDRVHLANIMRK..IRVMPDVIKV 697
TR5H_HUMAN      18 ATLIFSLKNE...VGGLIKALKIFQEKHVNLLHIESRKSK.....RRNSEFEIFVDC..DTNREQLNDIFHLLKS..HTNVLSVTPP 92
TYRA_BACSU     299 YDLYVDVPDH...PGVISEITAILAAERISITNIRIIETR....EDINGILRISFQ.....SDDDRKRAEQCIEA..RAEYETFYAD 371
TYRR_ECOLI       1 MRLEVFCEDR...LGLTRELLDLLVLRGIDLRGIEIDPIG....RIYLNFAELEFE........SFSSLMAEIRR..IAGVTDVRTV 70
```

Chapter 5: Figure 4. An alignment of representative ACT domains produced with the Belvu program (Erik Sonnhammer, unpublished). The Swiss-Prot identifier for each sequence is given. The known secondary structure taken from DSSP of PDB entry 1psd is shown for SERA_ECOLI.

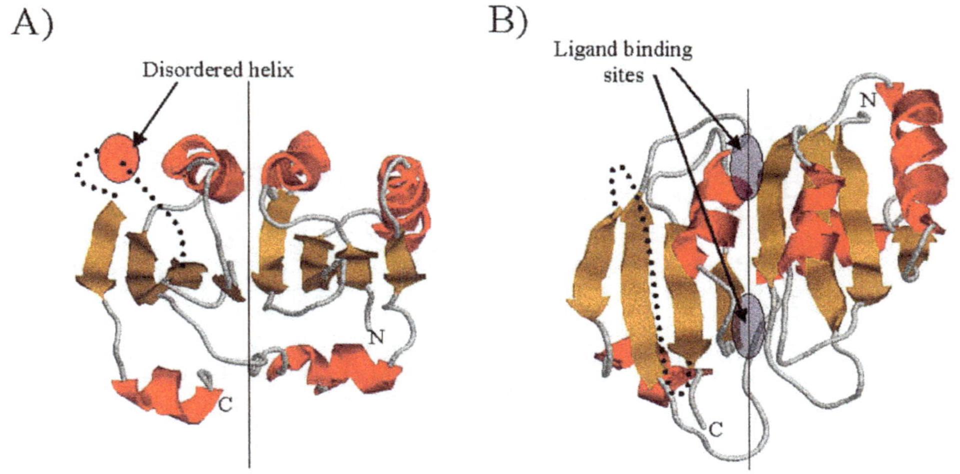

Chapter 5: Figure 5. The orientation of ACT domain dimers. The structures show the pseudo dimeric arrangement from the C-terminal ACT domains of threonine dehydratase (pdb entry 1thd). The site of ligand binding inferred from phosphoglycerate dehydrogenase is shown in part B. Part of this structure is disordered, this region is shown by a dotted line.

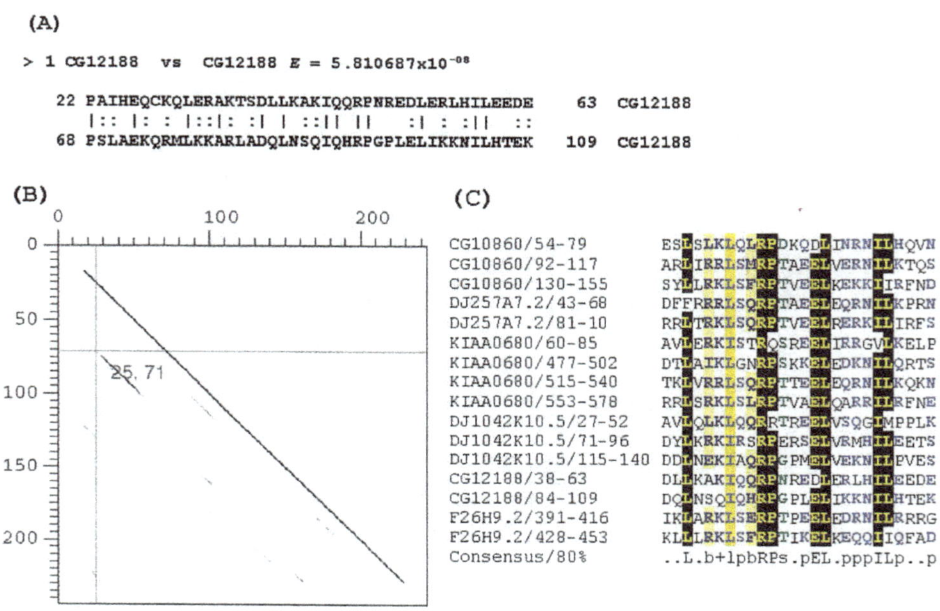

Chapter 5: Figure 6. Analysis of RPEL protein sequence repeats. A) The result of the best scoring self-self match of *Drosophila* protein CG12188 using prospero. B) A dot plot of the first 250 amino acids of CG12188 against itself. The strong diagonal line shows the 100% identical self match. Other off-diagonal lines represent potential repeats. C) An alignment of RPEL repeats from a variety of proteins. The sequence identifier is followed by the start and end points of the repeat. An alignment of RPEL repeats can be found in the Pfam database with accession number PF02755.

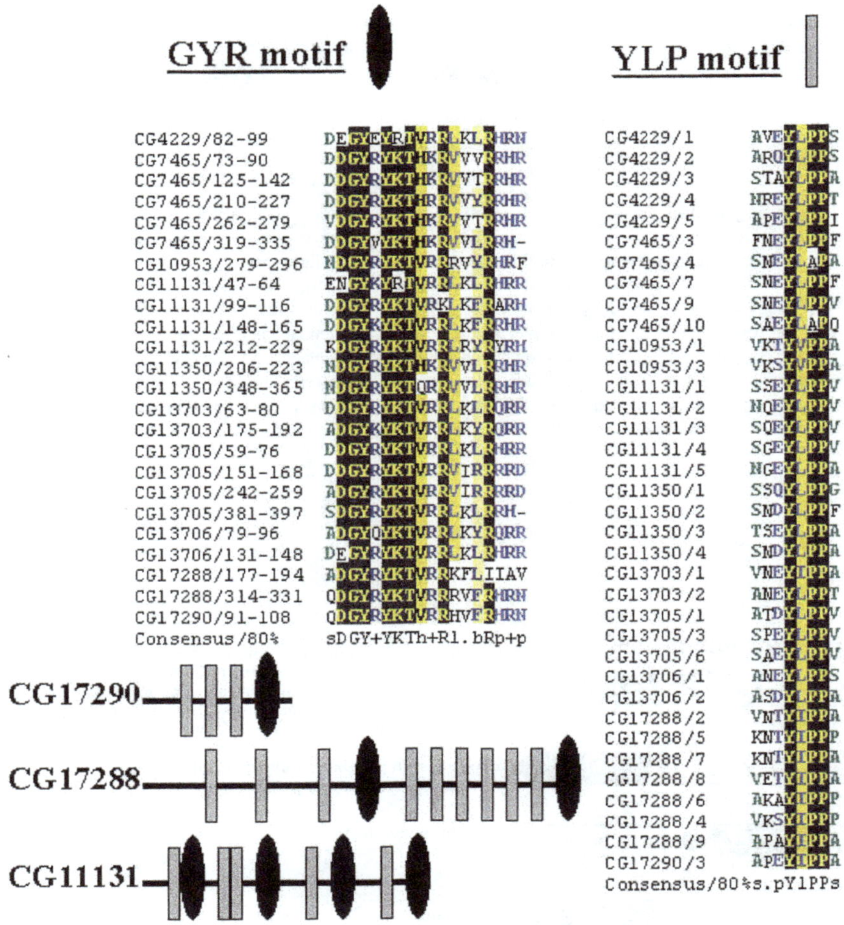

Chapter 5: Figure 7. Alignment of GYR and YLP motifs in *Drosophila* proteins. The motifs are shown graphically for three of these proteins. Alignments of GYR and YLP motifs can be found in the Pfam database with accession numbers PF02756 and PF02757 respectively.

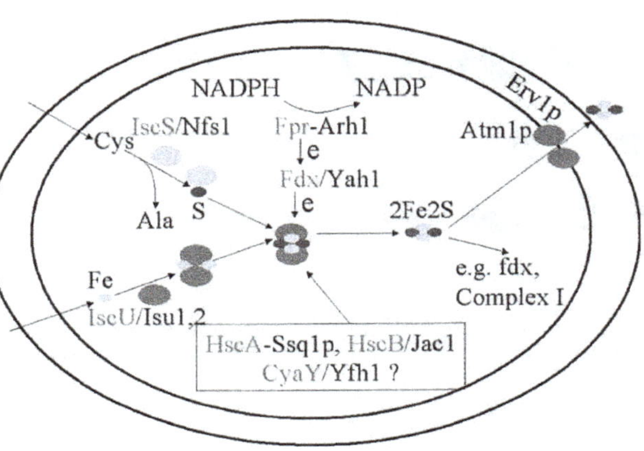

Chapter 6: Figure 3. Schematic representation of Fe-S cluster assembly in Proteobacteria and Fe-S cluster assembly and export from the mitochondria of *S.cerevisiae*. Gene displacements are marked with a "-", the Proteobacteria specific proteins are marked in gray. The genes of the proteobacteria and *S.cerevisiae* that are orthologous are separated by a "/". The genes for HscA, HscB and CyaY have a high genomic association are therefore grouped. The picture is largely based on (Lill and Kispal, 2000). The exact order of events in Fe-S cluster assembly on IscU (*e.g.* does the binding and reduction of sulfur take place before or after the binding of iron) has not been resolved (Urbina *et al.*, 2001).

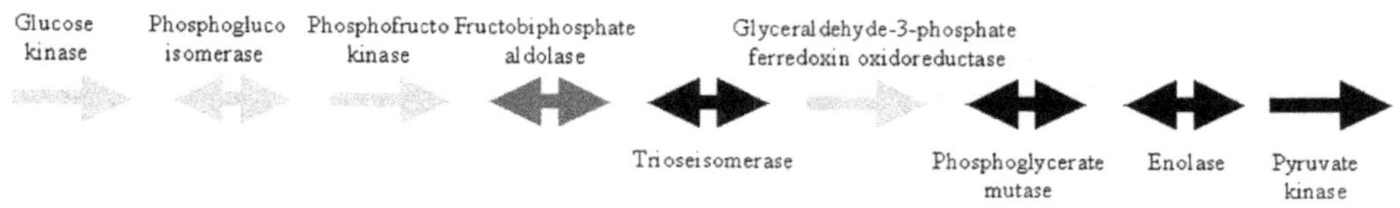

Chapter 6: Figure 5. Modular gene displacement in the glycolysis of *Pyrococcus furiosus*. Enzymes in the *P. furiosus* glycolysis that are orthologous to the corresponding *E. coli* enzyme are marked in black, enzymes that are paralogous to the *E.coli* ones in dark-gray and enzymes that are not homologous to the *E.coli* ones are marked in light-gray. The *P. furiosis* glycolysis contains nine enzymes, in contrast to the ten of *E.coli*, because *P. furiosis* Glyceraldehyde phosphate dehydrogenase (GAPOR, Oost *et al.*, 1998) combines glyceraldehyde-3-phosphate dehydrogenase and 3-phosphoglycerate kinase of the *E.coli* glycolysis. Notice that the displacements are glycolysis specific (in contrast to gluconeogenesis) as they tend to catalyze unidirectional reactions and are clustered in the top part of the pathway. For references to the non-homologous displacements see (Verhees *et al.*, 2001). Most of the enzymes of the glycolysis in *P. furiosus* that are orthologous to the ones in *E.coli* have high levels of sequence identity to these enzymes, and are correctly annotated in the databases. The one exception, in a distant ortholog (Dandekar *et al.*, 1999) of the Bacterial Phosphoglycerate mutase in *P. furiosus* does indeed have Phosphoglycerate mutase activity (Corne Verhees, pers. Comm.).

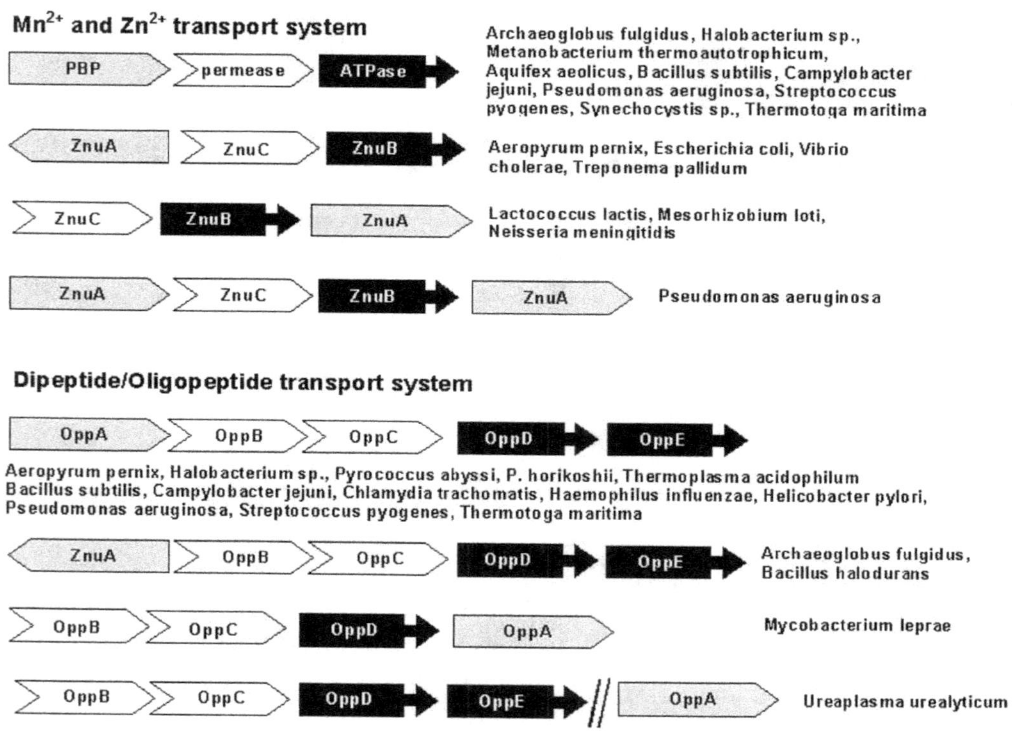

Chapter 7: Figure 1. Diversity of operon organization of common ABC-type transporters. The order of the shapes and the direction of the arrows reflect the gene order and orientation in given organisms. The shapes are drawn approximately to scale of the encoded proteins.

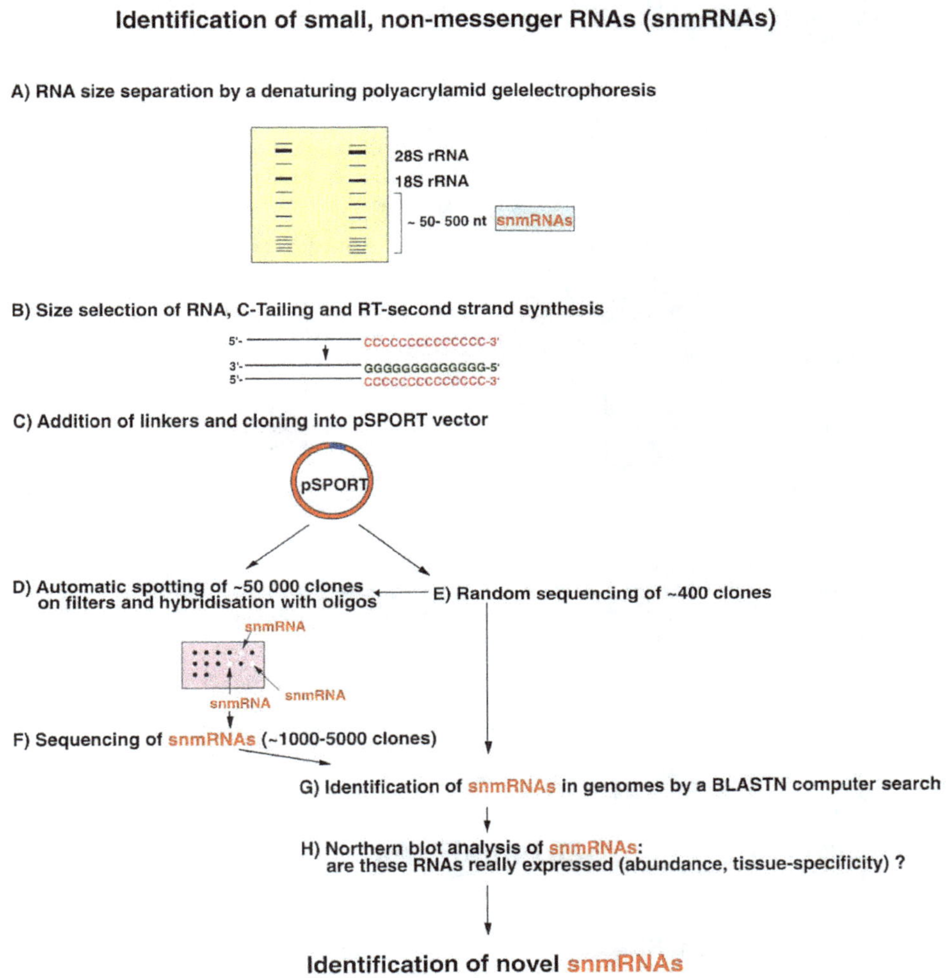

Chapter 9: Figure 1. Identification scheme of candidates for small non-messenger RNA (snmRNAs) in model organisms. Gel-purified bulk small RNA is C-tailed and reverse transcribed. Resulting cDNAs are cloned into the pSPORT plasmid vector. After identification and exclusion of clones derived from known RNAs by hybridization of arrays, remaining cDNAs are sequenced. Expression of candidates for snmRNAs is confirmed by Northern blot analysis.

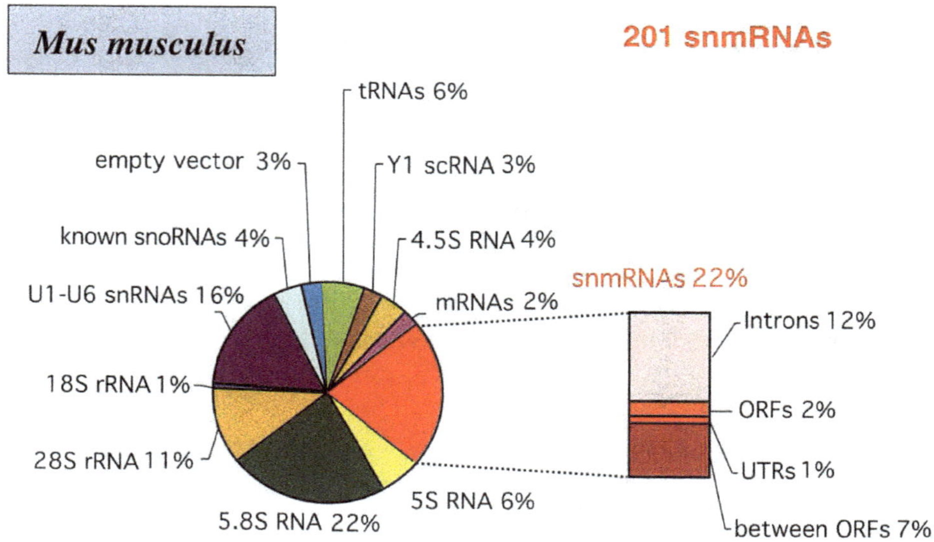

Chapter 9: Figure 2. Sequence analysis of a mouse brain cDNA library containing candidates for snmRNAs (Hüttenhofer *et al*., 2001). Indicated is the total number of novel snmRNA candidates identified (upper right), as well as the number of all RNAs species identified in our screen (in %). The location of genes on the mouse genome encoding novel snmRNA candidates is shown on the right.

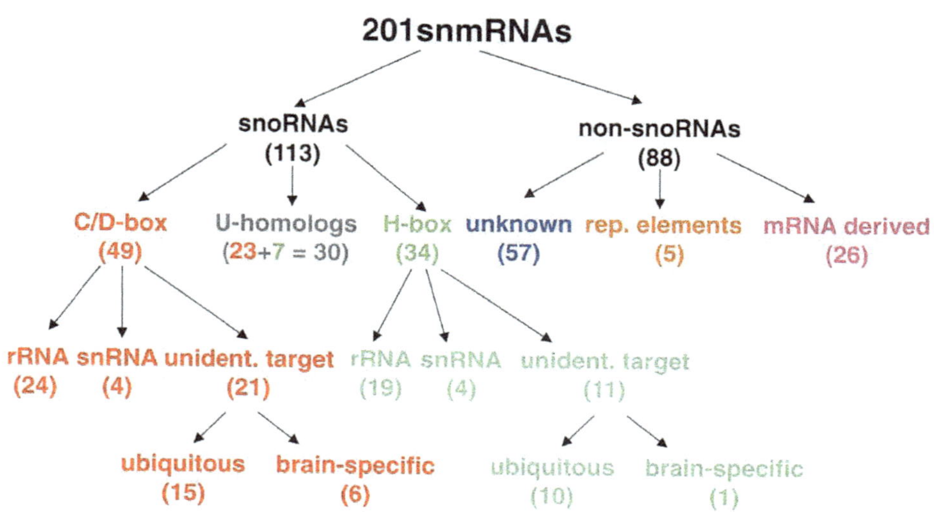

Chapter 9: Figure 3. Grouping and functional assignment of 201 novel snmRNAs identified in mouse brain.

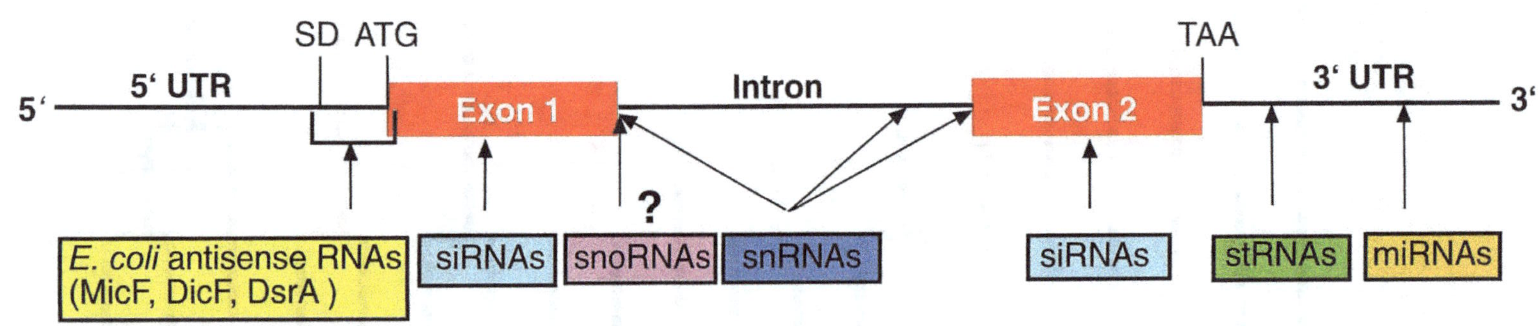

Chapter 9: Figure 4. Targeting of bacterial or eukaryal mRNAs or pre-mRNAs by snmRNAs thereby regulating gene expression. It should be noted that binding of a snoRNA to the splice-site of a eukaryal pre-mRNA or involvement in RNA editing has recently been proposed by our group (Cavaillé *et al.*, 2000) as two of several alternatives for a functional role. They remain to be experimentally sustained or disproven.

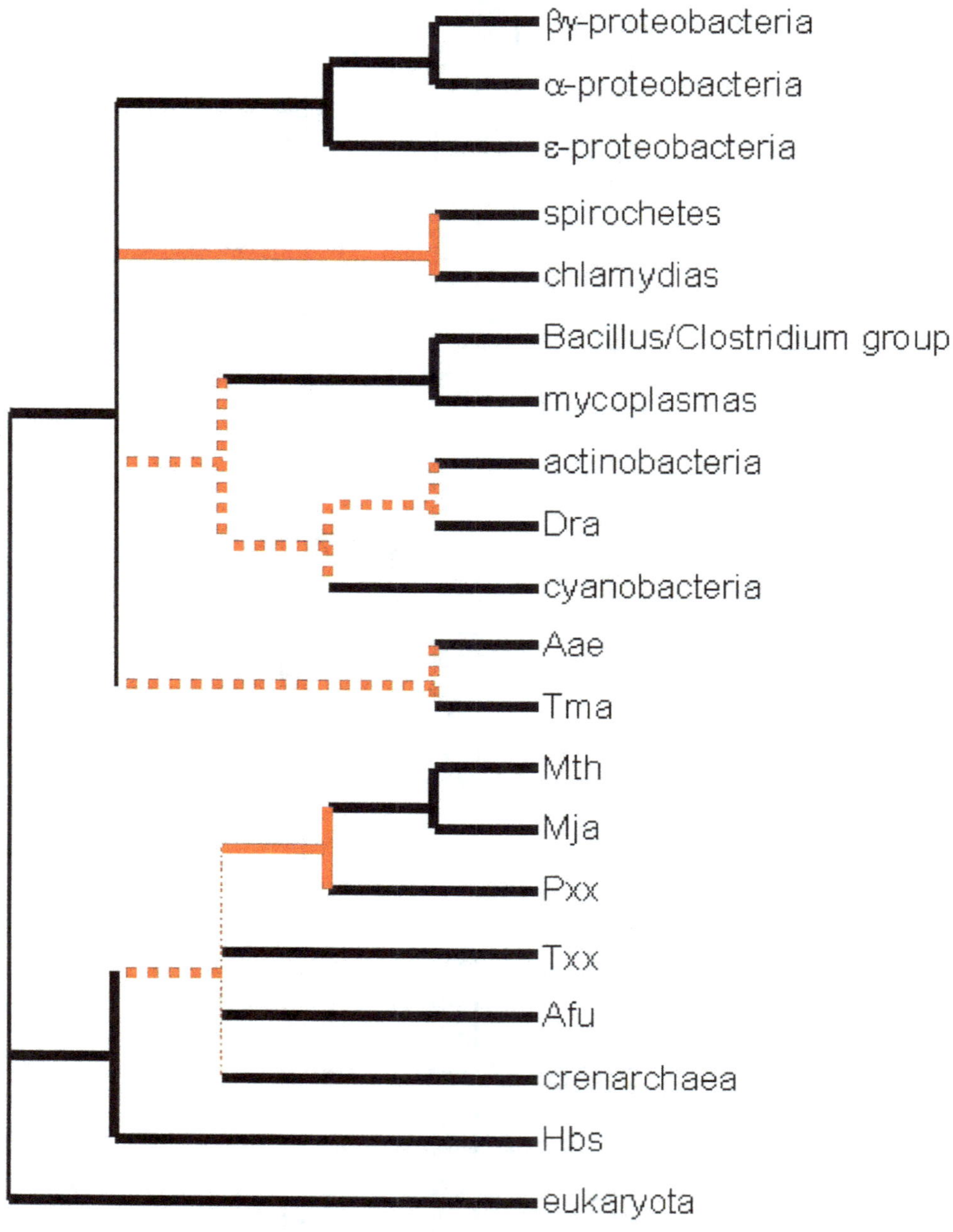

Chapter 10: Figure 2. A tentative consensus of genome-trees of prokaryotes. Black branches are conventional clades reproduced by genome-tree approaches; solid red branches are new clades that we consider to be firmly established by genome-trees; broken red branches are tentative clades suggested by genome-trees; thin lines are unresolved multifurcations. Species/clades abbreviation: the same as in Figure 1 and Txx, two *Thermoplasma* species (*T. acidophilum* and *T. volcanium*); Pxx, two *Pyrococcus* species (*P. horikoshii* and *P. abyssi*).

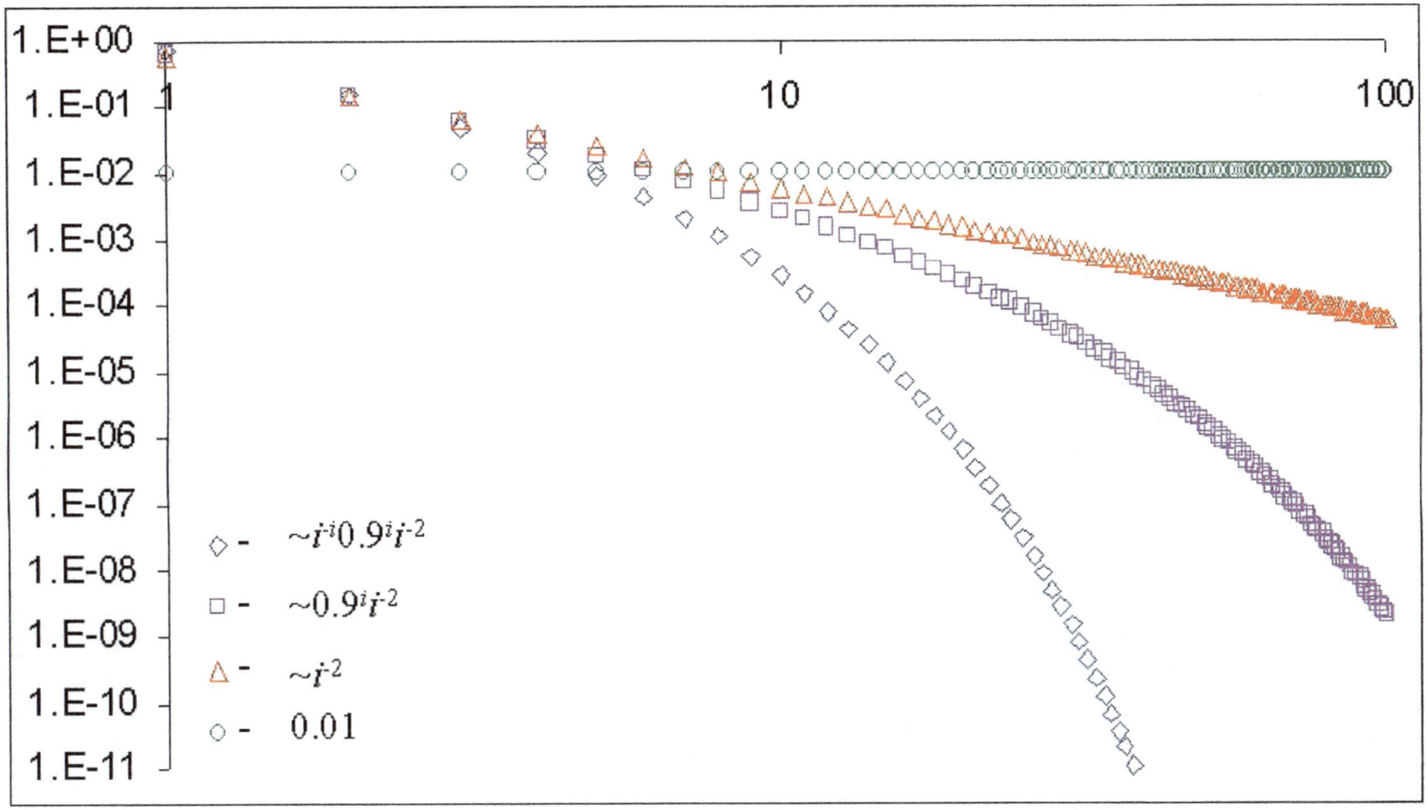

Chapter 11: Figure 4. Asymptotics of equilibrium distributions for balanced BDIMs of different orders.

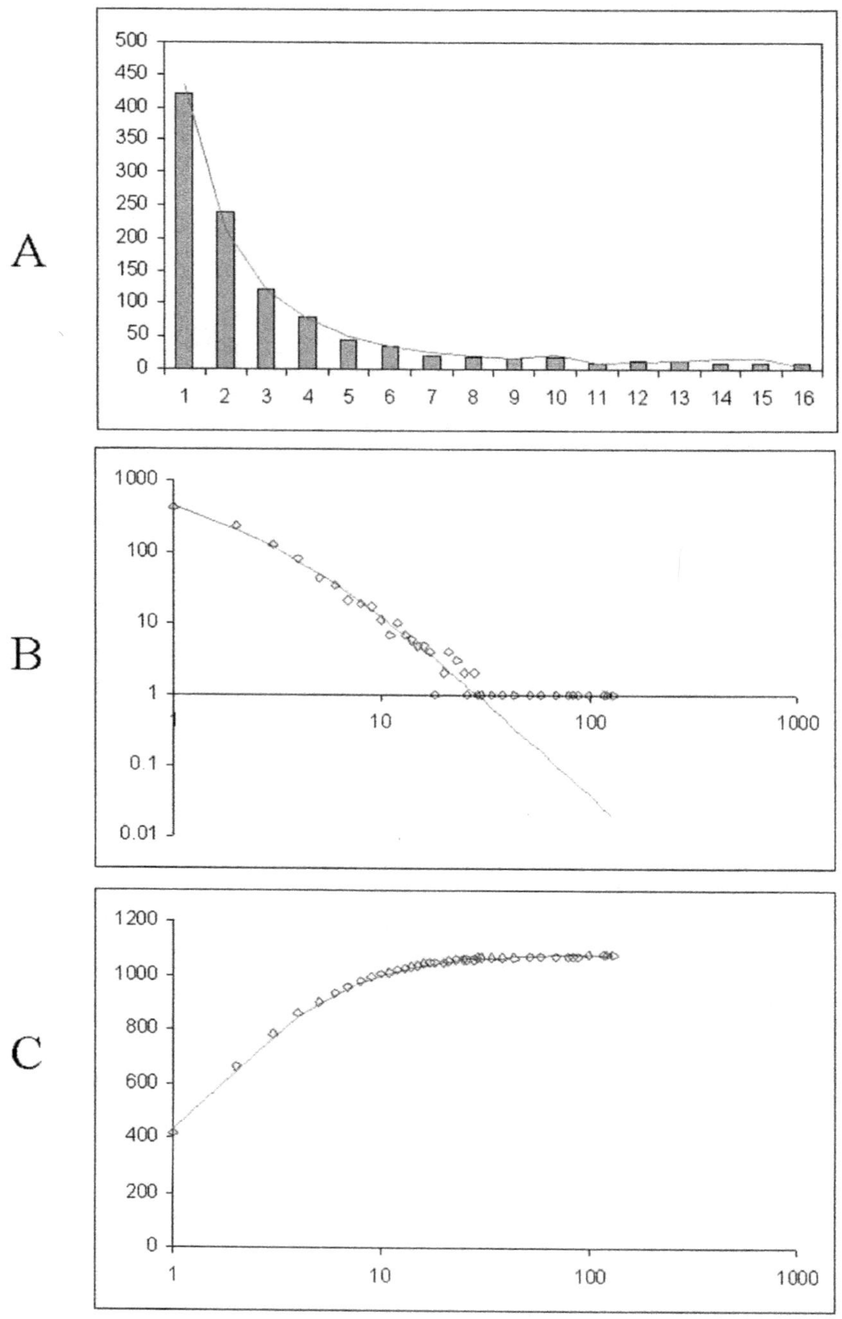

Sc

Chapter 11: Figure 5. Fit of empirical domain family size distributions to the second-order balanced linear BDIM. *Saccharomyces cerevisiae.*
A. Distribution of the size of domain families grouped into bins
B. Domain family size distribution in double logarithmic coordinates.
C. Cumulative distribution function of domain family size.
The line shows the prediction of the second-order balanced linear BDIM.

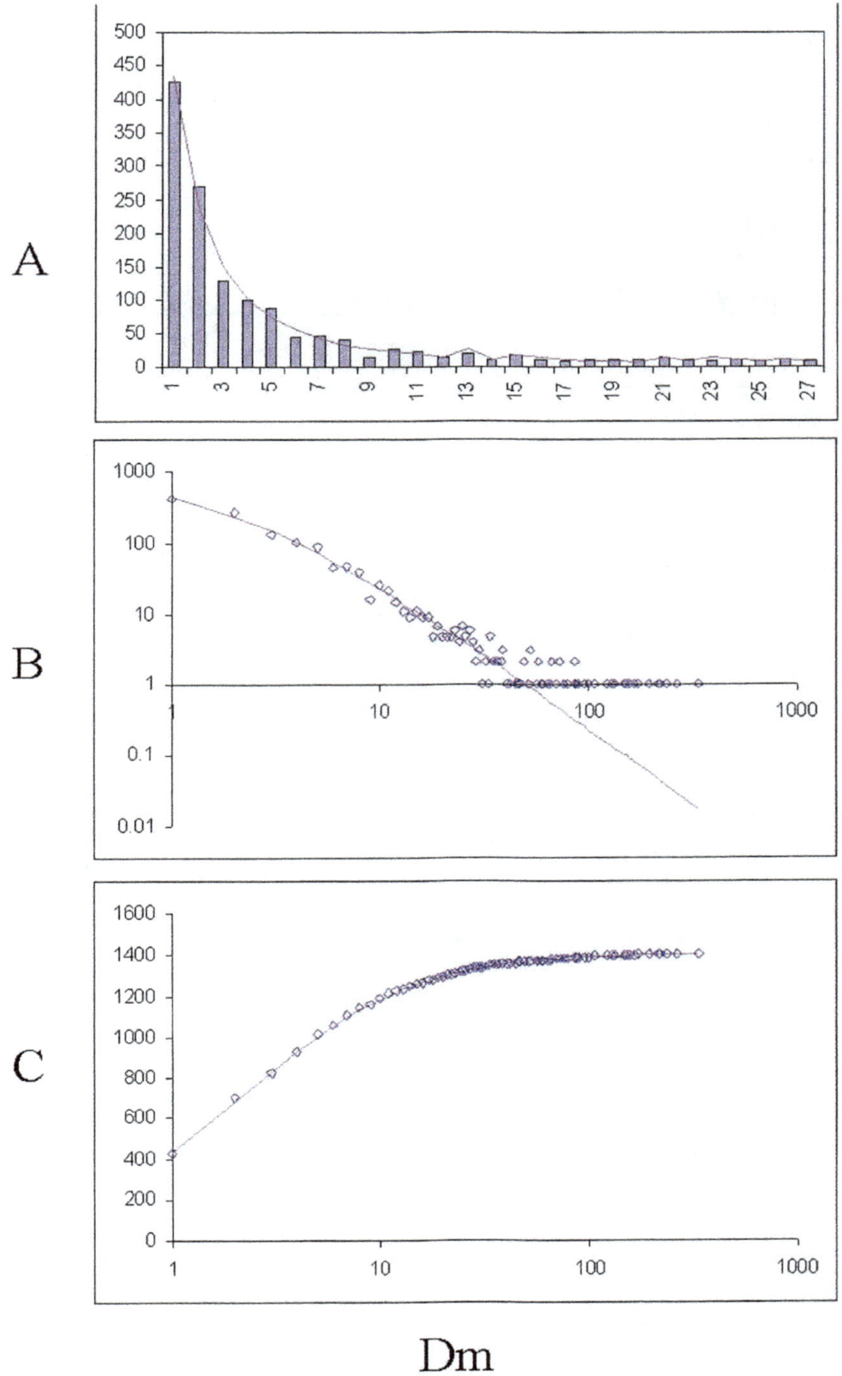

Chapter 11: Figure 6. Fit of empirical domain family size distributions to the second-order balanced linear BDIM. *Drosophila melanogaster*.

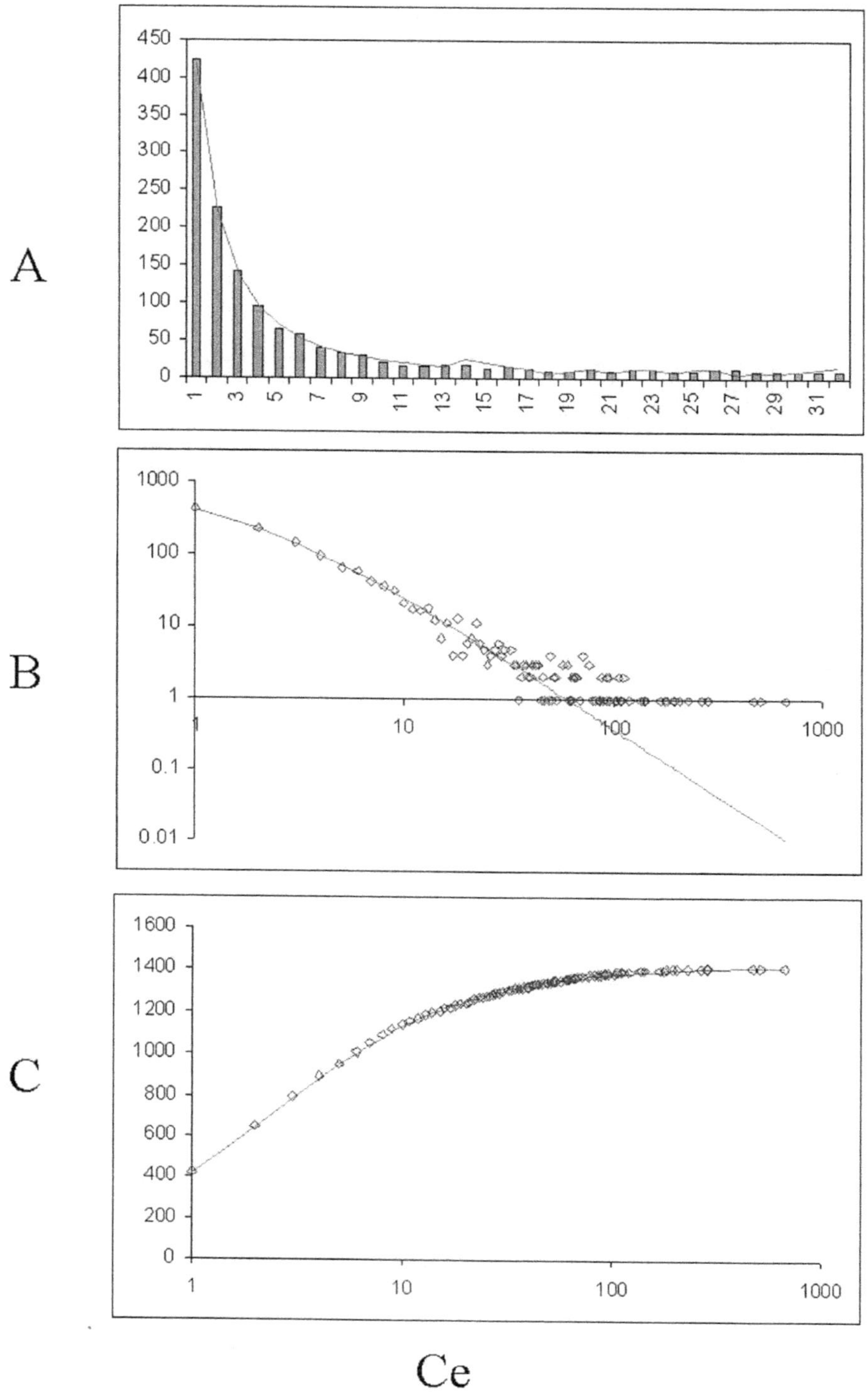

Chapter 11: Figure 7. Fit of empirical domain family size distributions to the second-order balanced linear BDIM. *Caenorhabditis elegans*.

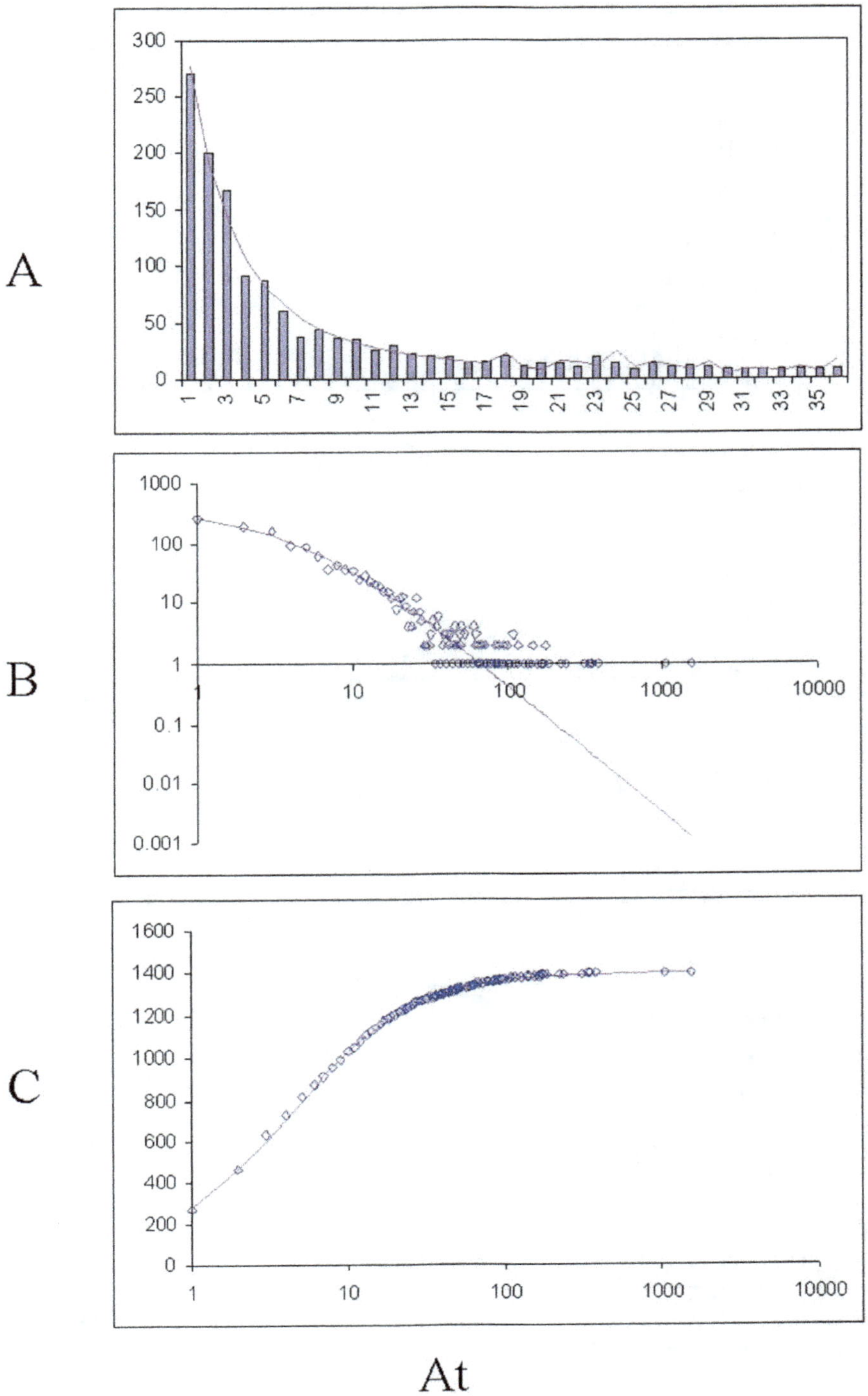

Chapter 11: Figure 8. Fit of empirical domain family size distributions to the second-order balanced linear BDIM. *Arabidopsis thaliana*.

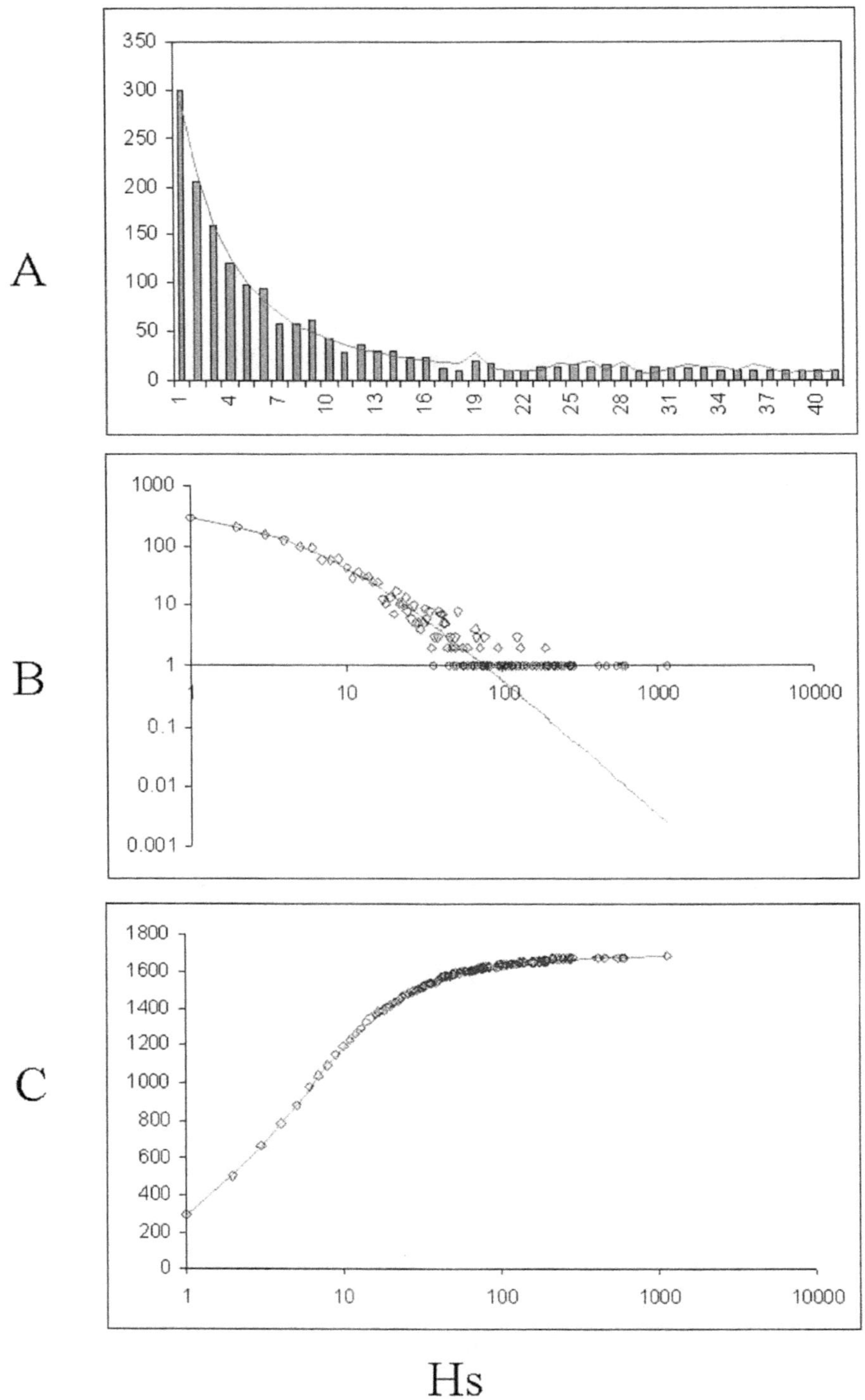

Chapter 11: Figure 9. Fit of empirical domain family size distributions to the second-order balanced linear BDIM. *Homo sapiens*.

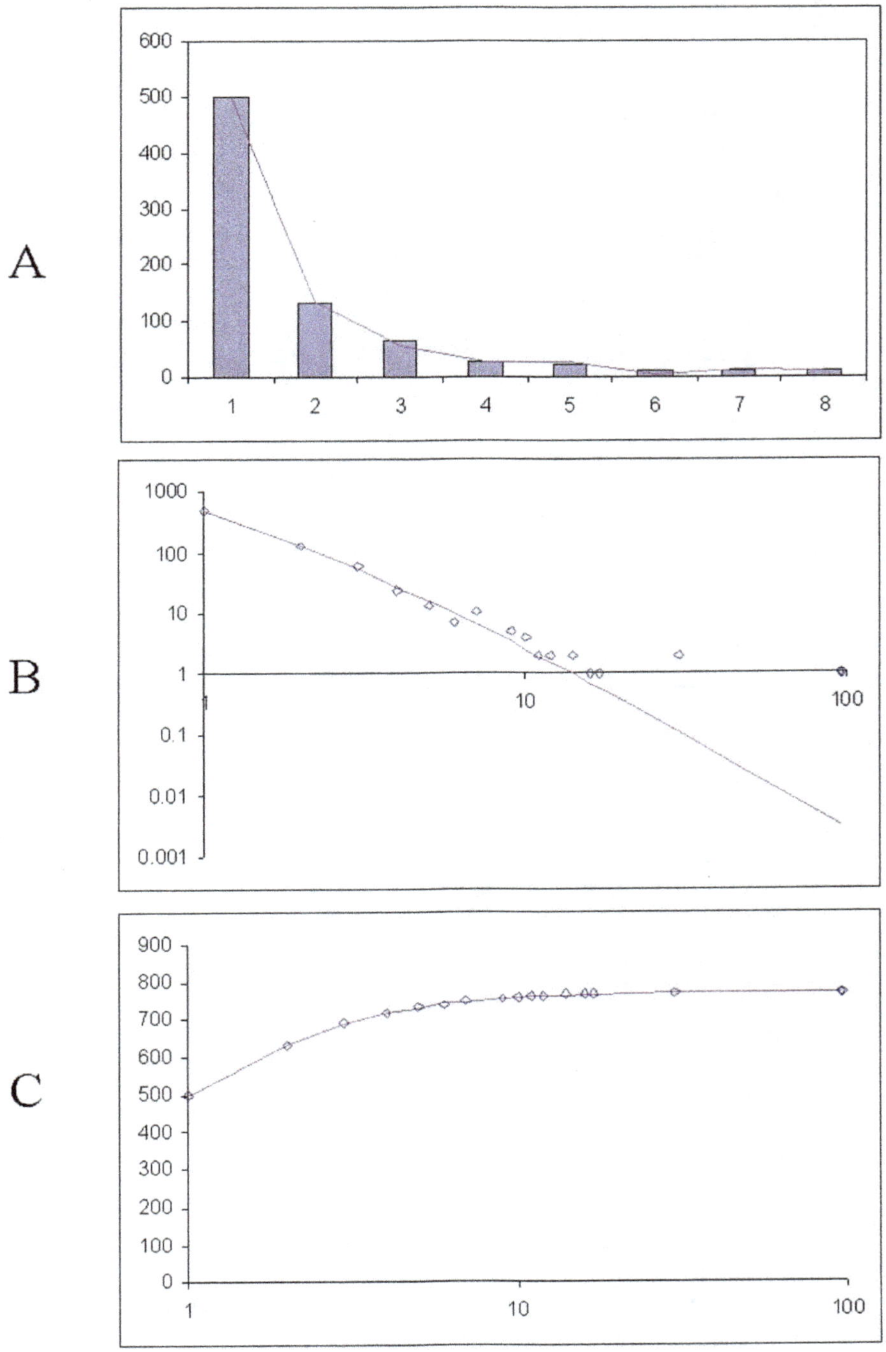

Thema

Chapter 11: Figure 10. Fit of empirical domain family size distributions to the second-order balanced linear BDIM. *Thermotoga maritima*.

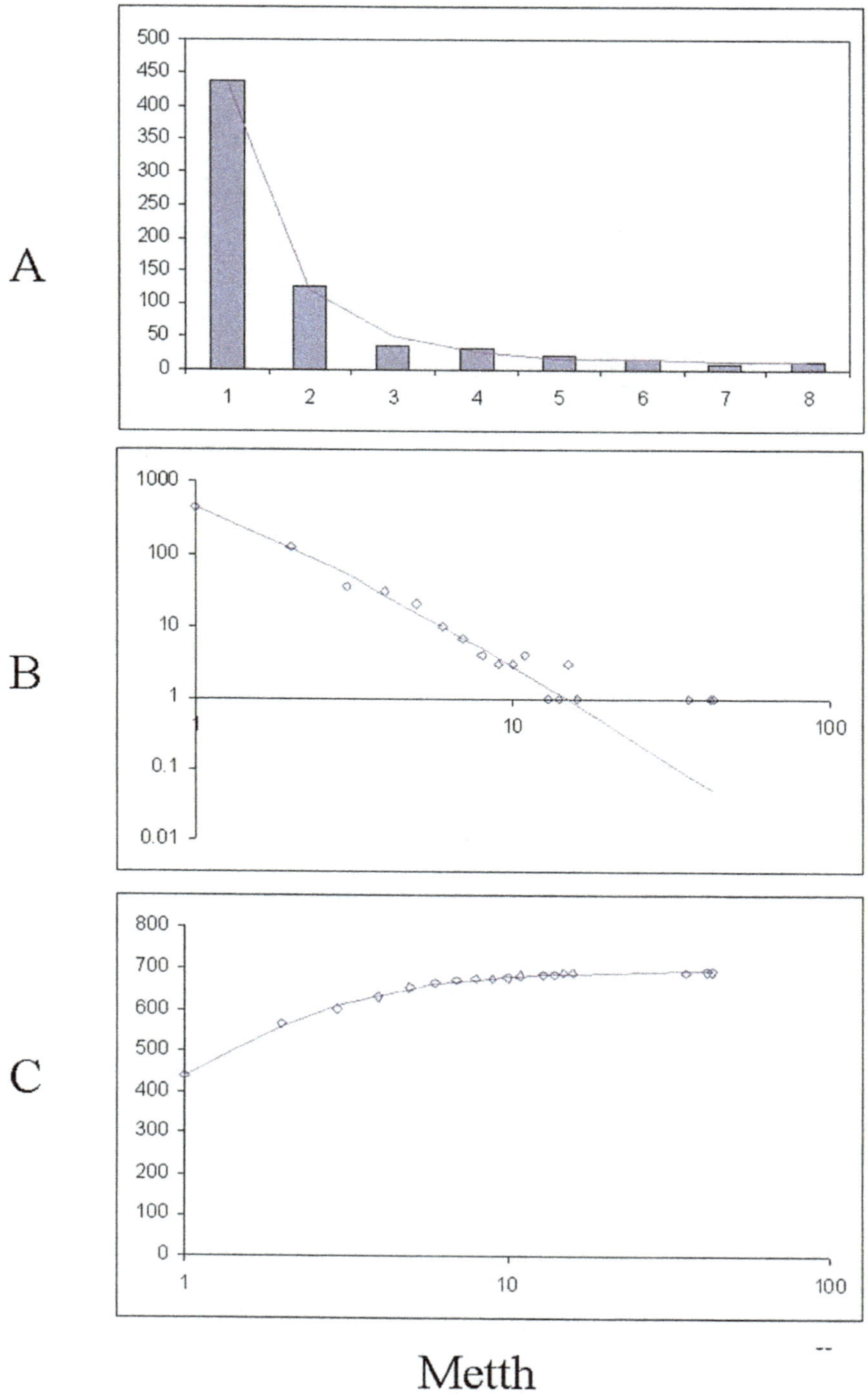

Chapter 11: Figure 11. Fit of empirical domain family size distributions to the second-order balanced linear BDIM. *Methanothermobacter thermautotrophicus*.

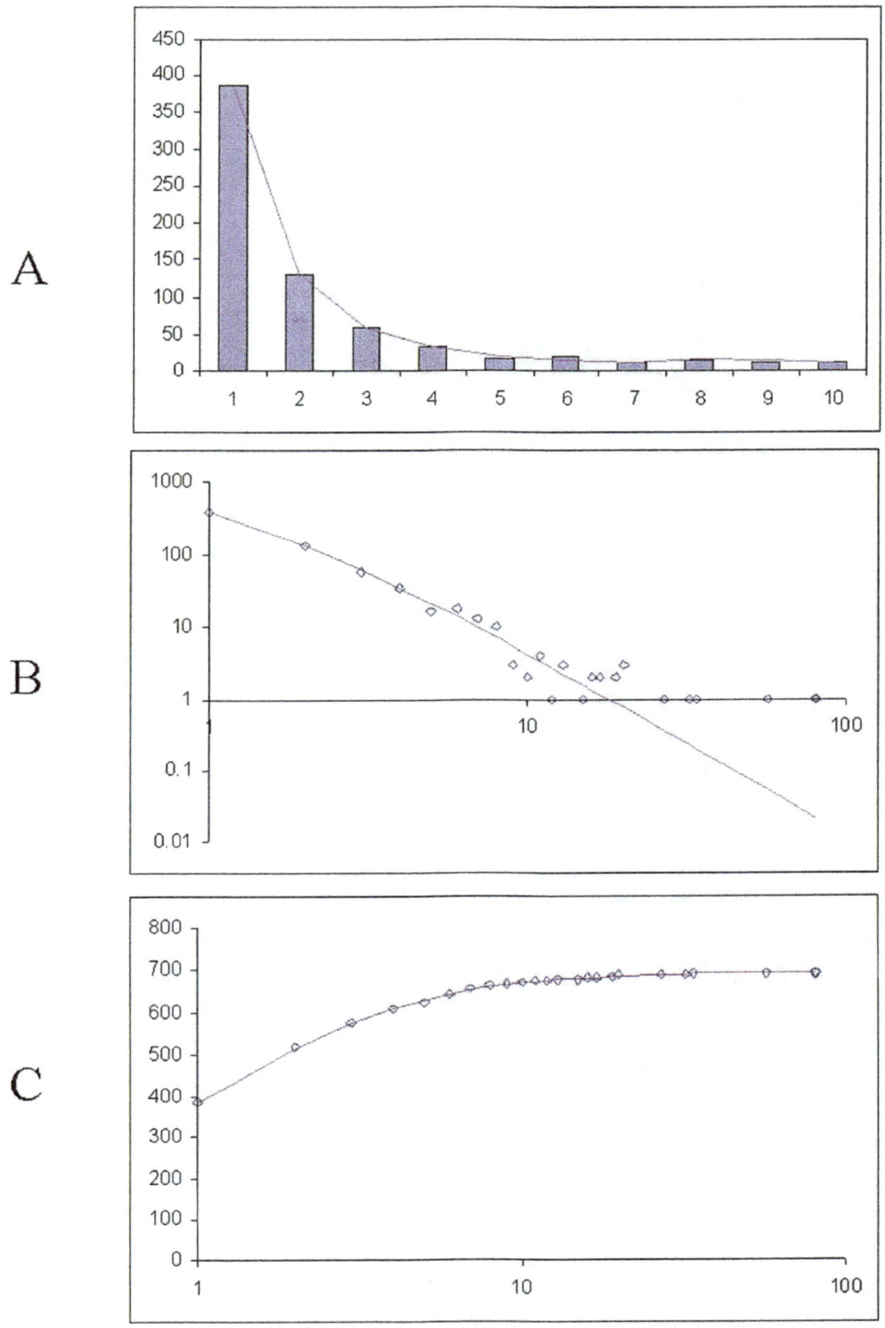

Sulso

Chapter 11: Figure 12. Fit of empirical domain family size distributions to the second-order balanced linear BDIM. *Sulfolobus solfataricus*.

A

B

C

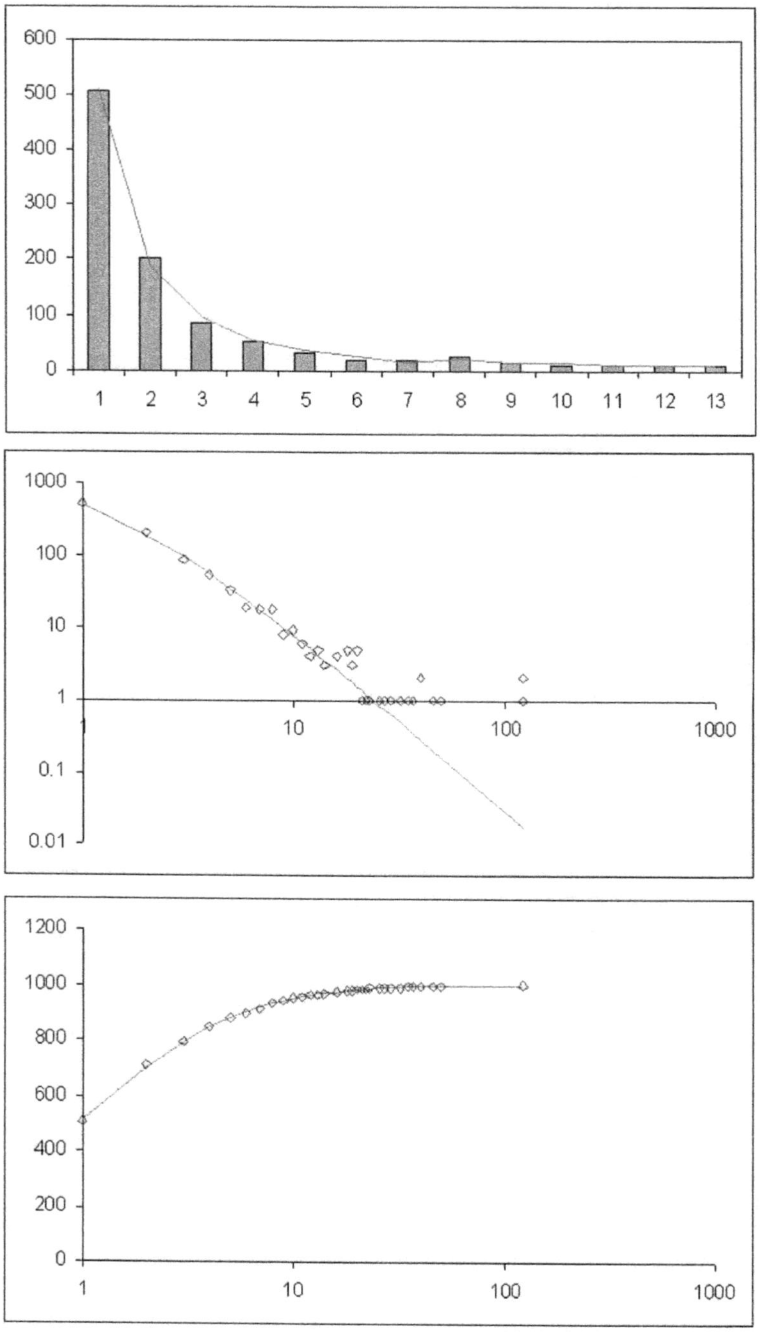

Bacsu

Chapter 11: Figure 13. Fit of empirical domain family size distributions to the second-order balanced linear BDIM. *Bacillus subtilis*.

A

B

C

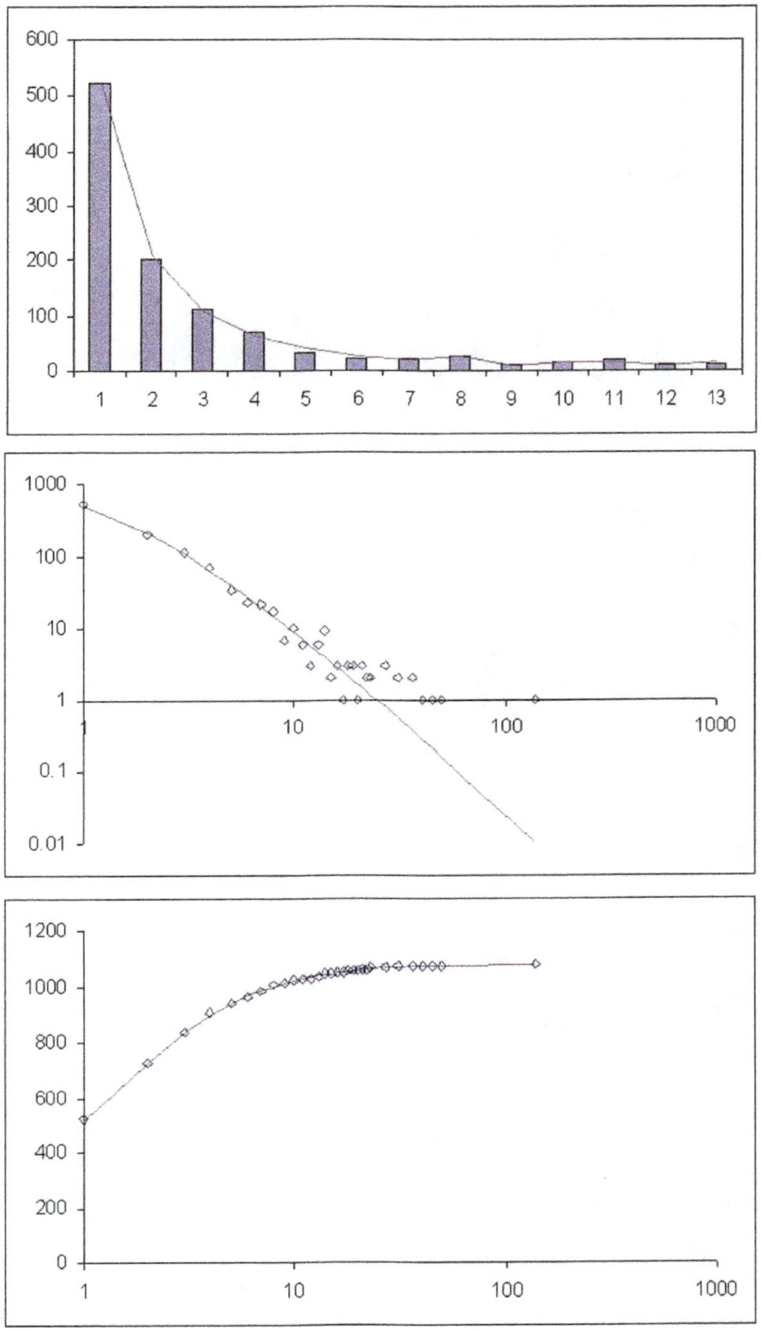

Escco

Chapter 11: Figure 14. Fit of empirical domain family size distributions to the second-order balanced linear BDIM. *Escherichia coli*.

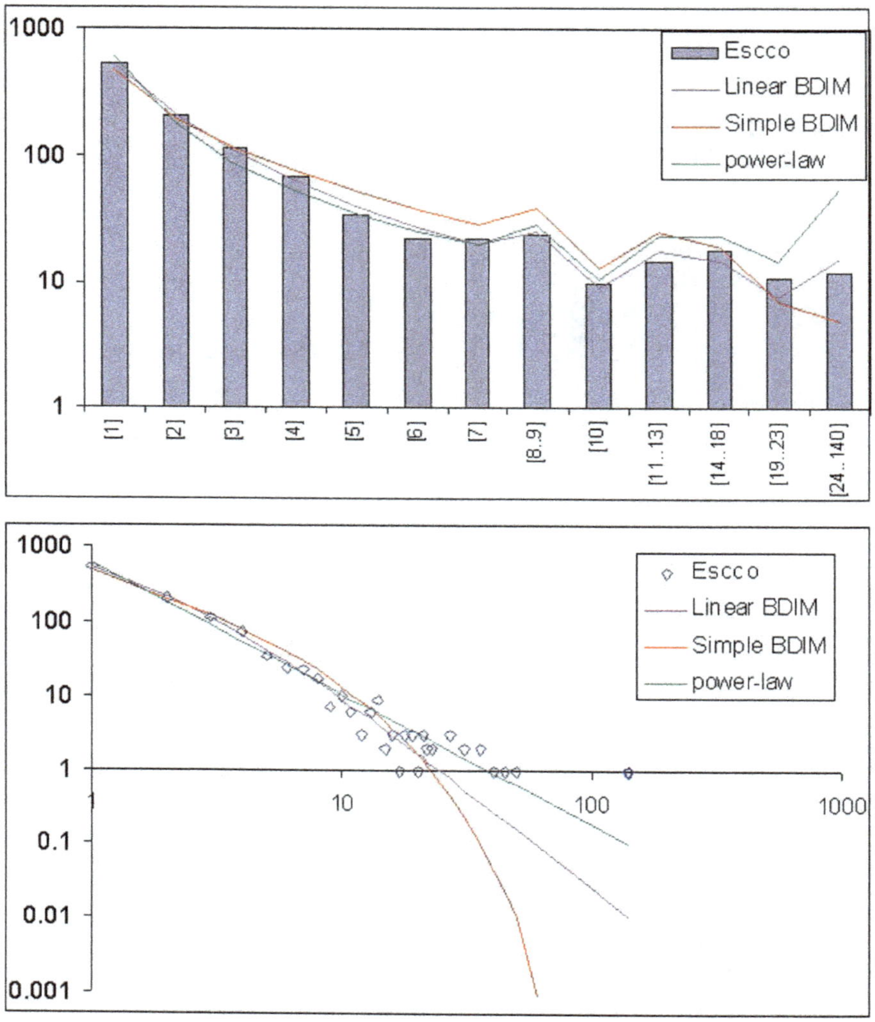

Chapter 11: Figure 15. Comparison of different approximations of the empirical domain family size distribution. *Escherichia coli.*
Magenta line: second-order balanced linear BDIM
Red line: simple BDIM
Cyan line: power law

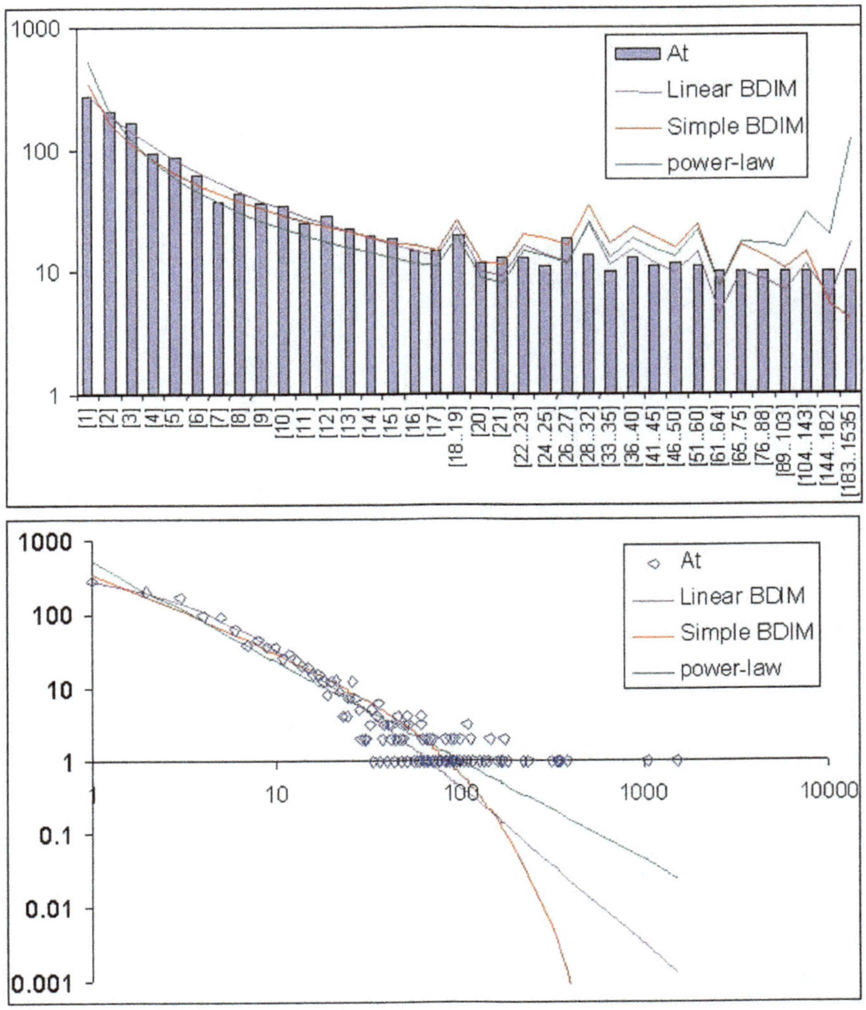

Chapter 11: Figure 16. Comparison of different approximations of the empirical domain family size distribution.
Arabidopsis thaliana.
Magenta line: second-order balanced linear BDIM
Red line: simple BDIM

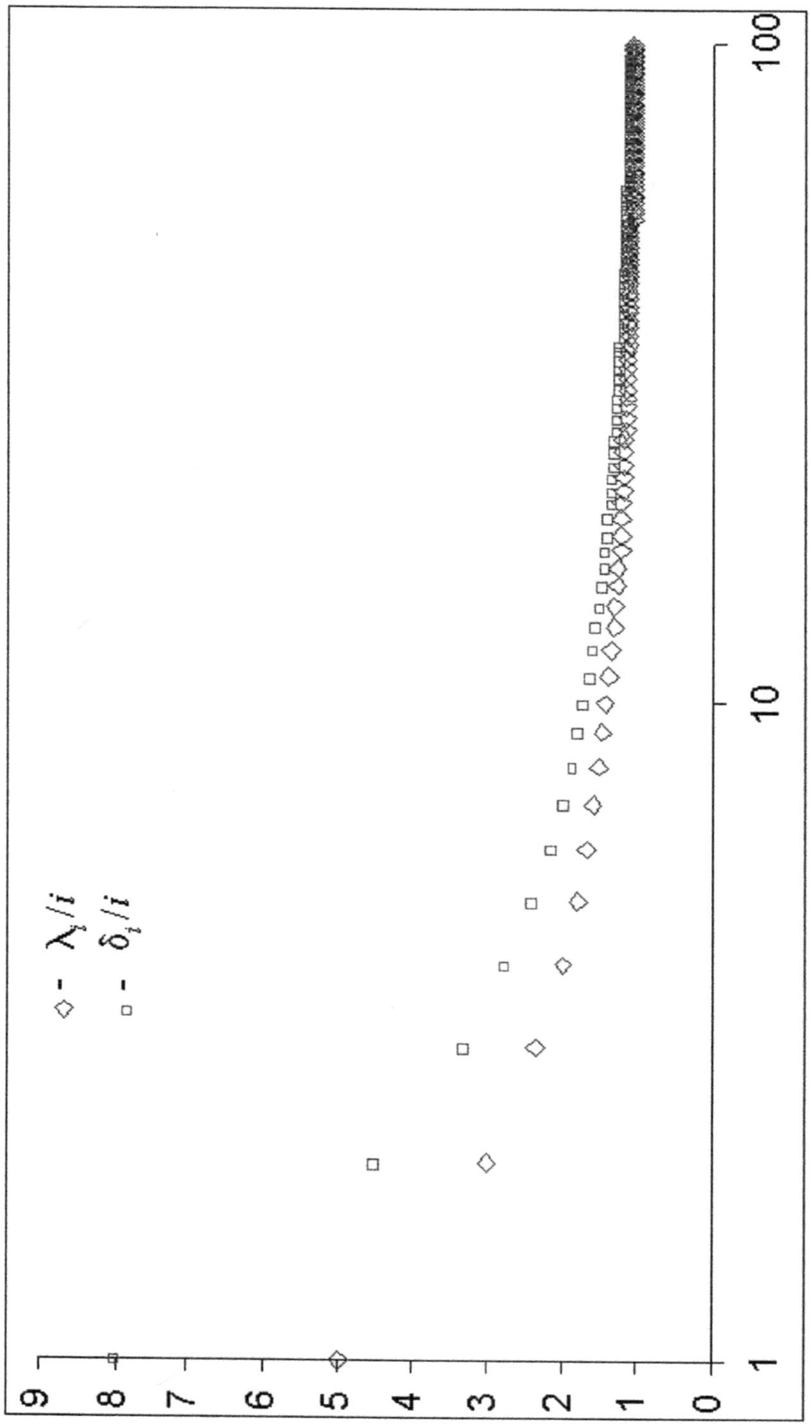

Chapter 11: Figure 17. Per domain birth and death rates for the second-order balanced BDIM.

www.ingramcontent.com/pod-product-compliance
Lightning Source LLC
LaVergne TN
LVHW082115200926
848484LV00008B/1951
* 9 7 8 0 9 5 4 2 4 6 4 4 0 *